Theory of Interacting Fermi Systems

Advanced Book Classics
David Pines, Series Editor

Anderson, P. W., *Basic Notions of Condensed Matter Physics*
Bethe, H. and Jackiw, R., *Intermediate Quantum Mechanics, Third Edition*
Nozières, P., *Theory of Interacting Fermi Systems*
Pines, D., *The Many-Body Problem*
Quigg, C., *Gauge Theories of the Strong, Weak, and Electromagnetic Interactions*

Theory of Interacting Fermi Systems

Philippe Nozières
Collège de France
Paris

Translation by
D. Hone
University of Pennsylvania

The Advanced Book Program

Addison-Wesley
Reading, Massachusetts

Many of the designations used by manufacturers and sellers to distinguish their products are claimed as trademarks. Where those designations appear in this book and Addison-Wesley was aware of a trademark claim, the designations have been printed in initial capital letters.

ISBN 0-201-32824-0

Addison-Wesley is an imprint of Addison Wesley Longman, Inc.

Cover design by Suzanne Heiser

1 2 3 4 5 6 7 8 9-MA-0100999897
First printing, November 1997

Find us on the World Wide Web at
http://www.aw.com/gb/

Editor's Foreword

Addison-Wesley's *Frontiers in Physics* series has, since 1961, made it possible for leading physicists to communicate in coherent fashion their views of recent developments in the most exciting and active fields of physics—without having to devote the time and energy required to prepare a formal review or monograph. Indeed, throughout its nearly forty-year existence, the series has emphasized informality in both style and content, as well as pedagogical clarity. Over time, it was expected that these informal accounts would be replaced by more formal counterparts—textbooks or monographs—as the cutting-edge topics they treated gradually became integrated into the body of physics knowledge and reader interest dwindled. However, this has not proven to be the case for a number of the volumes in the series: Many works have remained in print on an on-demand basis, while others have such intrinsic value that the physics community has urged us to extend their life span.

The *Advanced Book Classics* series has been designed to meet this demand. It will keep in print those volumes in *Frontiers in Physics* or its sister series, *Lecture Notes and Supplements in Physics*, that continue to provide a unique account of a topic of lasting interest. And through a sizable printing, these classics will be made available at a comparatively modest cost to the reader.

Although the informal monograph *Theory of Interacting Fermi Systems* was written some thirty-five years ago, when the distinguished French theoretical physicist and Wolf Prize–winner Philippe Nozières was a young man giving his first lectures on this topic, it continues to be the authoritative account of the way in which Landau Fermi liquid theory emerges from a field-theoretic description of interacting fermions. Written with unusual clarity and attention to detail, it is must-reading for anyone interested in applying field-theoretic techniques to problems in condensed matter or nuclear physics. It gives me great pleasure to see this book back in print, and to welcome Professor Nozières to the *Advanced Book Classics* series.

David Pines
Urbana, Illinois
October 1997

Contents

Introduction

Before 1950 there was practically no many-body problem. Only some precursors had touched on the study of condensed systems. Their efforts remained isolated and elicited little response from the main body of physicists. In ten years this subject has been developed to such a point that it is now involved in every area of physics.

The first efforts in this direction were quite disconnected. Several approximate theories were proposed, each of which treated a very specialized subject. This diversity is manifest in the courses offered at the 1958 session of the Summer School of Theoretical Physics at Les Houches. The essential step has been to set up a unified formalism, based on the methods of quantum field theory. The same "language" now allows us to treat nuclear matter, liquid helium, or superconductors.

These new methods also possess a great virtue—they can be generalized to systems at finite temperatures; the many-body problem thus becomes allied with quantum statistical mechanics, representing the latter's low-temperature limit. These new paths seem extremely promising.

There is no question of covering so vast a subject within the framework of this study; a severe limitation is imposed. A priori it seems natural to illustrate these theories by a sufficiently large selection of practical examples without being too concerned with details. By placing the methods within the framework of ordinary physics, one undoubtedly facilitates their immediate application. In spite of these advantages I have prefered to take a "vertical" section, thoroughly treating a restricted subject. The goal sought is not to exhaust "the" many-body problem, but to demonstrate its power in a simple case. This choice answers a very subjective need. It makes the presentation quite dry and neglects certain important aspects of the subject (in particular, finite temperature properties). On the other hand, such an analysis usefully complements the more descriptive treatments.

The work is devoted to the general properties of infinite systems of *fermions at zero temperature*. A double objective is pursued: to analyze the mechanism of correlations and to set forth a solid formalism that can be directly generalized to the most complex cases. With few exceptions the nature of these fermions is not specified; the results therefore apply to nuclear matter, to He^3, and to electrons in solids.

Since this subject is at present enjoying great popularity, there exist a large number of presentations of it (indicated, in part, in the bibliography). These theories are equivalent; in principle, a "dictionary" would be sufficient to go from the diagrams proposed by X to those favored by Y. Because this variety leads to confusion, I have worked uniquely with the formalism of the Soviet school, which seems to me to be the clearest and simplest. At this time I would like to pay homage to the great physicist L. D. Landau; his phenomenological theory of Fermi liquids (Chapter 1) has brilliantly clarified this whole area of physics. This book, inspired by his work, must pay him tribute.

In spite of the formal aspect of this book, no pretense as to mathematical rigor is attempted in the proofs; in particular, questions of convergence have been treated very lightly. A foreign colleague further characterizes certain proofs as "wishful thinking." In my opinion an incomplete but simple argument often clarifies the physical phenomenon better than a rigorous proof. I excuse myself in advance for these gaps.

The presentation assumes a thorough knowledge of elementary quantum mechanics, such as is taught in the first year of graduate studies. With this as a basis, the formalism is developed point by point; in particular, perturbation techniques are analyzed in great detail. Since this book is supposed to be self-contained, references to the original articles have been collected at the end and grouped by subject. I have, nevertheless, tried to attribute the origins of the principal results to their respective authors. I have certainly forgotten some of them; in advance I beg the pardon of those whom I have involuntarily ignored.

The organization of the work reflects the various aspects of the fermion gas. The first chapter is devoted to the theory of "normal" Fermi liquids proposed by Landau. This phenomenological introduction already deals with most important concepts. Chapter 2 establishes contact with the principal experimental techniques: response to an external field and scattering of a beam of incident particles. In Chapter 3 we enter into the heart of the subject by formally introducing Green's functions. The physical meaning of the latter is discussed in Chapter 4, first for "normal" systems, then for "superfluids." Chapters 5 and 6 are devoted to the development, then to the exploitation, of perturbation techniques; among other things, this allows us to justify the Landau theory. Finally, Chapter 7 generalizes perturbation methods to superfluid systems.

This book is a product of a course taught in 1959–1960 at the University of Paris in the "Troisième Cycle" of Theoretical and Solid-State Physics.

I must thank all those who have assisted or encouraged me in this task. By assigning me a course, Professors M. Lévy and J. Friedel gave me the opportunity to explore this subject. The pertinent criticism of E. Abrahams, D. Pines, and P. R. Weiss has been valuable to me. A part of Chapter 6 was worked out in collaboration with J. M. Luttinger and J. Gavoret; Miss O. Betbeder-Matibet assisted in the correction of proof. Finally, J. Giraud assumed the thankless task of preparing the manuscript. May all of them find here the expression of my recognition.

At the time of publication of this book I find many defects in it. Is it really worthwhile to perfect a tool without making use of it? It will be up to the reader to judge.

P. Nozières

Paris, France
May 1962

I wish to thank Dr. D. Hone for accepting the burden of translating this book into English. It was a rare opportunity to have the translation made by a specialist in the many-body problem. Dr. Hone's comments on the subject have been very helpful; I wish to thank him for his assistance.

P. Nozières

Paris, France
July 1963

Theory of Interacting Fermi Systems

Chapter 1

The Landau Theory

1. The Notion of a Quasi Particle

Let us consider a uniform gas of fermions, containing N particles in a volume Ω which we assume to be very large. In this chapter we propose to study the macroscopic properties of this system by means of a semiphenomenological method due to Landau. For the moment we shall use an intuitive approach, reserving the proof of the validity of our assertions for Chap. 6. It is from this point of view that Landau attacked this problem. His hypotheses having since been shown to be rigorously exact, we can only pay him homage.

The study of the macroscopic properties of a system at zero temperature requires knowledge of the ground state and the low-lying excited states. Let us first consider the very simple case of an "ideal" gas—that is, a gas of noninteracting particles. The eigenstates of such a system are well known: they are antisymmetric combinations of plane waves, one for each particle. Each plane wave is characterized by its wave vector k. To define an eigenstate of the whole system, it is sufficient to indicate which plane waves are occupied by means of a distribution function $n(k)$. The ground state of the system corresponds to an isotropic distribution $n_0(k)$, of the form indicated in Fig. 1. The cutoff level k_F is called the Fermi level. It is given by

$$\frac{k_F^3}{3\pi^2} = \frac{N}{\Omega} \tag{1-1}$$

account being taken of spin degeneracy. If the distribution function is changed by an infinitesimal quantity $\delta n(k)$, the total energy of the system changes by an amount

$$\delta E = \sum_k \frac{\hbar^2 k^2}{2m} \delta n(k)$$

We thus see that the energy $\hbar^2 k^2/2m$ of a particle of wave vector k can be defined as the functional derivative $\delta E/\delta n(k)$—that is to say, the variation δE when $\delta n(k')$ is equal to the Kronecker δ function $\delta_{kk'}$. Let us point out that δn is necessarily positive for $k > k_F$ and negative for $k < k_F$.

These considerations are trivial for an ideal gas. They become much less obvious when one tries to extend them to a real gas. We shall try to pass from one case to the other by introducing the interaction progressively, in an adiabatic manner. We shall *assume* that the states of the ideal system are gradually transformed into states of the real system as the interaction increases; we can then study the time development of each state by means of a perturbation treatment. Note that this hypothesis does not exclude the possibility of *other* elementary excitations of the real system which disappear when the interaction is reduced to zero. Some "new" states of this type always appear; this is the case, for instance, for sound waves.

The hypothesis which we have just made is very restrictive. There are cases where one knows in advance that it is false. If there is an attraction between particles, however weak, the ground state of the real system is radically different from that of the ideal gas. We are thus limited to repulsive forces, with the further condition that they not be too strong. Even in this restricted case the situation is far from clear. We shall see that the states of the real system are in general unstable and are damped out after a certain time τ. If the adiabatic switching on of the interaction requires a time $\gtrsim \tau$, the original state will have decayed even before the full interaction has been turned on; the result manifestly makes no sense.

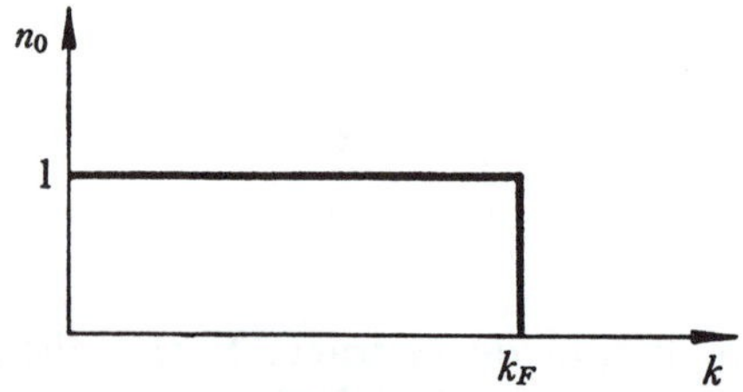

Figure 1

On the other hand, if one proceeds too fast, the final state is no longer an eigenstate. This dilemma disappears only if the lifetime τ is very long. We shall see in Chap. 4 that this limits us to low-lying excited states, very close to the ground state.

All these difficulties will arise in Chap. 5, where we shall try to connect the formalism of this chapter to the general theory of the many-body problem. For the moment, we content ourselves with asserting the continuity of the states as a function of the coupling constant. We furthermore assume that the ground state of the ideal system gives rise to the ground state of the real system.

Let us add an additional particle of wave vector k ($k > k_F$) to the "ideal" ground state, and then switch on the interaction. In this way we obtain an eigenstate of the real system; we shall say that we have added a *quasi particle* of wave vector k to the real ground state. We shall see later that the lifetime of the state thus defined is long only in the immediate neighborhood of $k = k_F$. The concept of a quasi particle is therefore valid only near the Fermi surface; this is an *essential* aspect of the problem which is important always to keep in mind.

Similarly, we define a *quasi hole* of wave vector k ($k < k_F$) by referring to the state of the ideal system in which we have removed a particle of wave vector k. This definition is easily extended to states containing several quasi particles or quasi holes, on condition, however, that these be all in the neighborhood of the Fermi surface. The same function $n(k)$ which characterizes the states of the ideal system thus allows us to characterize the real states; it now gives the distribution of *quasi particles*, no longer that of bare particles.

Note that quasi particles correspond to $k > k_F$, quasi holes to $k < k_F$; the distribution of quasi particles in the ground state, $n_0(k)$, is still given by Fig. 1. The notion of a Fermi surface remains. The excitation of the system is measured by

$$\delta n(k) = n(k) - n_0(k)$$

For the notion of the quasi particle to make sense, δn must be appreciable only in the neighborhood of $k = k_F$. [$n_0(k)$ is defined only in this region.] It is tempting to assert that the ground state is made up of N quasi particles located in the interior of the Fermi surface; this makes no sense, since most of these quasi particles are poorly defined. In summary, a quasi particle is just an elementary excitation in the neighborhood of $k = k_F$; it gives no information about the ground state.

The energy of the real system is a "functional" of $n(k)$, which we write $E[n(k)]$. For an ideal gas, this functional reduces to the sum of the energies of each particle. In the real case it is, in general, extremely complex. If we alter $n_0(k)$ by an amount $\delta n(k)$, the variation of the energy, to the first order in δn, is given by

$$\delta E = \sum_k \varepsilon_k \, \delta n(k) \tag{1-2}$$

where $\varepsilon = \delta E/\delta n(k)$ is the first functional derivative of E with respect to $n(k)$. For $k > k_F$, ε_k is the variation in the energy when a quasi particle of wave vector k is added. ε_k is thus the *energy* of the quasi particle. Let us at once make a *fundamental* observation: ε_k is defined as a derivative oi the energy with respect to the distribution function. This in no way predicts the value of the total energy. In particular, the energy of the whole system is *not* equal to the sum of the energies of the quasi particles.

For $k = k_F$, ε_k is the energy acquired in adding one particle at the Fermi surface. The state thus obtained is just the ground state of the $(N + 1)$-particle system. We can therefore write

$$\varepsilon_{k_F} = E_0(N + 1) - E_0(N) = \mu$$

The quantity $\mu = \partial E_0/\partial N$ is usually called the "chemical potential." We shall return later to this theorem.

Equation (1–2) implies that the energy of the quasi particles is an additive quantity. This result is only approximate and is valid if terms of order $(\delta n)^2$ can be neglected—that is to say, if the number of quasf particles added or removed is small compared with N. Actually, this conclusion is physically obvious; if there are few quasi particles, there is a large probability that they will be far from one another and thus that they will not interact; their energies simply add. In this argument we have set aside the possibility of *bound* states (such as the exciton, made up of an electron and a "hole" revolving around one another); in any case such bound states have no equivalent in the ideal gas and, as a consequence, remain beyond the scope of the present discussion. In practice, we shall see that at low temperatures there are always few quasi particles. (1–2) is thus generally valid, and the definition of ε_k is unambiguous.

There exists a whole class of gross phenomena for which the knowledge of ε_k is sufficient. But, in general, we need more precise details—for example, the variation of ε_k with the distribution function. We must then push (1–2) one step further by writing

$$\delta E = \sum_k \varepsilon_k^0 \, \delta n(k) + \tfrac{1}{2} \sum_k \sum_{k'} f(k, k') \, \delta n(k) \, \delta n(k') \tag{1-3}$$

$f(k,k')$ is the second functional derivative of E. By construction we have

$$f(k, k') = f(k', k)$$

ε_k° is the energy of the quasi particle k when it alone is present. When it is surrounded by a gas of other quasi particles of density $\delta n(k')$, its energy becomes

$$\varepsilon_k = \varepsilon_k^0 + \sum_{k'} f(k, k')\, \delta n(k') \qquad (1\text{–}4)$$

$f(k,k')$ thus characterizes the variation of ε_k with the distribution function $n(k)$. Note that since the summation over k introduces a factor Ω, $f(k,k')$ must be of order $1/\Omega$.

It is evident that the quasi particles are fermions, since only one can be put into a given level. This single hypothesis is sufficient for applying the classical methods of statistical mechanics (for example, for calculating the number W of possible configurations, the entropy $S = \varkappa \log W_{\max}$). We thus find that the probability $f(\varepsilon)$ of the existence of a quasi particle of energy ε is equal to

$$f(\varepsilon) = \frac{1}{\exp\left[(\varepsilon - \mu)/\varkappa T\right] + 1} \qquad (1\text{–}5)$$

where μ is the chemical potential, which is adjusted so as to normalize the total number of quasi particles to N. When $T \to 0$, $f(\varepsilon)$ tends toward the step function of Fig. 1. Note that at appreciable temperatures ε begins to depend on T, through the distribution function $n(k)$. The relation (1–5) then becomes much less trivial.

Until now we have ignored the spin of the particles. It is very easy to include it in our formulation. In fact, spin plays the same role as momentum in the classification of the levels of the ideal gas. A quasi particle is therefore characterized by its wave vector k and its spin σ. Hereafter we shall often simplify the notation by replacing the pair (k,σ) by the single symbol $\mathbf{k}$. The state of the system is characterized by a distribution function $n(\mathbf{k})$, the energy being given by

$$E = E_0 + \sum_{\mathbf{k}} \varepsilon(\mathbf{k})\, \delta n(\mathbf{k}) + \tfrac{1}{2} \sum_{\mathbf{k}\mathbf{k}'} f(\mathbf{k}, \mathbf{k}')\, \delta n(\mathbf{k})\, \delta n(\mathbf{k}') \qquad (1\text{–}6)$$

Let us assume the system to be isotropic (in particular, without a magnetic field). For reasons of symmetry, the energy $\varepsilon(\mathbf{k})$ then cannot depend on the spin σ. Similarly, the interaction between two quasi particles depends only on the *relative* orientation of their spins σ and σ'.

We can therefore write

$$f(\mathbf{k}, \mathbf{k}') = f(k\sigma, k'\sigma') = f_0(k, k') + f_e(k, k')\,\delta_{\sigma,\sigma'} \qquad (1\text{–}7)$$

The term f_e, which appears only when the spins are parallel, expresses the exchange interaction between the two quasi particles.

2. The Properties of Quasi Particles. Macroscopic Applications

a. Velocity; effective mass; specific heat

Until now we have defined a quasi particle by its momentum $\hbar k$, its spin σ, and its energy ε_k. In order to find its velocity v_k, we form a wave packet, and we calculate the corresponding group velocity. The standard result is:

$$v_{k\alpha} = \frac{1}{\hbar}\frac{\partial \varepsilon_k}{\partial k_\alpha} \qquad (1\text{–}8)$$

(the index α refers to one of the three components of the vector v_k). For an isotropic system, v_k and k are collinear; we can then write

$$v_k = \hbar k/m^* \qquad (1\text{–}9)$$

m^* is the effective mass of the quasi particle. In principle m^* depends on k; we shall be interested only in its value at $k = k_F$.

Starting with m^*, we can easily calculate the density of quasi-particle states per unit energy in the neighborhood of μ. The result is the same as for an ideal gas, with the sole difference that m is replaced by m^*,

$$\left(\frac{d\nu}{d\varepsilon}\right)_\mu = \frac{\Omega\, k_F\, m^*}{\pi^2\, \hbar^2} \qquad (1\text{–}10)$$

Let us turn now to the calculation of the specific heat C_v. By definition we have

$$\delta E = C_v\, \delta T$$

The variation δT has the effect of modifying $n(\mathbf{k})$ by an amount δn which can be deduced from (1–5). Rigorously it would be necessary to take account of the fact that ε_k depends on n and therefore on T; this effect is negligible at low temperatures. Knowing δn, we can find δE by means of the relation

$$\delta E = \sum_{\mathbf{k}} \varepsilon_k \, \delta n(\mathbf{k})$$

The calculations cause no difficulties. If we limit ourselves to terms linear in T, we find that the specific heat is given by

$$C_v = \frac{\pi^2 \varkappa^2 T}{3\,\Omega} \left(\frac{d\nu}{d\varepsilon}\right)_\mu \tag{1-11}$$

(where we recall that $\varkappa$ is the Boltzmann constant).

The specific heat [Eq. (1–11)] takes into account only the thermal excitation of quasi particles. To be complete, we must add to it the corrections arising from the thermal excitation of more complex levels, which have no equivalents in the ideal gas (phonons, bound states, etc.). In general, the contribution from these "pathological" states to the specific heat is negligible at low temperatures. (We know, for example, that the phonons contribute a T^3 term.) It is therefore reasonable to use measurements of the specific heat to calculate m^*.

b. Compressibility of the fermion gas

The ground-state energy of the system E_0, is a function of the particle number N and the volume Ω. For a macroscopic system, we can write

$$E_0 = \Omega\, f(N/\Omega) \tag{1-12}$$

where $N/\Omega = \rho$ is the particle density. The pressure can be found from E_0 by the relation

$$P = -\,\partial E_0/\partial\Omega \tag{1-13}$$

The compressibility of the system is then defined as

$$\chi = -\frac{1}{\Omega}\frac{\partial\Omega}{\partial P}$$

An elementary calculation shows that

$$1/\chi = \rho^2 f''(\rho) \tag{1-14}$$

On the other hand, we have seen that the chemical potential μ is related to E_0 by

$$\mu = \partial E_0/\partial N = f'(\rho) \tag{1–15}$$

Comparing (1–14) and (1–15), we see that

$$\frac{1}{\chi} = N\rho \left.\frac{d\mu}{dN}\right|_{\Omega} \tag{1–16}$$

The calculation of the compressibility is thus found to reduce to that of $d\mu/\partial N$.

The compressiblity χ is directly related to the velocity of sound propagation at low frequency (by "low frequency" we mean a period much longer than the collision time of the quasi particles). Under these conditions, the restoring force can be obtained from χ by a purely macroscopic argument. It is found that the velocity of sound, C, is given by

$$C^2 = \frac{1}{\chi m \rho} = \frac{N}{m}\frac{d\mu}{dN} \tag{1–17}$$

Let us now turn to the calculation of $d\mu/\partial N$. A variation δN in the number of particles is equivalent to a variation δk_F in the Fermi wave vector, deduced from (1–1),

$$\delta k_F = \frac{\pi^2}{\Omega k_F^2}\,\delta N \tag{1–18}$$

The corresponding variation $\delta n(\mathbf{k})$ in the distribution function thus takes the following form (assuming $\delta N > 0$):

$$\delta n(\mathbf{k}) \begin{cases} = 1 & \text{if } k_F < |k| < k_F + \delta k_F \\ = 0 & \text{otherwise} \end{cases} \tag{1–19}$$

Note: To generalize this calculation to an anisotropic system, it is preferable to proceed in the opposite direction and to take μ as the independent variable. In each direction, k_F is a function of μ; if $dk_F/d\mu$ is known in all directions, we can follow the deformation of the Fermi surface as a function of μ. Since N is proportional to the volume enclosed by this surface, we can derive $Nd/d\mu$ and then the compressibility.

When k_F varies, μ changes for two reasons:

Because ε_k depends on k, resulting in a correction $(\partial\varepsilon/\partial|k|)\,\partial k_F$.

Because ε_k depends on the distribution function, which itself changes when k_F varies.

Adding these two contributions, we find

$$\delta\mu = \hbar\, v_k\, \delta k_F + \sum_{\mathbf{k}'} f(\mathbf{k}, \mathbf{k}')\, \delta n(\mathbf{k}') \tag{1-20}$$

Let us use (1–19) and transform the sum over $\mathbf{k}'$ into an integral. For an isotropic system, $f(\mathbf{k},\mathbf{k}')$ depends only on the angle θ between the directions k and k' (let us recall that k and k' are both on the Fermi surface). We thus obtain

$$\frac{d\mu}{dk_F} = \hbar v_k + \sum_{\sigma'} \int d\gamma' \frac{\Omega k_F^2}{8\pi^3} f(\sigma, \sigma', \theta) \tag{1-21}$$

where $d\gamma'$ is the element of solid angle. From this we find

$$\frac{d\mu}{dN} = \frac{\pi^2\hbar^2}{\Omega k_F m^*} + \sum_{\sigma'} \int d\gamma' \frac{f(\sigma, \sigma', \theta)}{8\pi} \tag{1-22}$$

The velocity of sound is given by

$$C^2 = \frac{\hbar^2 k_F^2}{3mm^*} + \sum_{\sigma'} \int d\gamma' \frac{Nf(\sigma, \sigma', \theta)}{8\pi m} \tag{1-23}$$

This result is *rigorous*. If the interaction between quasi particles were neglected, the second term would disappear and we should have an erroneous result.

Here we see the first example of an important phenomenon: although in the expansion (1–6) the interaction term seems to be negligible, in practice it gives a contribution as important as the linear term. The latter, apparently of first order, actually gives a total contribution which is of second order. It is to the great credit of Landau that he noted that, as a consequence, it was necessary to carry the expansion one step further.

In the weak coupling limit, $f \to 0$, and $m^* \to$ m. The velocity of sound then tends to $\hbar k_F/m\sqrt{3}$. This is a well-known result.

c. *Current carried by a quasi particle*

During the adiabatic switching on of the interaction, the total number of particles remains constant. As a consequence, a quasi particle contains *one* bare particle, distributed among a large number of configurations. If these particles are electrons, we shall say that the quasi particle has a charge e (we shall see in the following that the effect of screening is to push this charge to the boundaries of the system; the charge of a quasi particle is not localized).

Let J_k be the current carried by the quasi particle $\mathbf{k}$. J_k is a particle current (for electrons, the electric current is eJ_k). It is tempting to say that J_k is equal to the velocity v_k of the quasi particle; this is *false* for a system of interacting particles. In fact, we then neglect the "backflow" of the medium around the quasi particles. This effect is illustrated in Fig. 2; the quasi particle moves forward with a velocity v_k; the neighboring particles move away from it, which produces a current in the reverse direction, with a roughly dipolar distribution. The current J_k is the sum of two terms: the current v_k of the quasi particle and the backflow of the medium.

To calculate J_k, we first need a precise definition of the current. In an arbitrary state $|\varphi\rangle$ the current J is given by

$$J = \langle \varphi | \sum_i p_i/m | \varphi \rangle \tag{1-24}$$

where p_i is the momentum of the ith particle and m its mass (bare, of course). To measure J, we put ourselves in a reference frame moving with respect to the system with the uniform velocity $\hbar q/m$. Let us emphasize that we are in no way deforming our system; we are simply *looking* at it from a moving reference frame. The Hamiltonian in the rest frame can be written

$$H = \sum_i \frac{p_i^2}{2m} + V \tag{1-25}$$

Let us assume that V depends only on the positions and the relative velocities of the particles; it is not modified by a translation. In the moving

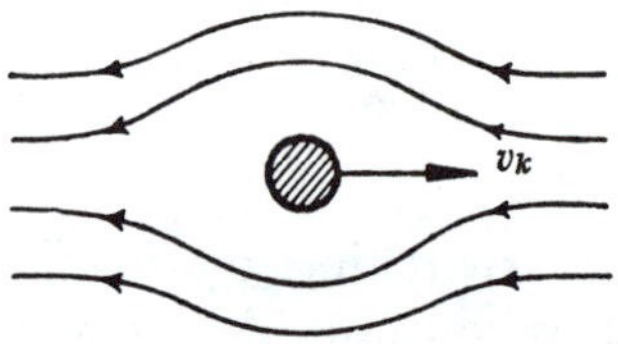

Figure 2

system only the kinetic energy changes; the apparent Hamiltonian becomes

$$H_q = \sum_i \frac{(p_i - \hbar q)^2}{2m} + V$$

$$= H - \hbar q \cdot \sum_i \frac{p_i}{m} + N \frac{\hbar^2 q^2}{2m} \tag{1-26}$$

Let us take the average value of (1–26) in the state $|\varphi\rangle$, and let E_q be the energy of the system as seen from the moving reference frame. When q tends to 0, we find

$$\partial E_q / \partial q_\alpha = -\hbar < \varphi \mid \sum_i p_{i\alpha}/m \mid \varphi > = -\hbar J_\alpha \tag{1-27}$$

(where α refers to one of the three coordinates). (1–27) will constitute our definition of current.

The ground state is invariant with respect to reflection; $\partial E_q / \partial q_\alpha$ is then zero, as well as J_α. Let us now consider the state containing a quasi particle $\mathbf{k}$; according to (1–27), the current J_k carried by the quasi particle is given by

$$J_{k\alpha} = -\frac{1}{\hbar} \frac{\partial \varepsilon_k}{\partial q_\alpha} \tag{1-28}$$

$\partial \varepsilon_k / \partial q_\alpha$ expresses the variation in the energy ε_k, when the coordinate system is displaced with the velocity $\hbar q/m$, or, what is equivalent, when *all* the particles are displaced with the velocity $-\hbar q/m$. This amounts to displacing the distribution in reciprocal space by an amount $-q$. The situation is illustrated by Fig. 3, where the equilibrium distribution is indicated by a solid line and the distribution after translation by a dotted line.

ε_k varies for two reasons:

(a) Because the wave vector k varies by an amount $-q$.

(b) Because the distribution of the quasi particles has changed, with creation in region A and destruction in region B.

Let $\delta n(\mathbf{k}')$ be the corresponding variation in the distribution function. The total change in ε_k is

$$\delta \varepsilon_k = -\hbar q_\alpha v_{k\alpha} + \sum_{\mathbf{k}'} f(\mathbf{k}, \mathbf{k}')\, \delta n(\mathbf{k}') \tag{1-29}$$

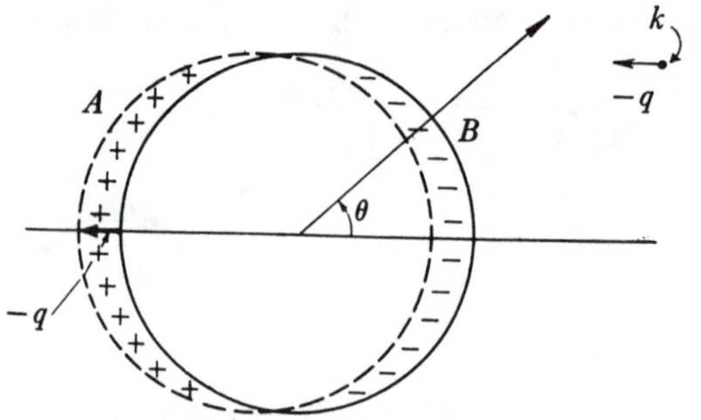

Figure 3

We finally obtain

$$J_{k\alpha} = v_{k\alpha} - \sum_{\mathbf{k}'} f(\mathbf{k}, \mathbf{k}') \frac{\delta n(\mathbf{k}')}{\hbar q_\alpha} \tag{1–30}$$

The second term of (1–30) is precisely the backflow that we sought to determine.

Let us now calculate $\delta n(\mathbf{k})$ for a fixed direction of k, making an angle θ with q. $\delta n(\mathbf{k})$ is given by Fig. 4. All the properties of interest are continuous functions of $|k|$. We can therefore replace δn by a Dirac δ function and write

$$\begin{aligned} \sum_{\mathbf{k}'} \delta n(\mathbf{k}') &= \frac{\Omega}{8\pi^3} \sum_{\sigma'} \int d^3 k' (-q \cos \theta)\, \delta(|k'| - k_F) \\ &= -\frac{\Omega}{8\pi^3} \sum_{\sigma'} \int d^3 k'\, \hbar q \cdot v_{k'}\, \delta(\varepsilon_{k'} - \mu) \end{aligned} \tag{1–31}$$

(1–30) can now be written as

$$J_{k\alpha} = v_{k\alpha} + \frac{\Omega}{8\pi^3} \sum_{\sigma'} \int d^3 k'\, v_{k'\alpha}\, \delta(\varepsilon_{k'} - \mu)\, f(\mathbf{k}, \mathbf{k}') \tag{1–32}$$

which we can put in the form

$$\mathbf{J}_k = \mathbf{v}_k + \sum_{\mathbf{k}'} f(\mathbf{k}, \mathbf{k}')\, \mathbf{v}_{k'}\, \delta(\varepsilon_{k'} - \mu) \tag{1–33}$$

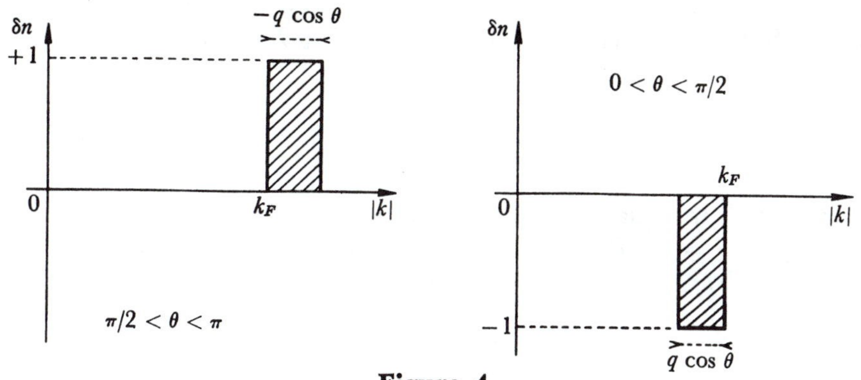

Figure 4

(1–33) is a very formal expression, since the Dirac δ function makes sense only when we go over to the integral (1–32).

(1–33) is a *rigorous* and very general result which remains valid for anisotropic systems and even for real solids (with a periodic potential). The phenomenon of backflow manifests itself very clearly. We cannot overemphasize its importance.

Let us now concern ourselves with the particular case of a translationally invariant system. In this case the total current is a constant of the motion, which commutes with the interaction V and which, as a consequence, does not change when V is switched on adiabatically. Let us consider, in particular, the state containing one quasi particle **k**; the total current J_k is the same as for the ideal system.

$$J_k = \hbar k/m \tag{1–34}$$

This result is a direct consequence of Galilean invariance. Comparing (1–34) with the general result (1–32), we obtain an identity which, for an isotropic system, takes the very simple form

$$\frac{1}{m} = \frac{1}{m^*} + \frac{\Omega k_F}{8\pi^3\hbar^2}\sum_{\sigma'}\int d\gamma' f(\sigma, \sigma', \theta)\cos\theta \tag{1–35}$$

(where θ is the angle between k and k'). (1–35) should be compared with (1–22). Both are valuable for determining f.

(1–35) was first established by Landau, using a slightly different approach. Landau *assumes* that the *total* current J is given by

$$J = \sum_{\mathbf{k}} n(\mathbf{k})\, v_k \tag{1-36}$$

where $n(\mathbf{k})$ is the quasi particle distribution function. In the ground state $n = n_0$ and $J = 0$. If n_0 is varied by an amount δn, J is given, to first order in δn, by

$$\begin{cases} J = \sum_{\mathbf{k}} \delta n(\mathbf{k})\, v_k + \sum_{\mathbf{k}} n_0(\mathbf{k})\, \delta v_k \\ \delta v_{k\alpha} = \dfrac{1}{\hbar}\dfrac{\partial}{\partial k_\alpha}\left[\sum_{\mathbf{k}'} f(\mathbf{k}, \mathbf{k}')\, \delta n(\mathbf{k}')\right] \end{cases} \tag{1-37}$$

Integrating the second term of J by parts, we obtain

$$J_\alpha = \sum_{\mathbf{k}} \delta n(\mathbf{k})\, v_{k\alpha} - \sum_{\mathbf{k}\,\mathbf{k}'} \frac{\partial n_0}{\partial k_\alpha}\frac{1}{\hbar} f(\mathbf{k}, \mathbf{k}')\, \delta n(\mathbf{k}') \tag{1-38}$$

If we note that

$$\partial n_0/\partial k_\alpha = -\, \hbar v_{k\alpha}\, \delta(\varepsilon_k - \mu) \tag{1-39}$$

we see that

$$J = \sum_{\mathbf{k}} \delta n(\mathbf{k})\, J_k \tag{1-40}$$

(1–40) is equivalent to (1–33), since to first order in δn the quasi particle currents are additive.

Landau's approach thus gives the correct result. However, it has the inconvenience of involving the distribution function $n(\mathbf{k})$, which makes no sense; only $\partial n(\mathbf{k})$ is defined unambiguously. Still, his approach remains very suggestive.

d. Spin-dependent properties: Pauli paramagnetism

In the presence of a magnetic field H, a free particle has its energy displaced by $\beta\sigma_z H$, where $\beta = e\hbar/mc$, $\sigma_z = \pm\frac{1}{2}$ is the component of spin along H. In the case of a real gas, we must add to this displacement a correction, arising from the fact that the magnetic field modifies the distribution function $n(\mathbf{k})$, and therefore the quasi particle energies. Let

$\delta n(\mathbf{k})$ be the variation in n caused by the magnetic field. The variation in the energy of the quasi particle $\mathbf{k}$ can then be written

$$\delta\varepsilon(\mathbf{k}) = -\beta\,\sigma_z H + \sum_{\mathbf{k}'} f(\mathbf{k}, \mathbf{k}')\,\delta n(\mathbf{k}') \tag{1-41}$$

We shall see that $\delta\varepsilon(\mathbf{k})$ has the following form,

$$\delta\varepsilon(\mathbf{k}) = -\eta\sigma_z H \tag{1-42}$$

where η is a positive constant. This expression is odd in σ_z. It follows that the distribution function takes the form indicated in Fig. 5, where δk_F is given by·

$$\delta k_F = \left(\frac{d\varepsilon}{d\,|\,k\,|}\right)^{-1} |\,\delta\varepsilon\,| = \frac{m^*}{\hbar^2 k_F}\frac{\eta H}{2} \tag{1-43}$$

Note that the average value of k_F remains constant to the order considered (in other words, the variation $\delta\mu$ in the chemical potential is of order H^2 and therefore negligible).

By using (1–42) and (1–43), we can write Eq. (1–41) in the form

$$\delta\varepsilon(\mathbf{k}) = -\beta\sigma_z H + \frac{\Omega m^* k_F}{8\pi^3\hbar^2}\frac{\eta H}{2}\int d\gamma' \left\{ f(\theta, \sigma, +\tfrac{1}{2}) - f(\theta, \sigma, -\tfrac{1}{2}) \right\} \tag{1-44}$$

Let us now refer to the expression (1–7) for f. We see that only the second term contributes. The relation (1–44) finally reduces to

$$\eta = \beta - \frac{\Omega m^* k_F}{8\pi^3\hbar^2}\,\eta \int d\gamma'\, f_e(\theta) \tag{1-45}$$

which gives η directly.

Let us now turn to the calculation of the magnetic moment M, and thus of the susceptibility per unit volume, χ_M. We have

$$M = \Omega\chi_M H = \sum_{\mathbf{k}} \beta\sigma_z\,\delta n(\mathbf{k}) \tag{1-46}$$

from which we find, referring to Fig. 5,

$$M = \beta\Omega\,\frac{4\pi k_F^2\,\delta k_F}{(2\pi)^3} = \frac{m^* k_F \Omega}{4\pi^2\hbar^2}\,\beta\eta H \tag{1-47}$$

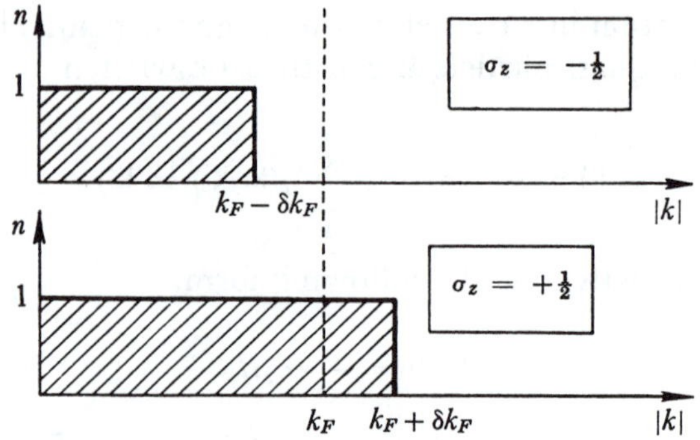

Figure 5

Combining (1–45) and (1–47), we finally obtain

$$\frac{1}{\chi_M} = \frac{1}{\beta^2}\left\{\frac{4\pi^2\hbar^2}{m^*k_F} + \frac{\Omega}{2\pi}\int d\gamma'\, f_e(\theta)\right\} \qquad \textit{(1–48)}$$

which allows us to express the magnetic susceptibility as a function of m^* and f_e.

Before leaving this question, let us point out that, in the absence of an exchange interaction ($f_e = 0$), χ_M can be very simply expressed as a function of the coefficient α in the specific heat

$$C_v = \alpha T$$

Namely, we have

$$\chi_M = \frac{3\alpha\beta^2}{4\pi^2\varkappa^2}$$

In practice, the exchange interaction is *always* present for a real gas, so that the above equation is *not valid.*

e. Stability of the ground state

We assumed, at the beginning of this chapter, that the distribution of the ground state was the same as that of an ideal gas—that is to say, an isotropic step function. This assertion will be valid only if the state so defined is stable, corresponding to a minimum of the free energy,

$$F = E - \mu N$$

This leads us to study the stability of the Fermi surface. Let us define a direction in k space by its polar angles θ and φ, and let us displace the Fermi surface of spin σ in this direction by an infinitesimal amount $u(\theta, \varphi, \sigma)$. According to (1–3), the resulting variation in free energy is given by

$$\delta F = \frac{\Omega}{8\pi^3} \sum_{\sigma} \int d\gamma \int_{k_F}^{k_F+u} (\varepsilon_k - \mu)\, k^2\, dk + \frac{1}{2} \cdot \frac{\Omega^2}{64\pi^6} \sum_{\sigma\sigma'} \iint d\gamma\, d\gamma' \int_{k_F}^{k_F+u} k^2\, dk \int_{k_F}^{k_F+u'} k'^2\, dk'\, f(\mathbf{k}, \mathbf{k}') \tag{1–49}$$

Note that the first term in (1–49) is actually of order u^2, since $(\varepsilon_k - \mu)$ is zero for $k = k_F$. The integration over k and k' is easy; it leads to

$$\delta F = \frac{\Omega k_F^3 \hbar^2}{16\pi^3 m^*} \sum_{\sigma} \int d\gamma\, u^2(\theta, \varphi, \sigma) + \frac{\Omega^2 k_F^4}{128\pi^6} \sum_{\sigma\sigma'} \iint d\gamma\, d\gamma'\, u(\theta, \varphi, \sigma)\, u(\theta', \varphi', \sigma') \times f(\xi, \sigma, \sigma') \tag{1–50}$$

where ξ is the angle between the directions (θ,φ) and (θ',φ').

In order to analyze this expression, we decompose u and f into spherical harmonics. For simplicity we assume u and f to be independent of the spin variables; thus we set

$$\begin{cases} u = \sum_{lm} u_{lm} Y_{lm}(\theta, \varphi) \\ f(\xi) = \sum_{l} f_l P_l(\cos \xi) \end{cases} \tag{1–51a}$$

The coefficients f_l are obtained by the well-known formula

$$f_l = \frac{2l + 1}{4\pi} \int f(\xi)\, P_l(\cos \xi)\, d\gamma \tag{1–51b}$$

Inserting these expansions into the expression for δF and using the orthonormality relations for the spherical harmonics, we obtain the relation

$$\delta F = \frac{\Omega k_F^3}{(2\pi)^3} \frac{\hbar^2}{m^*} \sum_{lm} | u_{lm} |^2 \frac{4\pi}{2l + 1} \frac{(l + m)\,!}{(l - m)\,!} \left[1 + \frac{m^*}{2l + 1} \frac{\Omega k_F f_l}{\pi^2 \hbar^2}\right] \tag{1–52}$$

For δF to be always positive, it is necessary and sufficient that all of the following conditions be satisfied:

$$1 + \frac{m^*}{2l+1}\frac{\Omega k_F f_l}{\pi^2\hbar^2} > 0 \qquad (1\text{–}53)$$

(For $l = 0$ and $l = 1$ these conditions imply, respectively, that C^2 and m are positive.) If $f(\mathbf{k},\mathbf{k}')$ depends on spin, we must also concern ourselves with the separation of the Fermi surfaces of the two spins; this brings in a new set of stability conditions, analogous to (1–53). The first of these conditions ($l = 0$) ensures that the magnetic susceptibility is positive —that is to say, that the system is not ferromagnetic.

3. Transport Properties of the Quasi Particles: Collective Modes

a. Nonuniform distributions of quasi particles

Until now we have been interested only in homogeneous systems, whose wave functions are translationally invariant; the state is completely characterized by the distribution function $\delta n(\mathbf{k})$. Let us now consider an inhomogeneous excited state; the properties of the system vary from point to point. We shall assume that this deformation occurs on a *macroscopic* scale; in other words, the state remains homogeneous over a microscopic distance, such as the average interparticle spacing or the range of the forces. We can then define a local distribution function $\delta n(\mathbf{k},r)$, giving the quasi-particle distribution in a unit volume centered at the point r. (The discrete values of $\mathbf{k}$ refer to this same unit volume.)

Again the energy is a functional of the distribution function $\delta n(\mathbf{k},r)$. By analogy with (1–3), we shall write

$$\delta E = \sum_{\mathbf{k}} \int d^3r\, \varepsilon_0(\mathbf{k}, r)\, \delta n(\mathbf{k}, r) + \tfrac{1}{2}\sum_{\mathbf{k}\mathbf{k}'} \iint d^3r\, d^3r'\, f(\mathbf{k}r, \mathbf{k}'r')\, \delta n(\mathbf{k}, r)\, \delta n(\mathbf{k}', r') \qquad (1\text{–}54)$$

where f is now defined for unit volume. The ground state being translationally invariant, the energy $\varepsilon_0(\mathbf{k},r)$ is independent of r, equal to ε_k^0, whereas the interaction $f(\mathbf{k},r;\mathbf{k}',r')$ depends only on the distance $(r - r')$.

At this stage, we must distinguish between two cases. Let us first consider short-ranged forces; the interaction f decreases rapidly as soon as $|r - r'|$ is greater than the range. In the interval where f is appreciable,

$\delta n(\mathbf{k},r)$ and $\delta n(\mathbf{k},r')$ are practically equal. We can then write

$$\left\{\begin{aligned} \delta E &= \sum_{\mathbf{k}} \int d^3r\, \varepsilon_k^0\, \delta n(\mathbf{k}, r) + \tfrac{1}{2} \sum_{\mathbf{k},\mathbf{k}'} \int d^3r\, f(\mathbf{k}, \mathbf{k}')\, \delta n(\mathbf{k}, r)\, \delta n(\mathbf{k}', r) \\ f(\mathbf{k}, \mathbf{k}') &= \int d^3r'\, f(\mathbf{k}r, \mathbf{k}'r') \end{aligned}\right. \tag{1-55}$$

If a is the range of the interaction and Δr the scale of the inhomogeneities, (1–55) involves an error of order $a/\Delta r$. For macroscopic phenomena, this error is negligible.

(1–55) can be written in the form

$$\left\{\begin{aligned} \delta E &= \int d^3r\, \delta E(r) \\ \delta E(r) &= \sum_{\mathbf{k}} \varepsilon_k^0\, \delta n(\mathbf{k}, r) + \tfrac{1}{2} \sum_{\mathbf{k},\mathbf{k}'} f(\mathbf{k}, \mathbf{k}')\, \delta n(\mathbf{k}, r)\, \delta n(\mathbf{k}', r) \end{aligned}\right. \tag{1-56}$$

At each point the system is thus quasi-homogeneous and is described by Eq. (1–3), with a *local* distribution function $\delta n(\mathbf{k},r)$. We can define an energy density $\delta E(r)$, the total energy being obtained by integrating δE over all space. Note that any two different regions of the system are completely independent.

The situation is different for long-range forces, such as the Coulomb interaction. In this case very widely separated quasi particles can interact, and we can no longer reduce (1–54) to (1–55). In order to avoid this difficulty, Silin has proposed the following device. We decompose the Coulomb interaction into two parts:

(1) Electrostatic interaction between the *average* densities at the points r and r'. The corresponding contribution to (1–54) can be written as

$$\tfrac{1}{2} \iint d^3r\, d^3r' \frac{e^2}{|r - r'|} \sum_{\mathbf{k}} \delta n(\mathbf{k}, r) \sum_{\mathbf{k}'} \delta n(\mathbf{k}', r') \tag{1-57}$$

(where e is the charge of the particles). This term is of infinite range; we shall treat it explicitly by introducing an electric field due to space charge, $\mathscr{E}_H$, given by

$$\operatorname{div} \mathscr{E}_H = 4\pi e \sum_{\mathbf{k}} \delta n(\mathbf{k}, r) \tag{1-58}$$

We shall consider $\mathscr{E}_H$ as an external force, tending to drive the quasi particles. We are thus led to solve simultaneously the transport equation (see the following paragraphs) and Poisson's equation (1–58).

(2) To the "Hartree" contribution (1–57), we must add corrections expressing the correlations between particles. These correlation corrections are short-range, the range being of the order of the screening length. We can thus treat them by Landau's method and express them in the form of an interaction $f(\mathbf{k},\mathbf{k}')$. In other words, density fluctuations are negligible when viewed from a distance; at large separations only the interactions between the average space charges remain. At short distances fluctuations give rise to important corrections, described by $f(\mathbf{k},\mathbf{k}')$.

The Landau theory thus applies to any inhomogeneous system, whatever the nature of the forces may be, on condition that the deformation occur on a macroscopic scale.

b. Boltzmann equation and applications

Let us consider a quasi particle $\mathbf{k}$ located at the point r. It has an energy

$$\varepsilon(\mathbf{k}, r) = \varepsilon_k^0 + \sum_{\mathbf{k}'} f(\mathbf{k}, \mathbf{k}')\, \delta n(\mathbf{k}', r) \tag{1-59}$$

In order to study the *scattering* of these particles, Landau proposed considering $\varepsilon(\mathbf{k},r)$ as the *classical* Hamiltonian of a quasi particle, thereby neglecting the interaction between several quasi particles. In this scheme the interaction has only an *average* effect on the energy which is absorbed into the definition (1–59) of $\varepsilon(\mathbf{k},r)$. This is a very bold hypothesis. We shall see in Chap. 6 that it is rigorously correct; on a macroscopic scale, fluctuations are negligible.

Hereafter, each quasi particle is an isolated entity, analogous to the molecule of a dilute gas. Its velocity v has components

$$v_{k\alpha} = \frac{1}{\hbar}\frac{\partial \varepsilon}{\partial k_\alpha}$$

Furthermore, the variation of ε with r is equivalent to a force $\mathscr{F}$ of components

$$\mathscr{F}_\alpha = -\,\partial\varepsilon/\partial r_\alpha \tag{1-60}$$

$\mathscr{F}$ is a *diffusion* force which tends to push particles towards regions of minimum energy.

To study the time development of the distribution function $n(\mathbf{k},r)$, we count the number of particles which enter and the number which leave an element of volume in phase space $(\mathbf{k},r)$. The calculation is a standard one and leads to the well-known result

$$\frac{\partial n\,(\mathbf{k}, r, t)}{\partial t} + \frac{1}{\hbar}\left\{\frac{\partial n}{\partial r_\alpha}\frac{\partial \varepsilon}{\partial k_\alpha} - \frac{\partial n}{\partial k_\alpha}\frac{\partial \varepsilon}{\partial r_\alpha}\right\} = 0 \qquad \textit{(1-61)}$$

(1–61) is the Boltzmann equation, which regulates the "flow" of quasi particles in phase space.

In the form (1–61), our description of transport phenomena still lacks an essential element—collisions between quasi particles. In fact, the Landau theory systematically ignores the existence of *real* transitions from one state to another. In certain cases, such as thermal conductivity and viscosity, the collisions play a fundamental role, since they limit the response of the system to the external excitation. We must therefore take them into account. We shall rewrite (1–61) in the form

$$\frac{\partial n}{\partial t} + \frac{1}{\hbar}\left\{\frac{\partial n}{\partial r_\alpha}\frac{\partial \varepsilon}{\partial k_\alpha} - \frac{\partial n}{\partial k_\alpha}\frac{\partial \varepsilon}{\partial r_\alpha}\right\} = I(n) \qquad \textit{(1-62)}$$

where $I(n)$ is the "collision integral."

At low temperatures, collisions are rare (this is what justifies the use of the Landau theory). The system of quasi particles thus has all the properties of a *dilute gas*, with, however, a new feature: the diffusion force $-\ \partial\varepsilon/\partial r$.

Let us now assume the system to be subjected to an external force. This force acts on the quasi particles and, as a consequence, modifies the Boltzmann equation. In general, the real force on the quasi particle is different from that on the bare particle and remains unknown (this is a manifestation of more or less complicated screening.) Fortunately, electromagnetic forces do not have this problem. Let us consider a gas of particles of charge e and apply an external field $\mathscr{E}$. According to the discussion given above, we must add to $\mathscr{E}$ the space-charge field $\mathscr{E}_H$. The local field is thus

$$\mathcal{E}_L = \mathcal{E} + \mathcal{E}_H \qquad \textit{(1-63)}$$

The force applied to each quasi particle is

$$\mathscr{F}_{\text{ext}} = e\,\mathcal{E}_L \qquad \textit{(1-64)}$$

(let us recall that the charge of a quasi particle is e, identical to that of a bare particle). The effect of screening is contained in the space-charge field $\mathscr{E}_H$ and does not enter into (1–64). This argument remains very qualitative; we shall prove (1–64) rigorously in Chap. 6. In any case, we must add to the Boltzmann equation the driving term due to the external force $\mathscr{F}$.

(1–62) is a very general equation, which allows rigorous treatment of any inhomogeneity on a macroscopic scale. In general, we are restricted by our ignorance of the collision integral $I(n)$. In the following paragraphs, we shall treat some applications for which $I(n)$ does not appear. Let us point out, without going into more detail, that Abrikosov and Khalatnikov have analyzed the problem of collisions and have calculated the thermal conductivity and the viscosity of a Fermi liquid (He^3, in their case).

In its present form (1–62) suffers from the inconvenience already mentioned above; it involves the distribution function $n(\mathbf{k})$ throughout k space, including regions in which it makes no sense. Actually this difficulty is only an apparent one. Let us write

$$n(\mathbf{k}) = n_0(\mathbf{k}) + \delta n(\mathbf{k})$$

and consider first an isolated system, without external forces. Only δn depends on the time t and the position r. To first order in δn Eq. (1–62) can then be written as

$$\frac{\partial\, \delta n}{\partial t} + \frac{1}{\hbar}\left\{\frac{\partial\, \delta n}{\partial r_\alpha}\frac{\partial \varepsilon_k^0}{\partial k_\alpha} - \frac{\partial n_0}{\partial k_\alpha}\frac{\partial \varepsilon}{\partial r_\alpha}\right\} = I(n) \qquad (1\text{–}65)$$

According to (1–59) we have

$$\frac{\partial \varepsilon}{\partial r_\alpha} = \sum_{\mathbf{k}'} f(\mathbf{k}, \mathbf{k}')\, \frac{\partial\, \delta n(\mathbf{k}')}{\partial r_\alpha} \qquad (1\text{–}66)$$

The linearized Boltzmann equation then takes the final form

$$\frac{\partial\, \delta n(\mathbf{k}, r)}{\partial t} + v_{k\alpha}\frac{\partial\, \delta n}{\partial r_\alpha} + v_{k\alpha}\, \delta(\varepsilon_k - \mu) \sum_{\mathbf{k}'} f(\mathbf{k}, \mathbf{k}')\, \frac{\partial\, \delta n(\mathbf{k}', r)}{\partial r_\alpha} = I(n) \qquad (1\text{–}67)$$

Only the quantities δn and $\delta(\varepsilon_k - \mu)$ are involved. We are automatically limited to the immediate neighborhood of the Fermi surface. This conclusion remains true if we introduce an external force $\mathscr{F}_{\text{ext}}$ (the driving term due to $\mathscr{F}_{\text{ext}}$ contains a factor $\partial n_0/\partial k_\alpha$).

*c. **Definition of fluxes—continuity equation***

At each point the quasi particle density varies by

$$\delta n(r) = \sum_{\mathbf{k}} \delta n(\mathbf{k}, r) \tag{1-68}$$

The density of bare particles varies according to the same law, since the quasi particles and the bare particles have the same Fermi surface—that is to say, the same density. Furthermore, the total current density is given by (1–40) and (1–33),

$$J_\alpha(r) = \sum_{\mathbf{k}} \delta n(\mathbf{k}, r) \left\{ v_{k\alpha} + \sum_{\mathbf{k}'} f(\mathbf{k}, \mathbf{k}')\, v_{k'\alpha} \delta(\varepsilon_{k'} - \mu) \right\} \tag{1-69}$$

Conservation of the total number of particles implies the continuity equation,

$$\frac{\partial \delta n(r)}{\partial t} + \operatorname{div} J = 0 \tag{1-70}$$

We are going to verify this relation.

In order to do this, let us sum (1–67) over all values of **k**. The collision integral gives no contribution, since each collision conserves the total number of particles. Let us interchange the indices **k** and **k**′ in the interaction term. We thus obtain

$$\frac{\partial \delta n(r)}{\partial t} + \sum_{\mathbf{k}} \frac{\partial \delta n(\mathbf{k}, r)}{\partial r_\alpha} \left\{ v_{k\alpha} + \sum_{\mathbf{k}'} f(\mathbf{k}, \mathbf{k}')\, v_{k'\alpha} \delta(\varepsilon_{k'} - \mu) \right\} = 0 \tag{1-71}$$

Comparing this with (1–69), we see that (1–71) is just the desired continuity equation.

Let us now consider the energy density E, or, more exactly, the variation δE from the ground-state energy E_0. To first order we have

$$\delta E(r) = \sum_{\mathbf{k}} \varepsilon_k^0\, \delta n(\mathbf{k}, r)$$

Energy flows within the system; at each point one can define an energy flux $Q(r)$, related to δE by the continuity equation

$$\frac{\partial \delta E}{\partial t} + \operatorname{div} Q = 0 \tag{1–72}$$

We propose to find an expression for Q. In order to do this, let us multiply (1–67) by ε_k° and sum over $\mathbf{k}$. The collision integral gives no contribution, since collisions conserve energy. We thus obtain

$$\frac{\partial \delta E}{\partial t} + \frac{\partial}{\partial r_\alpha}\left\{\sum_{\mathbf{k}} v_{k\alpha}\, \delta n(\mathbf{k}, r)\, \varepsilon_k^0 + \sum_{\mathbf{k}} v_{k\alpha}\, \delta(\varepsilon_k - \mu)\, \varepsilon_k^0 \sum_{\mathbf{k}'} f(\mathbf{k}, \mathbf{k}')\, \delta n(\mathbf{k}', r)\right\} = 0 \tag{1–73}$$

The energy flux can then be written

$$Q(r) = \sum_{\mathbf{k}} Q_k\, \delta n(\mathbf{k}, r) \tag{1–74}$$

where the flux Q_k of a single quasi particle is given by

$$Q_{k\alpha} = v_{k\alpha}\varepsilon_k^0 + \sum_{\mathbf{k}'} f(\mathbf{k}, \mathbf{k}')\, v_{k'\alpha}\, \delta(\varepsilon_{k'} - \mu)\, \varepsilon_{k'}^0 \tag{1–75}$$

In principle, Q_k is different from $J_k \varepsilon_k^\circ$. If, however, we remain in the immediate vicinity of the Fermi surface, we have $\varepsilon_k^\circ \approx \mu$, so that

$$Q_k \approx \mu\, J_k \tag{1–76}$$

which seems quite obvious. It would be interesting to make a more careful analysis and to study the transport of free energy, $F = E - \mu N$. This is difficult when one is in the vicinity of the Fermi surface; we shall therefore put this problem aside.

An analogous method allows us to calculate the "momentum-flux" tensor. We leave to the reader the problem of finding an expression for it.

d. Response of an electron gas to an electric field

Let $\mathscr{E}$ be the external field applied to the system. The local field $\mathscr{E}_L$ is given by (1–63). We shall assume, for simplicity, that $\mathscr{E}$ is periodic in space and time, having the form

$$\mathcal{E}(r, t) = \mathcal{E} \exp[i(qr - \omega t)]$$

The deformation $\delta n(\mathbf{k},r)$ varies according to the same law. For Landau's theory to be applicable, the wave vector q must be much less than k_F and the frequency ω much less than μ—in other words, we limit ourselves to *macroscopic* perturbations. We shall assume, in addition, that the frequency ω is much greater than the collision frequency ν, so that we can neglect the collision integral $I(n)$. At low temperatures, ν varies as T^2; it is therefore perfectly possible to realize the conditions

$$\nu \ll \omega \ll \mu \tag{1-77}$$

In this restricted region, the Boltzmann equation (1–62) becomes (after linearization)

$$(q\cdot v_k - \omega)\,\delta n(\mathbf{k}) + q\cdot v_k\,\delta(\varepsilon_k - \mu)\sum_{\mathbf{k}'} f(\mathbf{k},\mathbf{k}')\,\delta n(\mathbf{k}') + ie\mathcal{E}_L\cdot v_k\,\delta(\varepsilon_k - \mu) = 0 \tag{1-78}$$

(1–78) is an integral equation whose solution is in general difficult. This is a new feature, arising from the interaction term $f(\mathbf{k},\mathbf{k}')$, which does not appear in the usual treatments of the Boltzmann equation.

Let us assume that the solution $\delta n(\mathbf{k})$ is known. The electric current density $I(r,t)$ is then given by

$$I(r,t) = e\sum_{\mathbf{k}} \delta n(\mathbf{k})\; J_k \exp[i(qr - \omega t)] \tag{1-79}$$

I is related to $\mathcal{E}_L$ by an equation of the form

$$I_\alpha = \sigma_{\alpha\beta}\mathcal{E}_{L\beta}$$

where $\sigma_{\alpha\beta}(q,\omega)$ is the conductivity tensor. In principle, (1–78) and (1–79) allow us to calculate σ for small values of q and ω, whatever the ratio qv_F/ω may be. We shall discuss this question in detail in Chap. 6. For the moment, we treat only some simple cases.

Let us first assume that $q = 0$—the perturbation is uniform throughout the system. The solution of (1–78) is then trivial,

$$\delta n(\mathbf{k})\Big|_{q=0} = \frac{ie\mathcal{E}_L\cdot v_k}{\omega}\,\delta(\varepsilon_k - \mu) \tag{1-80}$$

Inserting this into (1–79), and assuming that $J_k = \hbar k/m$ (true for translationally invariant systems), we obtain the conductivity,

$$\sigma_{\alpha\beta}(0, \omega) = \mathrm{i}\,\frac{Ne^2}{m\omega}\,\delta_{\alpha\beta} \tag{1–81}$$

Note that it is m, and not m^*, which enters (1–81). This result is well known and can be proved directly. It is a direct consequence of translational invariance.

Let us now consider the other limit, $\omega/q \to 0$. A priori, (1–78) does not seem to have a simple solution. However, let us define

$$\overline{\delta n}(\mathbf{k}) = \delta n(\mathbf{k}) + \delta(\varepsilon_k - \mu) \sum_{\mathbf{k}'} f(\mathbf{k}, \mathbf{k}')\, \delta n(\mathbf{k}') \tag{1–82}$$

The transport equation becomes

$$q \cdot v_k\, \overline{\delta n}(\mathbf{k}) - \omega\, \delta n(\mathbf{k}) + \mathrm{i}e\mathcal{E}_L \cdot v_k\, \delta(\varepsilon_k - \mu) = 0 \tag{1–83}$$

On the other hand, expression (1–33) for J_k allows us to write

$$\begin{aligned} \sum_{\mathbf{k}} \delta n(\mathbf{k})\, J_k &= \sum_{\mathbf{k}} \delta n(\mathbf{k}) \left\{ v_k + \sum_{\mathbf{k}'} f(\mathbf{k}, \mathbf{k}')\, v_{k'}\, \delta(\varepsilon_{k'} - \mu) \right\} \\ &= \sum_{\mathbf{k}} \overline{\delta n}(\mathbf{k})\, v_k \end{aligned} \tag{1–84}$$

The total current density I can then be put into the form

$$I(r, t) = e \sum_{\mathbf{k}} \overline{\delta n}(\mathbf{k})\, v_k \exp[\mathrm{i}(qr - \omega t)] \tag{1–85}$$

Thus, to obtain I, we can work directly with $\overline{\delta n}$; we have a choice between the two expressions (1–79) and (1–85).

When $\omega \to 0$, Eqs. (1–83) and (1–85) are required. From (1–83) we obtain

$$\overline{\delta n}(\mathbf{k}) \Big|_{\omega=0} = -\,\mathrm{i}e\,\frac{\mathcal{E}_L . v_k\, \delta(\varepsilon_k - \mu)}{q \cdot v_k} \tag{1–86}$$

When $\mathscr{E}$ is parallel to q (longitudinal field), (1–86) takes the very simple form

$$\overline{\delta n} = -\frac{ie\,|\,\mathcal{E}_L\,|}{|\,q\,|}\,\delta(\varepsilon_k - \mu)$$

For reasons of symmetry, I is zero. Physically, the field $\mathscr{E}$ is equivalent to a static space charge. The gas polarizes so as to screen this external charge. In equilibrium the charge density is fixed, and the current I is therefore zero.

When $\mathscr{E}$ is perpendicular to q, the situation is altogether different. In this case the second member of (1–86) is singular when $q{\cdot}v_k = 0$; there is then a resonance between $\mathscr{E}$ and the system and hence real transitions and dissipation of energy. To avoid this problem, we must switch on the field adiabatically, introducing into $\mathscr{E}$ a factor $e^{\eta t}$, where η is a positive infinitesimal (see Chap. 2). This has the effect of modifying the denominator of (1–86), which becomes

$$\frac{1}{q\cdot v_k - i\eta} = P(1/q\cdot v_k) + i\pi\delta(q\cdot v_k) \tag{1–87}$$

Again, the principal part gives no contribution to the current, for reasons of symmetry. Only the δ function enters, giving a conductivity

$$\sigma_{\alpha\alpha}(q, 0) = \frac{3\pi N e^2}{4\hbar q k_F} \qquad (q_\alpha = 0) \tag{1–88}$$

This result is real, and therefore purely dissipative. Physically, these conditions are realized in the anomalous skin effect. Note that the mass does not enter (1–88); the skin effect is therefore the same, whether or not the Coulomb interaction between electrons is taken into account. This is a result of great practical importance.

Let us emphasize that these results are rigorous, to within an error of order q/k_F or ω/E_F. These examples demonstrate the power of the Landau theory, which, with a very simple formalism, describes a large number of phenomena.

Before leaving this problem, let us point out that δn and $\overline{\delta n}$ have a very simple physical interpretation. These two quantities always contain a factor $\delta(\varepsilon_k - \mu)$. We can therefore set

$$\begin{cases} \delta n(\mathbf{k}) = -\dfrac{\partial n_0}{\partial\,|\,k\,|}\,u(\mathbf{k}) = \delta(\varepsilon_k - \mu)\,\hbar\,|\,v_k\,|\,u(\mathbf{k}) \\ \overline{\delta n}(\mathbf{k}) = \delta(\varepsilon_k - \mu)\,\hbar\,|\,v_k\,|\,\overline{u}(\mathbf{k}) \end{cases} \tag{1–89}$$

Actually, $\delta(\varepsilon_k - \mu)$ is just a convenient approximation for a unit step function (see Fig. 4). The correction $\delta n(\mathbf{k})$ reduces to a *deformation of the Fermi surface*, n always remaining equal to either 0 or 1. Let S_F° be the Fermi surface at equilibrium, characterized by the relation

$$\varepsilon_k^0 = \mu$$

and S_F the Fermi surface in the presence of the perturbation $\delta n(\mathbf{k})$. It is clear from (1–89) that $u(\mathbf{k})$ is just the displacement of the Fermi surface of spin σ in the direction k with respect to S_F°.

It remains to interpret $\bar{u}$. For this purpose we note that the quasi-particle energy is actually

$$\varepsilon(\mathbf{k}, r) = \varepsilon_k^0 + \sum_{\mathbf{k}'} f(\mathbf{k}, \mathbf{k}')\, \delta n(\mathbf{k}', r) \tag{1–90}$$

For each spin direction, let us define a *local* Fermi surface S_F^L by the condition

$$\varepsilon(\mathbf{k}, r) = \mu$$

In any direction k, the displacement $u_L(\mathbf{k})$ which takes S_F° into S_F^L is of magnitude

$$u_L(\mathbf{k}) = -\frac{\varepsilon(\mathbf{k}, r) - \varepsilon_k^0}{\hbar \mid v_k \mid} \tag{1–91}$$

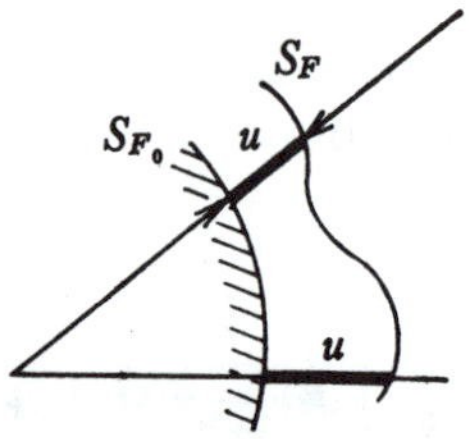

Figure 6

Comparing Eqs. (1–82), (1–89), (1–90), and (1–91), we can easily see that

$$\bar{u}(\mathbf{k}) = u(\mathbf{k}) - u_L(\mathbf{k}) \qquad \textit{(1–92)}$$

$\bar{u}(\mathbf{k})$ is thus the displacement of the Fermi surface of spin σ with respect to its *local* equilibrium position.

This very ingenious argument is due to Heine, who emphasizes, moreover, that Eq. (1–83) then becomes very natural; the diffusion terms are affected only by the *local* energy and therefore bring in $\overline{\delta n}$ instead of δn.

e. Collective oscillations

Let us now consider a neutral Fermi liquid—He^3, for example—in the absence of any external force. Let $\delta n(\mathbf{k})$ be the deformation of the gas of quasi particles; we assume it to be periodic, of wave vector q and frequency ω. We again choose ω much greater than the collision frequency ν of the quasi particles. The transport equation is then obtained by suppressing the driving field in (1–78),

$$(q \cdot v_k - \omega)\, \delta n(\mathbf{k}) + q \cdot v_k\, \delta(\varepsilon_k - \mu) \sum_{\mathbf{k}'} f(\mathbf{k}, \mathbf{k}')\, \delta n(\mathbf{k}') = 0 \qquad \textit{(1–93)}$$

(1–93) is a *homogeneous* integral equation, which has a solution only for certain *eigenvalues* of the parameter ω/q. These solutions correspond to oscillations of the system in the absence of any external force—thus to *free* oscillations, as opposed to the *forced* oscillations studied in the preceding section. Such free oscillations are called the *collective modes* of the system.

At this point, it is important to realize the fundamental difference between high-frequency ($\omega \gg \nu$) and low-frequency ($\omega \ll \nu$) oscillations. In the calculation of the velocity of ordinary sound ($\omega \ll \nu$), we did not explicitly mention collision phenomena. However, they are essential to the macroscopic method used above. The relaxation time that the system takes to return to equilibrium is of order ν^{-1}. For a phenomenon of period $\gg \nu^{-1}$, we can consider the gas of quasi particles always to be in equilibrium and to be deformed in an adiabatic manner. The internal mechanism of relaxation is then of no importance; this is why we did not speak of it. However, it is clear that an adiabatic approximation implies a quasi-instantaneous relaxation toward equilibrium, which depends on the existence of collisions. In the inverse case $\omega \gg \nu$, on the other hand, we do not speak about collisions because, practically, they play no role.

We have already indicated that the collision time ν^{-1} of the quasi particles varies as $1/T^2$. Therefore, whatever the frequency of the phenomenon considered may be, we can always expect that ω will become $\gg \nu$ if we go to a sufficiently low temperature. In practice, ν^{-1} always remains very small. For He^3, for example, $\nu^{-1} \sim 10^{-12}$ sec for $T = 1°K$. In the present state of low-temperature techniques, the phenomena which we are going to discuss will be of interest only at very high frequencies, of the order of 10^6 megacycles, which is well beyond the range of radio frequencies which we know how to produce. These questions thus are of a rather speculative nature. However, it is not impossible that some day we shall be able to observe these collective modes of a Fermi gas. As far as He^3 is concerned, Abrikosov and Khalatnikov have made a very ingenious proposal, based on the use of the Brillouin effect. We know that, if we analyze the light scattered by a liquid (Rayleigh scattering), we find, in addition to the incident frequency, two satellite lines, whose frequencies are shifted by

$$\frac{\Delta\omega}{\omega} = \pm \frac{2u}{c} \sin\frac{\theta}{2}$$

where u and c are the velocities of sound and light, respectively, and θ is the scattering angle. Applying this technique to He^3 at low temperatures, we might hope to demonstrate the collective oscillations of the system. An order-of-magnitude calculation shows that this effect occurs just at the limit of possible observations.

Let us return to Eq. (1–93). Again $\delta n(\mathbf{k})$ is going to contain a factor $\delta(\varepsilon_k - \mu)$; the collective mode is just an oscillation of the Fermi surface, similar to that of a liquid drop (which is compressible, since the volume enclosed by S_F can vary). It is convenient to introduce the displacement u of the Fermi surface, defined by (1–89). Equation (1–93) then becomes

$$(q \cdot v_k - \omega)\, u(\mathbf{k}) + q \cdot v_k \sum_{\mathbf{k}'} f(\mathbf{k}, \mathbf{k}')\, \delta(\varepsilon'_k - \mu)\, u(\mathbf{k}') = 0 \qquad (1\text{–}94)$$

Let us choose polar coordinates (θ,φ), with the polar axis parallel to q. Let us introduce, in addition, the phase velocity ω/q of the collective mode and refer it to the Fermi velocity by setting

$$s = \frac{\omega}{qv_F} = \frac{\omega m^*}{\hbar q k_F} \qquad (1\text{–}95)$$

With these conventions (1–94) becomes

$$(\cos\theta - s)\, u(\theta, \varphi, \sigma) + \cos\theta \sum_{\sigma'} \int d\gamma' F(\xi, \sigma, \sigma')\, u(\theta', \varphi', \sigma') = 0 \qquad (1\text{–}96)$$

where $d\gamma'$ is the element of solid angle, ξ the angle between k and k', and we have set

$$F(\mathbf{k}, \mathbf{k}') = \frac{m^* k_F}{\hbar^2 (2\pi)^3} f(\mathbf{k}, \mathbf{k}') \qquad (1\text{–}97)$$

Equation (1–96) is very important, since it describes *all* low-frequency collective modes.

A priori, the possible solutions are numerous. We first distinguish two main categories, according to whether the Fermi surfaces of the two spins oscillate in phase (normal waves) or 180° out of phase (spin waves). In each category, there can exist several types of polarization; longitudinal, transverse, and even more complex types. A complete study requires an expansion of u in spherical harmonics, of the type (1–51). The calculations are tedious, so we shall study only some representative examples.

Let us assume that $s < 1$; the factor $(\cos\theta - s)$ can vanish; a resonance is then produced between the wave and the individual quasi particles, giving rise to a dissipation of energy. The solution for s contains an imaginary part $-is_1$, corresponding to a damping of the wave. In general this damping is so important that it divests the collective mode of any physical meaning. Therefore, let us consider only undamped modes, such that $s > 1$. Furthermore, we exclude any pure imaginary solution of the type is_1, which would lead to an instability of the system.

Let us first consider normal waves, independent of spin. Only the symmetric kernel

$$F_s(\xi) = \tfrac{1}{2}\{F(\xi, +\tfrac{1}{2}, +\tfrac{1}{2}) + F(\xi, +\tfrac{1}{2}, -\tfrac{1}{2})\} \qquad (1\text{–}98)$$

is involved. (1–96) can then be written

$$\left(\frac{s}{\cos\theta} - 1\right) u(\theta, \varphi) = \int 2F_s(\xi)\, u(\theta', \varphi')\, d\gamma' \qquad (1\text{–}99)$$

Let us study the solution of this equation in some simple cases.

(1) $F_s(\xi) = F_{s0}$, independent of ξ. We then find

$$\begin{cases} u(\theta, \varphi) = \dfrac{\cos\theta}{s - \cos\theta} \times C^{te} \\ \dfrac{s}{2}\ln\left(\dfrac{s+1}{s-1}\right) - 1 = \dfrac{1}{8\pi F_{s0}} \end{cases} \tag{1-100}$$

The first of these relations shows that the Fermi surface is deformed into the shape of an egg, the small end pointing in the direction of q. This mode is longitudinal; it is usually called "zero sound." The second part of (1–100) is the dispersion relation, giving s. It always has a solution for $F_{s0} > 0$. For F_{s0} large we have $s = (8\pi F_{s0}/3)^{\frac{1}{2}}$. On the other hand, for $F_{s0} \to 0$, $s \to 1$. This last result is completely general, valid for any form whatever of $F(\xi)$: in the weak-coupling limit, zero sound propagates with the Fermi velocity. The expression for u shows that the deformation of the Fermi surface then reduces to a small, very localized protuberance in the direction of q. All the quasi particles in this region propagate in approximately the same direction, with velocity v_F; we can see that, for weak coupling, the collective mode to which it gives rise propagates with the same velocity.

It is interesting to compare these results with those concerning ordinary sound ($\omega \ll \nu$). The latter corresponds to an oscillation of a "slab" of gas as a whole, at the frequency ω. This motion is comprised of two components: oscillation of the velocity, and 90° out of phase with this, oscillation of the pressure. The macroscopic velocity has the effect of displacing the Fermi distribution without deforming it; the pressure compresses it, while retaining its spherical shape. The displacement of a surface element thus has a complex amplitude $(\alpha + i\beta\cos\theta)$, very different from that given by (1–100). Let us point out, moreover, that in the weak-coupling limit the velocities of propagation are different: $v_F/\sqrt{3}$ for ordinary sound, v_F for zero sound.

Let us analyze the dispersion equation (1–100) more in detail. We assume F_{s0} to be real. By studying the analytic properties of (1–100), we can verify that there exist two types of roots:

(a) Two equal and opposite real roots if $0 < F_{s0}$

(b) Two equal and opposite pure imaginary roots if $-1 < 1/(8\pi F_{s0}) < 0$

The condition of stability of the system is thus

$$8\pi F_{s0} > -1 \tag{1-101}$$

Using the definition (1–97), and recalling that f is defined for unit volume, we see that (1–101) is just the first stability condition (1–53). Our description is therefore consistent, and it shows that instabilities give rise to collective oscillations of exponentially growing amplitudes.

(2) The solution which we have just studied corresponds to longitudinal waves, having cylindrical symmetry. In order to give an example of a transverse wave, let us assume that

$$F_s(\xi) = F_{s0} + F_{s1} \cos \xi \qquad \textit{(1–102)}$$

and look for a solution proportional to $e^{i\varphi}$ (*circularly* polarized wave). The calculations are elementary and give the following results:

$$\begin{cases} u(\theta, \varphi) = \dfrac{\sin\theta\cos\theta}{s - \cos\theta} e^{i\varphi} \times C^{te} \\ 2s^2 - \dfrac{4}{3} + s(1 - s^2) \ln\left(\dfrac{s+1}{s-1}\right) = \dfrac{1}{2\pi F_{s1}} \end{cases} \qquad \textit{(1–103)}$$

If we solve the dispersion equation graphically, we find that it has a solution only if $F_{s1} > 3/4\pi$. The interaction between quasi particles must attain a certain strength in order to permit propagation of this circular mode. This is a general result: all the modes predicted from a study of the symmetries of the problem do not necessarily exist; it is still necessary for the interaction to satisfy certain conditions.

We can repeat the same discussion for oscillations of the spin-wave type, corresponding to a deformation

$$u(\theta, \varphi, \sigma) = u(\theta, \varphi) \times \text{sign of } \sigma \qquad \textit{(1–104)}$$

Only the antisymmetric kernel enters in this case,

$$F_a(\xi) = \tfrac{1}{2}\{F(\xi, +\tfrac{1}{2}, +\tfrac{1}{2}) - F(\xi, +\tfrac{1}{2}, -\tfrac{1}{2})\} \qquad \textit{(1–105)}$$

The equation of propagation is obtained by replacing F_s by F_a in (1–99) All the preceding discussion can be transposed without modification; for each type of polarization we have three possibilities:

(a) A real solution $s > 1$, giving a collective mode
(b) No solution (or, rather, a damped solution)
(c) A pure imaginary solution, producing an instability.

For He^3, for example, it seems that the longitudinal mode does not exist.

In practice, the solution of (1–99) often poses delicate mathematical problems. Nevertheless, we must emphasize the simplicity of this description of the collective modes.

4. Conclusions

It is essential to be well aware of the limitations of this theory:

We have assumed a unique correspondence between the excited states of the ideal system and those of the real system; this limits us to "normal" systems.

We have assumed T to be small enough for the quasi particles to have a sufficiently long lifetime to be well defined. More precisely, the existence of a mean free path leads to some uncertainty in the momentum of the quasi particle, which we assume small compared with the width $\varkappa T/v_F$ of the thermally excited region in k space. It is due to this hypothesis that we have been able to assert that ε_k and $f(\mathbf{k},\mathbf{k}')$ are real. This is an approximation, certainly valid at low temperatures. If one wants to extend this formalism to other than thermal phenomena, it is, in general, necessary to introduce quasi particles with large values of k, whose energies have large imaginary parts.

The assumption that the quasi particles are well defined does not completely take care of the problem. There exist other excitations (collective oscillations, bound states, etc.) which must be studied separately. In general, these additional states of the system play a secondary role at low temperatures. For example, the "phonons" have a T^3 specific heat. When this term starts to be of importance, the quasi-particle lifetime has diminished to a point where the whole model is of dubious value. We must therefore pay attention to maintaining the consistency of the approximations throughout the calculation.

Within this restricted framework, the Landau method offers a simple, suggestive, and very powerful method for solving a large variety of problems. This theory rests on a certain number of more or less obvious hypotheses: structure of the elementary excitation spectrum, nature of the current, transport properties of the quasi particles, etc. These hypotheses are all very natural ones but are a little lacking in rigor. In Chap. 6 we firmly establish the foundations of this chapter within the framework of perturbation theory.

How can we use this formalism? First, from a qualitative point of view, for understanding and interpreting a phenomenon. We can also attempt to relate different properties, by experimentally determining the unknown functions $F_s(\xi)$ and $F_a(\xi)$. Let us expand these in Legendre polynomial series [see (1–51)], with coefficients F_{sl} and F_{al}. Combining definition (1–97) with (1–23), (1–35), and (1–48), we see that

$$\left\{\begin{aligned} 8\pi F_{s0} &= \frac{3mm^*c^2}{\hbar^2 k_F^2} - 1 \\ 8\pi F_{s1} &= \frac{m^*}{m} - 1 \\ 8\pi F_{a0} &= \frac{\beta^2 m^* k_F}{4\pi^2\hbar^2\chi_M} - 1 \end{aligned}\right. \tag{1-106}$$

k_F can be deduced from the density, m^* from the specific heat, and c and χ_M can be measured directly. The three coefficients F_{s0}, F_{s1}, and F_{a0} are therefore experimentally accessible (and are known for He^3). Provided with this fragmentary information, we can try to calculate the velocity of the collective modes or the conductivity for an electron gas.

In practice, the method has already been applied to numerous problems. We have mentioned several times the properties of He^3: thermodynamic properties, transport coefficients, absorption of sound, and Brillouin effect—all studied by Abrikosov and Khalatnikov. Silin has extended this formalism to a degenerate electron gas in a magnetic field and has applied his results to electrons in metals: plasma oscillations, zero sound, spin waves, and transverse oscillations (propagation of light and cyclotron resonance). We can thus treat a large number of problems, such as the anomalous skin effect, optical properties, and the characteristic energy losses of fast electrons. Abrikosov and Dzialoschinskii have generalized the method to the case of a ferromagnetic system, for which the Fermi surfaces of the two spins do not coincide. They have thus been able to show that the frequency of the spin waves must be proportional to q^2 rather than to q, which agrees with the usual results in the theory of ferromagnetism. For more details we refer the reader to the original articles indicated in the Bibliography.

Chapter 2

Response of the System to External Excitations

1. Response to a Time-Dependent Perturbative Potential

a. Generalities

The majority of the macroscopic properties of a system can be related to the following general problem: An external perturbation varying in both space and time is imposed on the system (e.g., an electric field, magnetic field, etc.), and we ask what the *response* is to this perturbation (electric polarization and current, magnetization, etc.). The problem is always the same; only the form of the perturbation is changed. For a weak perturbative potential H_1, the response is in general *linear*, proportional to H_1. In order to characterize it, it is then sufficient to find the coefficient of proportionality.

$$\frac{\text{response}}{\text{excitation}} = \text{admittance}$$

(the term "admittance" arises by analogy with the properties of an electrical circuit). In general, these admittances depend on wavelength and frequency. They are complex, their phase representing the delay of the response after excitation. We mention as an example the dielectric constant of the system, which we shall study in detail in the next section.

If the excitation becomes too strong, the response *saturates*. The study of these nonlinear effects is difficult, as Fourier expansion of the excitation can no longer be used. In what follows we limit ourselves exclusively to linear phenomena for which the notion of admittance is valid. Furthermore, we shall consider only perturbations characterized by a potential (electric, magnetic, etc.). We leave aside, for instance, the question of thermal conductivity (that is to say, the response to a temperature gradient), which lies outside the area to which we limited ourselves at the start of this volume. This problem, as well as that of diffusion, can be treated by a simple generalization of the methods which we are going to discuss (for more detail, see the original articles by Kubo listed in the Bibliography).

Let us therefore consider a system at absolute zero, with a Hamiltonian H_0. H_0 includes the interaction between the particles of the system. Let us apply it to a perturbation whose intensity varies with time, characterized by an interaction potential $AF(t)$. A is a Hermitian operator which operates on the wave function of the system. Its time dependence is absorbed completely in the real scalar factor $F(t)$. (Let us remark that we have thus limited ourselves to a restricted class of potentials: those whose "shape" is constant and for which only the intensity varies with time.) The wave function $|\varphi_s(t)\rangle$ satisfies the Schrödinger equation

$$\mathrm{i}\frac{\partial|\varphi_s\rangle}{\partial t} = |H_0 + AF(t)|\varphi_s\rangle \qquad (2\text{–}1)$$

(where we have taken units such that $\hbar = 1$). In order completely to define $|\varphi_s\rangle$, it is sufficient to impose one boundary condition: we shall assume that, for $t = -\infty$, $F(t) = 0$, the system being in its ground state $|\varphi_{s0}\rangle$.

Until now we have used the Schrödinger representation, in which the operators A, $H_0, \ldots$ are independent of time. It is convenient to switch to a new representation, defined by the transformation

$$|\varphi(t)\rangle = e^{\mathrm{i}H_0 t}|\varphi_s(t)\rangle \qquad (2\text{–}2)$$

In order to retain the same average values, it is necessary simultaneously to modify the definition of operators, which become

$$\mathbf{A}(t) = e^{\mathrm{i}H_0 t} A e^{-\mathrm{i}H_0 t} \qquad (2\text{–}3)$$

[This is just the Heisenberg representation for the isolated system.] After this transformation, Eq. (2–1) is written as

$$\mathrm{i}\frac{\partial|\varphi\rangle}{\partial t} = \mathbf{A}(t)\, F(t)\,|\varphi\rangle \qquad (2\text{–}4)$$

The "response" of the system is nothing more than the average value of a certain operator B, which can be the current, the magnetic moment, etc. Thus we want to calculate

$$< \varphi \mid \mathbf{B}(t) \mid \varphi > = \langle B \rangle \tag{2-5}$$

Now, since we are limiting ourselves to the *linear* response, it is sufficient to calculate $|\varphi\rangle$ to first order in A by a classical perturbation calculation. Let us write $|\varphi\rangle$ in the form of an expansion

$$\mid \varphi > = \mid \varphi_0 > + \mid \varphi_1 > + \ldots$$

We see at once that

$$\frac{\mathrm{i}\,\partial \mid \varphi_1 >}{\partial t} = \mathbf{A}(t)\, F(t) \mid \varphi_0 > \tag{2-6}$$

This equation is easily integrated because of the boundary condition indicated above, and the result is

$$\mid \varphi_1 > = -\mathrm{i} \int_{-\infty}^{t} \mathbf{A}(t')\, F(t')\, \mathrm{d}t' \mid \varphi_0 > \tag{2-7}$$

(where $|\varphi_0\rangle$ is independent of time because of the choice of representation). To first order, $\langle B\rangle$ is given by

$$\langle B \rangle = B_0 + < \varphi_0 \mid \mathbf{B}(t) \mid \varphi_1 > + < \varphi_1 \mid \mathbf{B}(t) \mid \varphi_0 > \tag{2-8}$$

where B_0 is the average value of B in the ground state $|\varphi_0\rangle$. By inserting (2–7) into (2–8) we obtain the final expression for $\langle B\rangle$,

$$\langle B \rangle - B_0 = \mathrm{i} \int_{-\infty}^{t} < \varphi_0 \mid [\mathbf{A}(t'), \mathbf{B}(t)] \mid \varphi_0 > F(t')\, \mathrm{d}t' \tag{2-9}$$

b. The response function of the system

If the perturbation is reduced to an impulse given to the system at time $t = 0$, that is to say $F(t) = \delta(t)$, the response becomes

$$\langle B \rangle - B_0 = \varphi_{BA}(t) = \begin{cases} 0 \text{ for } t < 0 \\ \mathrm{i} < \varphi_0 \mid [\mathbf{A}, \mathbf{B}(t)] \mid \varphi_0 > \text{ for } t > 0 \end{cases} \tag{2-10}$$

We shall call $\varphi_{BA}(t)$ the "response function" of the system: it describes the deformation at a time t *after* the excitation impulse. Let us remark that φ_{BA} is zero for $t < 0$, which is in conformity with the principle of causality. We see that φ_{BA} is completely determined if the commutator $[\mathbf{A},\mathbf{B}(t)]$ is known.

Every function $F(t)$ can be considered as a succession of impulses, of appropriately chosen amplitude. More precisely, we have

$$F(t) = \int_{-\infty}^{+\infty} F(t')\,\delta(t - t')\,dt'$$

Since the response of the system is linear, in order to obtain the response to the excitation $AF(t)$ it is sufficient to add the responses to each of the elementary impulses,

$$\langle B \rangle - B_0 = \int_{-\infty}^{+\infty} F(t')\,\varphi_{BA}(t - t')\,dt' \qquad (2\text{–}11)$$

Let us insert into (2–11) the expression (2–10) for $\varphi_{BA}(t)$, and let us remark that

$$< \varphi_0 \,|\, [\mathbf{A}(t'), \mathbf{B}(t)] \,|\, \varphi_0 >$$

in fact depends only on $(t - t')$: we recover Eq. (2–9).

Until now we have placed no restriction on $F(t)$. Actually, if the perturbation $AF(t)$ induces *real* transitions of the system, that is to say, produces a permanent deformation, it is necessary to limit ourselves to functions $F(t)$ which vanish sufficiently quickly as $t \to -\infty$ (for example, as $e^{\eta t}$). However weak the applied perturbation is, it will eventually "excite" the system if it is allowed to act for a sufficiently long time: the wave function is then modified to an appreciable extent, and the perturbation expansion used above no longer makes sense. To avoid this problem, it suffices to turn on the perturbation *adiabatically*, by introducing into $F(t)$ a factor $e^{\eta t}$. η must be small enough so that the system *follows* perfectly but large enough so that it does not have time to become excited. This lower bound on η tends to zero when the strength of the perturbation diminishes. To first order we can therefore assume η to be infinitesimally small; at the same time we must not forget it altogether, as it is precisely η which allows us to define resonance phenomena unambiguously.

Note: It is necessary to distinguish carefully between real transitions, conserving energy, which lead to a permanent and irreversible deformation, and virtual transitions, which give rise only to an "elastic" and reversible deformation of the wave function.

There exist cases for which this problem of excitation does not arise. Let us consider for example a "constant" perturbation, established adiabatically $[F(t) = e^{\eta t}]$, and spatially localized. As an application of these ideas, let us consider an electric field acting on an electron gas. The system remains at each instant in the ground state corresponding to the applied potential. This state being nondegenerate, there are no real transitions, nor is there a transfer of energy from the external field to the system. To put it simply, the spatial-distribution function of the electrons adjusts itself progressively so as to screen the locally induced charge. If the perturbation is removed adiabatically, the system returns to the ground state $|\varphi_0\rangle$. The situation is not the same for an electric field oscillating with frequency ω or for a constant uniform field. In the first case the field supplies quanta of energy $\hbar\omega$ to the system and produces excitation, characterized by the "loss angle," well known to electrical engineers. The second case poses a more subtle problem: the electrons are set in motion, and the current thus produced induces excitation by means of the Joule effect: the difficulty arises from the fact that the boundary conditions have changed. Then the solution of the problem requires completely different methods, which lie outside the scope of this account.

This discussion allows us to specify the limitations of the general theory formulated above: it is applicable only to weak perturbations, adiabatically turned on and not changing the boundary conditions of the system.

There are two especially important cases from the point of view of physics, corresponding, respectively, to

$$F(t) = \begin{cases} e^{\eta t} & \text{for} \quad t < 0 \\ 0 & \text{for} \quad t > 0 \end{cases}$$

$$F(t) = \begin{cases} 0 & \text{for} \quad t < 0 \\ 1 & \text{for} \quad t > 0 \end{cases}$$

The first case corresponds to the relaxation of the system when a perturbation turned on adiabatically is suddenly removed. The response is equal to

$$\langle B \rangle - B_0 = \Phi^R_{BA}(t)$$

where the "relaxation function" Φ^R_{BA} is given by

$$\Phi^R_{BA}(t) = \lim_{\eta \to +0} \int_t^\infty \varphi_{BA}(t') \exp(-\eta t')\, dt' \qquad \textit{(2–12)}$$

$\Phi^R_{BA}(0)$ describes the deformation *at equilibrium* produced by the perturbation. When $t \to +\infty$, it is reasonable to suppose that the system relaxes to its initial equilibrium condition and that $\langle B \rangle \to B_0$. In other words $\Phi_{BA}^R \to 0$ after a certain characteristic time τ_R. This assertion, as obvious as it seems, is very difficult to justify (it follows directly from the ergodic theorem). We content ourselves with emphasizing this difficulty, without exploring it more deeply.

The second case mentioned corresponds to the sudden turning on of a constant perturbation. The response of the system is then given by the "excitation function"

$$\Phi_{BA}^E = \int_0^t \varphi_{BA}(t')\, dt' \qquad (2\text{–}13)$$

We see that

$$\Phi_{BA}^E(t) + \Phi_{BA}^R(t) = \Phi_{BA}^R(0) \qquad (2\text{–}14)$$

The phenomena of relaxation and excitation are thus complementary to each other. When $t \to +\infty$, $\Phi_{BA}^E \to \Phi_{BA}^R(0)$, that is to say, to the equilibrium condition in the presence of the perturbation, under the restriction that Φ_{BA}^R must tend to zero: this is a conclusion which seems quite natural.

c. Response to a periodic excitation: admittance

Every function $F(t)$ can be expanded in a Fourier series

$$F(t) = \frac{1}{2\pi} \int_{-\infty}^{+\infty} F(\omega) \exp[(-i\omega + \eta)\, t]\, d\omega \qquad (2\text{–}15)$$

As $F(t)$ is a real function, we deduce that $F(-\omega) = [F(\omega)]^*$. Since we have limited ourselves to linear effects, we shall obtain the total response by adding the responses to each Fourier component. This leads us to study the perturbation

$$F(t) = \exp[(-i\omega + \eta)\, t]$$

(which makes sense physically only if we combine it with its complex conjugate of frequency $-\omega$). By putting this expression for F into the general equation (2–11), we obtain

$$\begin{cases} \langle B \rangle - B_0 = \chi_{BA}(\omega) \exp[(-i\omega + \eta)\, t] \\ \chi_{BA}(\omega) = \int_0^{\infty} \exp[(i\omega - \eta)t']\, \varphi_{BA}(t')\, dt' \end{cases} \qquad (2\text{–}16)$$

$\chi_{BA}(\omega)$ is a complex quantity, whose modulus and argument describe, respectively, the amplitude and the phase of the response. By analogy with the response of electric circuits to an alternating current, we call it the *admittance* of the system. Since we have assumed B to be Hermitian, $\langle B \rangle$ must be real, so

$$\chi_{BA}(\omega) = [\chi_{BA}(-\omega)]^* \tag{2-17}$$

Physical examples of admittance are numerous: dielectric constant, magnetic susceptibility, etc. We know that these quantities depend on frequency: we therefore say that the system is "dispersive." It is easy to express the response to any perturbation $F(t)$ whatsoever by means of the admittance $\chi_{BA}(\omega)$. We have

$$\langle B \rangle - B_0 = \frac{1}{2\pi} \int_{-\infty}^{+\infty} \exp[(-i\omega + \eta)\, t]\, F(\omega)\, \chi_{BA}(\omega)\, d\omega \tag{2-18}$$

Let us now turn to the explicit calculation of the quantities $\varphi_{BA}(t)$ and $\chi_{BA}(\omega)$. We shall utilize for this the *complete* set of eigenfunctions $|\varphi_n\rangle$ of the isolated system Hamiltonian H_0, of energy E_n. In order to simplify the writing of the equations, we shall put

$$E_n - E_0 = \omega_{n0}$$
$$\langle \varphi_0 | A | \varphi_n \rangle = A_{0n}$$

A glance at (2–10) shows that φ_{BA} is given by

$$\varphi_{BA}(t) = \begin{cases} 0 \text{ for } t < 0 \\ i \sum_n \{ A_{0n} B_{n0} \exp(i\omega_{n0} t) - B_{0n} A_{n0} \exp(-i\omega_{n0} t) \} \text{ for } t > 0 \end{cases} \tag{2-19}$$

It is apparent from this equation that φ_{BA} is a real function. If we insert (2–19) into Eq. (2–16), we obtain

$$\chi_{BA}(\omega) = \sum_n \left\{ \frac{B_{0n} A_{n0}}{\omega - \omega_{n0} + i\eta} - \frac{A_{0n} B_{n0}}{\omega + \omega_{n0} + i\eta} \right\} \tag{2-20}$$

We see from this expression that $\chi_{BA}(\omega)$ has singularities only for $\omega = \pm\, \omega_{n0} - i\eta$. In practice, the values of ω_{n0} form a continuous spectrum, and these singularities are transformed into a branch cut located at a distance $i\eta$ *below* the real axis. χ_{BA} suffers a discontinuity on crossing this branch cut.

We can therefore conclude that $\chi_{BA}(\omega)$ is an analytic function of the complex variable ω in the upper half-plane. This is just another way of expressing the principle of *causality*, according to which $\varphi_{BA}(t)$ is zero for negative times. In fact, if we invert the Fourier transform

$$\varphi_{BA}(t) = \frac{1}{2\pi}\int_{-\infty}^{+\infty} \exp(-i\omega t)\,\chi_{BA}(\omega)\,d\omega \tag{2-21}$$

we see that, for $t < 0$, it is necessary to close the contour of integration in the upper half-plane (see Fig. 1). As χ_{BA} is analytic there, the integral is zero. For $t > 0$, on the contrary, it would be necessary to close the contour in the lower half-plane, which cannot be done without crossing the cut.

When $\omega \to \infty$, the expression for χ_{BA} becomes very simple:

$$\chi_{BA}(\omega) \to \frac{1}{\omega} < \varphi_0 \,|\, [B, A] \,|\, \varphi_0 > \tag{2-22}$$

If B and A commute, it is necessary to carry the expansion one order higher,

$$\chi_{BA}(\omega) \to \sum_n \frac{\omega_{n0}}{\omega^2} \{ B_{0n}A_{n0} + A_{0n}B_{n0} \}$$

which can be written in the form

$$\chi_{BA}(\omega) \to -\frac{1}{\omega^2} < \varphi_0 \,|\, [[H_0, A], B] \,|\, \varphi_0 > \tag{2-23}$$

These commutators can frequently be evaluated directly, which makes

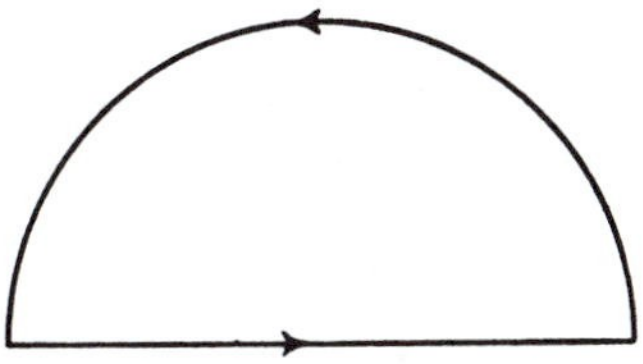

$t < 0$

Figure 1

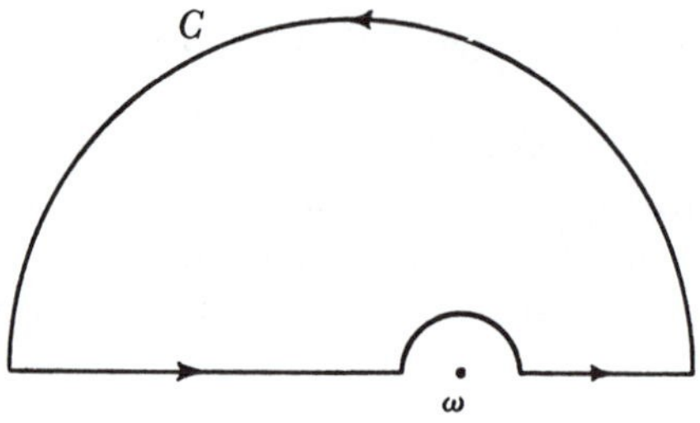

Figure 2

possible a simple calculation of the asymptotic forms of χ_{BA}. In all cases, χ_{BA} is zero at infinity.

The analytic properties of χ_{BA} have numerous interesting consequences. Let us consider, for example, the expression

$$\int_C \frac{\chi_{BA}(\omega')}{\omega - \omega'} \, d\omega' = 0$$

the contour C being defined by Fig. 2. The semicircle at infinity gives no contribution; the semicircle around ω contributes a term $- i\pi\chi_{BA}(\omega)$. The rest is by definition the principal part of the integral. This leads to the relation

$$\int_{-\infty}^{+\infty} \chi_{BA}(\omega') \, P\left[\frac{1}{\omega - \omega'}\right] d\omega' = i\pi\chi_{BA}(\omega) \tag{2-24}$$

If we separate the real and imaginary parts of this equation, we obtain the "Kramers-Kronig" relations, relating the real and imaginary parts of $\chi_{BA}(\omega)$. If we know one of them for all values of ω, we can calculate the other with these relations.

Developing another line of thought, let us study the quantity

$$\int_{-\infty}^{+\infty} \chi_{BA}(\omega) \, d\omega$$

We can close the contour of integration at infinity in the upper half-plane. In this region only the term proportional to $1/\omega$ in the expansion of χ_{BA} contributes: this leads to the relation

$$\int_{-\infty}^{+\infty} \chi_{BA}(\omega) \, d\omega = - i\pi < \varphi_0 \,|\, [B, A] \,|\, \varphi_0 > \tag{2-25}$$

The integral is zero if B and A commute. In this last case, we can also calculate the quantity

$$\int_{-\infty}^{+\infty} \chi_{BA}(\omega)\,\omega\,d\omega = i\pi < \varphi_0 \,|\, [[H_0, A], B] \,|\, \varphi_0 > \qquad (2\text{–}26)$$

These expressions, relating certain integrals to the asymptotic values of χ_{BA}, are known by the name of "*sum rules.*"

These few examples do not exhaust the numerous effects of the analytic properties of $\chi_{BA}(\omega)$.

2. Example: The Dielectric Constant of an Electron Gas

To illustrate the preceding discussion, we shall develop an example of particular practical importance—the longitudinal dielectric constant of a gas of particles that carry an electric charge *e*.

Let us introduce into this system an electric-charge distribution playing the role of a perturbation. To simplify, we suppose these charges to be periodic in time (frequency ω) and in space (wave vector q). The density of charge introduced is written as

$$Q(r, t) = er_q \cos(q \cdot r) \cos(\omega t) \qquad (2\text{–}27)$$

where r_q is a real number sufficiently small so that the response remains linear. This distribution produces a longitudinal electric field which can be calculated by using Poisson's equation and which is equal to

$$\mathcal{E}(r, t) = \frac{4\pi e \boldsymbol{q}}{q^2} r_q \sin(q \cdot r) \cos(\omega t) \qquad (2\text{–}28)$$

Under the influence of this electric field, the distribution of electrons readjusts itself in such a way as to screen the charges introduced. It is this shielding effect that we wish to study.

The vector defined by (2–28) is in fact the electric displacement $\mathscr{D}$. The local electric field $\mathscr{E}_L$ is given by Poisson's equation

$$\operatorname{div} \mathcal{E}_L = 4\pi(Q + Q_e)$$

where Q_e is the density of *electronic* charge resulting from polarization of the medium (we assume that at equilibrium the system is neutral, the charge of the electrons being compensated by a uniform positive charge). In general, Q_e is out of phase with Q. In order to take account of this dephasing, we resort to the usual artifice, and we study the response to the fictitious perturbation

$$Q(\omega) = er_q \cos(q \cdot r) \exp(-i\omega t)$$

The dielectric constant $\varepsilon(q,\omega)$ is then defined by

$$\varepsilon(q,\omega) = \frac{\mathfrak{D}(\omega)}{\mathcal{E}_L(\omega)} = \frac{Q(\omega)}{Q(\omega) + Q_e(\omega)}$$

which we can rewrite in the following form:

$$\frac{1}{\varepsilon(q, \omega)} - 1 = \frac{Q_e(\omega)}{Q(\omega)} \tag{2-29}$$

Q_e is equal to the *average value* of the operator $e\rho(r)$, where $\rho(r)$ is the density of electrons at the point r,

$$\rho(r) = \sum_i \delta(r - r_i)$$

(r_i being the position of the ith electron). It is convenient to express $\rho(r)$ in terms of its Fourier components,

$$\rho_q = \int \exp(-iq \cdot r)\, \rho(r)\, d^3r = \sum_i \exp(-iq . r_i)$$

For this purpose, let us enclose the system in a box of volume Ω, so as to quantize the values of q. It is then easy to verify that

$$\delta(r - r_i) = \frac{1}{\Omega} \sum_{q'} \exp[iq' \cdot (r - r_i)]$$

(to convince oneself of this, it is sufficient to integrate over r). It follows that Q_e can be written as

$$Q_e = \frac{e}{\Omega} \sum_{q'} \langle \rho_{q'} \rangle \exp(iq' \cdot r) \tag{2-30}$$

Let us limit ourselves to a homogeneous and isotropic system: it is clear that the polarization is going to contain the same factor $\cos(q \cdot r)$ as the external charge. Consequently, only the Fourier components $q' = \pm q$ will contribute to Q_e. We can therefore write

$$\begin{cases} Q_e = \dfrac{2e}{\Omega} \langle \rho_q \rangle \cos(q \cdot r) \\ \langle \rho_q \rangle = \langle \rho_{-q} \rangle \end{cases} \tag{2-31}$$

Equation (2–31) makes possible the simplification of the definition of $\varepsilon(q,\omega)$, which becomes

$$\frac{1}{\varepsilon} - 1 = \frac{2 \langle \rho_q \rangle}{\Omega r_q \exp(-i\omega t)} \tag{2-32}$$

We are thus led to the calculation of the average value of ρ_q in the presence of the perturbation. In order to apply the results of the preceding section, we need the interaction Hamiltonian between the electrons and the external charges. The latter create at each point an electrostatic potential

$$V(r, t) = \frac{4\pi e}{q^2} r_q \cos(q.r) \exp(-i\omega t) \tag{2-33}$$

The interaction energy can be written

$$\sum_i eV(r_i, t)$$

By returning to the definition of ρ_q given above, we see that the operator A defined in the first section is in this case equal to

$$A = \frac{4\pi e^2}{q^2} \frac{r_q}{2} (\rho_q + \rho_{-q}) \tag{2-34}$$

We are now ready to apply the general formulas (2–16). In the expression for $\varphi_{BA}(t)$, the commutator $[\rho_q, \rho_q(t)]$ has an average value of zero for obvious reasons of translational invariance. Only the term ρ_{-q} in A contributes. Under these conditions, Eqs. (2–16) and (2–20) give us

$$\langle \rho_q \rangle = \frac{4\pi e^2}{q^2} \frac{r_q}{2} \exp(-i\omega t) \sum_n \left\{ \frac{|(\rho_q)_{on}|^2}{\omega - \omega_{no} + i\eta} - \frac{|(\rho_{-q})_{on}|^2}{\omega + \omega_{no} + i\eta} \right\} \tag{2-35}$$

We can simplify expression (2–35) by noticing that, for symmetry reasons, to each state n of the first term of (2–35) there corresponds a state m of the second term, obtained by reflection, such that

$$\begin{cases} |(\rho_q)_{on}|^2 = |(\rho_{-q})_{om}|^2 \\ \omega_{no} = \omega_{mo} \end{cases}$$

We can therefore write

$$\langle \rho_q \rangle = \frac{4\pi e^2}{q^2}\frac{r_q}{2}\exp(-i\omega t)\sum_n |(\rho_q)_{on}|^2 \frac{2\omega_{no}}{(\omega - \omega_{no} + i\eta)(\omega + \omega_{no} + i\eta)} \tag{2–36}$$

(At the same time, this discussion shows that $\langle\rho_q\rangle = \langle\rho_{-q}\rangle$.) By comparing this result with the definition (2–32), we finally obtain the following expression for the dielectric constant:

$$\frac{1}{\varepsilon(q,\omega)} - 1 = \frac{4\pi e^2}{\Omega q^2}\sum_n \frac{2\omega_{no}|(\rho_q)_{on}|^2}{(\omega + i\eta)^2 - \omega_{no}^2} \tag{2–37}$$

In order to separate the real and imaginary parts of $1/\varepsilon$, we remark that

$$\lim_{\eta\to 0}\frac{1}{\omega \pm i\eta} = P\left(\frac{1}{\omega}\right) \mp i\pi\,\delta(\omega)$$

Upon comparison with (2–37) we obtain

$$\begin{cases} \mathrm{Re}(1/\varepsilon) = 1 + \dfrac{4\pi e^2}{\Omega q^2}\displaystyle\sum_n 2\omega_{no}|(\rho_q)_{on}|^2\, P\left(\frac{1}{\omega^2 - \omega_{no}^2}\right) \\ \mathrm{Im}(1/\varepsilon) = \dfrac{4i\pi^2 e^2}{\Omega q^2}\displaystyle\sum_n |(\rho_q)_{on}|^2\{\delta(\omega + \omega_{no}) - \delta(\omega - \omega_{no})\} \end{cases} \tag{2–38}$$

Re $(1/\varepsilon)$ and Im $(1/\varepsilon)$ are, respectively, even and odd functions of ω. The real part of $1/\varepsilon$ corresponds to an induced charge in phase with Q, that is to say, to a current of electrons 90° out of phase with the electric field: we thus have a polarization current, purely reactive, not giving rise to any transfer of energy. On the other hand, Im $(1/\varepsilon)$ corresponds to a current of electrons in phase with the electric field, which produces excitation by the Joule effect. The presence of the Dirac δ functions in Im $(1/\varepsilon)$ furthermore shows directly that this term describes *real* transitions, for which the energy of the system varies by $\pm\hbar\omega$. At zero temperature, the situation is simplified, as the energy can only increase ($\omega_{no} > 0$): the first term of Im $(1/\varepsilon)$ contributes only for $\omega < 0$, the second for $\omega > 0$.

When $\omega \to \infty$, $1/\varepsilon - 1$ is of order $1/\omega^2$. Its limiting value is given by Eq. (2–23), which is written here

$$\frac{1}{\varepsilon} - 1 \to -\frac{4\pi e^2}{\Omega q^2} < \varphi_0 \mid [[H_0, \rho_{-q}]\,\rho_q] \mid \varphi_0 > \tag{2–39}$$

This commutator can be calculated explicitly if we assume that the law of interaction between the particles of the system depends only on the distance between them. We can then write

$$H_0 = \sum_i \frac{p_i^2}{2m} + \frac{1}{2}\sum_{i \neq j} V(r_i - r_j)$$

(where m is the mass of an electron). ρ_q commutes with V: the only contribution arises from the kinetic energy, which gives

$$[H_0, \rho_{-q}] = \sum_i \left(-\frac{q \cdot p_i}{m} + \frac{q^2}{2m}\right) \exp(iq \cdot r_i)$$

$$[[H_0, \rho_{-q}], \rho_q] = \sum_i \left(-\frac{q^2}{m}\right) = -\frac{Nq^2}{m}$$

where N is the total number of electrons. (2–39) is therefore written as

$$\frac{1}{\varepsilon} - 1 \to \frac{4\pi N e^2}{m\Omega\omega^2} \tag{2–40}$$

In other words, at high frequencies, the response of the system becomes independent of the law of interaction. Only the inertia of the electrons and their density N/Ω come into play.

The quantity $1/\varepsilon - 1$ being an admittance in the general sense, it possesses all the properties described in the preceding section. In particular, it satisfies the Kramers-Kronig relations [which can be verified directly by means of the explicit expression (2–38)]. Furthermore, the sum rules (2–25) and (2–26) are written in this case as

$$\left\{\begin{aligned} &\int_{-\infty}^{+\infty} (1/\varepsilon - 1)\, d\omega = 0 \\ &\int_{-\infty}^{+\infty} (1/\varepsilon - 1)\, \omega\, d\omega = -\frac{4i\pi^2 N e^2}{m\Omega} \end{aligned}\right. \tag{2–41}$$

The second equation is intimately related to the "f sum rule" encountered in the study of the optical properties of atoms, molecules, or solids. It is sometimes useful for "normalizing" relative measurements of the dielectric constant.

We have defined ε by referring to the effect of shielding. In fact there exist many other physical problems where the dielectric constant comes into play. We shall see in the following section that it plays a major role in the diffusion of charged particles. Furthermore, the poles of $1/\varepsilon$ give the frequencies of the collective excitations of the system: in fact, we see from Eq. (2–29) that, if $1/\varepsilon \to \infty$, we can have an oscillation of electronic charge Q_e in the *absence* of excitation by an external charge Q. We are thus led from a problem of *forced* oscillations to a problem of *self*-excitations. The latter are merely longitudinal compressional waves, known as "plasma oscillations." For small values of q, one can show that the frequency of these oscillations is close to

$$\omega_p = \sqrt{\frac{4\pi N e^2}{m\Omega}}$$

We shall return to this question later.

Before leaving these questions of admittance, let us emphasize that this formalism has been applied to numerous practical problems: conductivity in a magnetic field, optical properties (which go back to the study of the conductivity tensor), magnetic susceptibility, etc. This method furnishes directly a number of useful relations: Kramers-Kronig relations, sum rules, etc. The reader will find a discussion of these applications, generalized to the case of nonzero temperature, in Kubo's articles cited in the Bibliography.

3. Scattering of an Incident Particle in the Born Approximation

At the beginning of this chapter we studied the response of the system to a macroscopic excitation. Now we deal with another type of property, relative to the excitation of the system by a beam of incident particles (photons, electrons, neutrons, etc.). We shall limit ourselves exclusively to problems which can be treated in the Born approximation (that is to say, where we can consider the interaction between the system and the incident beam to occur only to first order). In principle this approximation is valid only for sufficiently fast particles; actually, it turns out to be satisfactory for numerous practical problems: scattering of electrons in the 10-Kev range, Compton scattering of photons, etc. It is applicable also to the scattering of slow neutrons on condition that the nucleon-nucleon interaction be replaced by a pseudopotential of the type proposed by Fermi.

Let us thus consider an incident particle, of mass M. Let us assume that it interacts with each of the particles of the system through a potential $V(r)$ depending uniquely on the distance separating them. The Hamiltonian of the system plus particle can be written as

$$H = H_0 + \frac{P^2}{2M} + \sum_i V(R - r_i)$$

where R and P are the position and momentum of the incident particle. It is convenient to expand the potential in a Fourier series, which gives

$$H = H_0 + \frac{P^2}{2M} + \sum_q \frac{V_q \exp(iq \cdot R)}{\Omega} \rho_q \tag{2-42}$$

where the component V_q is defined by

$$V_q = \int \exp(iq \cdot R) \, V(R) \, d^3 R$$

A typical collision will have the effect of making the system go from its ground state $|\varphi_0\rangle$ to the excited state $|\varphi_n\rangle$ and the incident particle from the state P_0 to the state $(P_0 - q.)$ Thus there is transferred to the system an energy $\omega = \omega_{n0}$ and a momentum q. The particle is deflected through an angle θ, in the direction indicated by Fig. 3. The probability that a certain collision takes place is

$$W_{nq} = 2\pi \frac{|V_q|^2}{\Omega^2} |(\rho_q)_{on}|^2 \, \delta\left(\omega_{no} + \frac{q^2}{2M} - \frac{q \cdot P_0}{M}\right) \tag{2-43}$$

In practice we are interested only in the final state of the incident

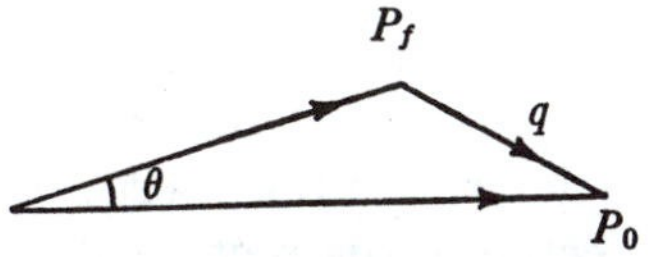

Figure 3

particle. The total probability that the latter will go into the state $|P_0 - q|$ is

$$W_q = \sum_n W_{nq} \tag{2-44}$$

In general, we express the experimental results in terms of an effective differential cross section $d^2\sigma/d\gamma\, d\omega$ for scattering into the solid angle $d\gamma$ with an energy transfer of an amount between ω and $\omega + d\omega$. These conditions restrict the momentum transfer q to the shaded region in Fig. 4. To obtain the cross section, it is sufficient to sum W_q over the values of q included in this region. We thus find

$$\frac{d^2\sigma}{d\gamma\, d\omega} = \frac{\Omega M}{P_0} \frac{d^2W}{d\gamma\, d\omega} = \frac{M^2\Omega^2}{(2\pi)^3} \frac{P_f}{P_0} W_q \tag{2-45}$$

where P_f is the magnitude of the final momentum.

By combining Eqs. (2–43) to (2–45), we see that we can write the differential cross section in the form

$$\frac{d^2\sigma}{d\gamma\, d\omega} = A S(q, \omega) \tag{2-46}$$

where the factors A and S are defined as follows:

$$A = \frac{\Omega M^2}{8\pi^3} \frac{P_f}{P_0} | V_q |^2$$
$$S(q, \omega) = \sum_n | (\rho_q)_{on} |^2 \delta(\omega_{no} - \omega) \frac{2\pi}{\Omega} \tag{2-47}$$

The value of q is known once the direction γ and the frequency ω have been fixed (see Fig. 4). The factor A contains all the elements relative to the incident particle: mass, initial and final momentum, law of interaction with the system. Conversely, S is entirely characteristic of the system itself. If the incident particle is changed, S remains the same; only A varies. We thus see the great generality of Eq. (2–47): it is sufficient to study the function $S(q,\omega)$ once and for all to describe scattering in the Born approximation for any particle whatsoever.

When we know how to determine the energy of the scattered particles, we obtain directly the effective cross section $d^2\sigma/(d\gamma\, d\omega)$ and formula (2–46) can be applied without further comment. Actually, we are often

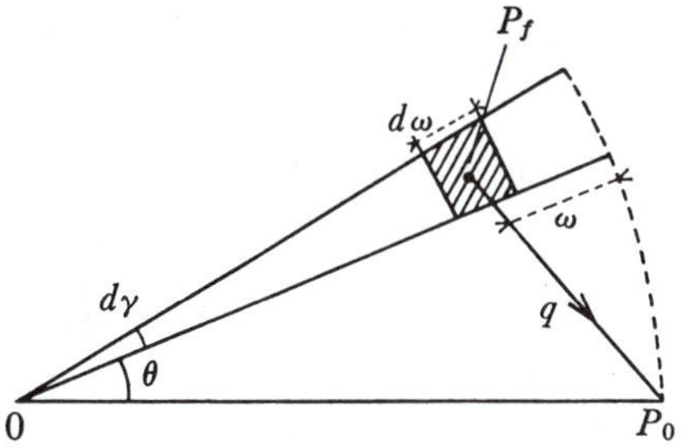

Figure 4

obliged to limit ourselves to a measurement of $d\sigma/d\gamma$, without determination of energy. Then it is necessary to integrate (2–46) over ω. In general this is not a simple operation, for q depends on ω (see Fig. 4). If, however, the velocity of the incident particle is raised, so that

$$\frac{|q|P_0}{M} \gg \langle \omega_{n0} \rangle \tag{2-48}$$

the problem is considerably simplified. In fact, in this case, conservation of energy requires q to be practically perpendicular to P_0, the angle of deflection θ remaining small. Under these conditions, q is practically independent of ω, and we can write

$$\frac{d\sigma}{d\gamma} = A \int_{-\infty}^{+\infty} S(q, \omega)\, d\omega = 2\pi A\, S(q) \tag{2-49}$$

valid if (2–48) is true.

It is important to understand that (2–46) is much more general than (2–49). For example (2–49) is not valid for the scattering of slow neutrons, which can nevertheless be described by a direct application of (2–46). On the other hand, electrons and photons in the 10-Mev range satisfy condition (2–48).

The function $S(q,\omega)$ is real; it is zero for negative values of ω. Let us calculate its Fourier transform with respect to ω,

$$S(q, t) = \frac{1}{2\pi} \int_{-\infty}^{+\infty} S(q, \omega) \exp(-i\omega t)\, d\omega \tag{2-50}$$

By comparison with (2–47), we see that

$$S(q, t) = \frac{1}{\Omega} \sum_n |(\rho_q)_{on}|^2 \exp(-i\omega_{no}t) = \frac{1}{\Omega} < \varphi_0 | \rho_q(t) \, \rho_{-q} | \varphi_0 > \qquad (2\text{–}51)$$

(we adopt the Heisenberg representation for the isolated system). In this expression the resemblance between S and the admittances defined above becomes evident. We shall return to this expression later.

Let us now take the Fourier transform with respect to q, by defining

$$S(r, t) = \frac{1}{\Omega} \sum_q S(q, t) \exp(iq \cdot r) \qquad (2\text{–}52)$$

Let us return to the definition of the ρ_q: we can write

$$S(r, t) =$$

$$\frac{1}{\Omega^2} \sum_q \int d^3 r' \int d^3 r'' \exp \left[iq \cdot (r - r' + r'') \right] < \varphi_0 | \rho(r', t) \, \rho(r'', 0) | \varphi_0 >$$

The summation over q is simple—it gives a δ function—and we finally obtain

$$S(r, t) = \frac{1}{\Omega} \int d^3 r' < \varphi_0 | \rho(r', t) \, \rho(r' - r, 0) | \varphi_0 > \qquad (2\text{–}53)$$

which, because of translational invariance, can be transformed into

$$S(r, t) = < \varphi_0 | \rho(r, t) \, \rho(0, 0) | \varphi_0 > \qquad (2\text{–}54)$$

[Let us point out that (2–53) remains valid for an inhomogeneous system, whereas (2–54) does not.]

The function $S(r,t)$ is called the "correlation function" between the density at (r,t) and that at $(0,0)$. It is intimately related to the probability of finding a particle at (r,t) when we know that there is one at $(0,0)$. This can be clearly seen if we neglect quantum effects: let us return to the definition of $\rho(r)$ and put it into (2–53):

$$S(r, t) = \frac{1}{\Omega} < \varphi_0 | \int d^3 r' \sum_{ij} \delta(r' - r_{it}) \, \delta(r' - r - r_{j0}) | \varphi_0 > \qquad (2\text{–}55)$$

where r_{it} and r_{i0} denote the position of the ith particle at the times t and

0. In the classical approximation, r_{it} and r_{j0} commute, which makes possible the performance of the integration over r',

$$S(r, t) = \frac{1}{\Omega} < \varphi_0 \mid \sum_{ij} \delta(r_{it} - r - r_{j0}) \mid \varphi_0 > \tag{2-56}$$

Since all the particles are equivalent, each value of j gives the same contribution. We can thus write

$$S(r, t) \sim N/\Omega < \rho(r + r_{j0}, t) > \tag{2-57}$$

(where N is the total number of particles). $\Omega S/N$ is then a real function, measuring the density at time t as seen from a point where there was a particle at time 0.

Let us set $t = 0$. We can write [see (2–49)]

$$\begin{cases} S(q, 0) = S(q) = \dfrac{1}{\Omega} < \varphi_0 \mid \rho_q \rho_{-q} \mid \varphi_0 > \\ S(r, 0) = S(r) = < \varphi_0 \mid \rho(r)\, \rho(0) \mid \varphi_0 > \end{cases} \tag{2-58}$$

$S(r)$ is a real function for which the interpretation (2–57) is rigorous, in contrast to the case of $t \neq 0$, where it was only approximate.

We turn now to a more detailed study of the properties of S. $S(q,\omega)$ is a real function. From this we deduce at once that

$$\begin{cases} S(q, t) = S(q, -t)^* \\ S(r, t) = S(-r, -t)^* \end{cases} \tag{2-59}$$

When r or t tends to infinity, it is physically obvious that correlations play no role. We see from (2–54) that $S(r,t)$ should tend to the square of the average density,

$$S(r, t) \to N^2/\Omega^2 \qquad \text{if } r \text{ or } t \to \infty \tag{2-60}$$

It is frequently convenient to subtract from $S(r,t)$ this limiting value by defining

$$S'(r, t) = S(r, t) - N^2/\Omega^2$$

S' tends to zero at infinity and uniquely describes the corrections to S due to the interaction between the particles of the system. The limiting value N^2/Ω^2 is of no physical interest whatever, as it gives a contribution to

$S(q,\omega)$ proportional to $\delta(\omega)\delta(q)$, corresponding to an elastic-scattering process without deflection: such a scattering cannot be separated from the incident beam.

In general, $S' \to 0$ past a certain average distance L, called the "correlation length," or past a certain average time T, called the "relaxation time." Direct calculation of S' is a very difficult problem, usually insoluble. On the other hand one can, by qualitative arguments, choose a simple analytic form for S', purely phenomenological, containing L and T as adjustable parameters. The general behavior of numerous phenomena can thus be described. In any case, the introduction of correlation functions leads to a concise and elegant formalism.

We have considered in this section only a restricted case, the scattering of particles interacting with the system by means of a potential $V(r)$. Actually, this method is applicable to any problem whatsoever for which only the *density* of the system is involved. Let us consider, for example, the Compton effect, which arises from the following term of the photon-electron interaction:

$$\frac{e^2}{2mc^2}\sum_i A^2(r_i)$$

$A(r_i)$ is the vector potential of the electromagnetic field at the point where the ith particle is found. This interaction can be written as

$$\frac{e^2}{2mc^2}\int A^2(r)\,\rho(r)\,d^3r$$

Only $\rho(r)$ is involved. Let q be the momentum transfer in the collision; the transition probability is proportional to

$$\sum_n |(\rho_q)_{0n}|^2 = \langle \rho_q \rho_{-q} \rangle$$

The scattering cross section for our system will be equal to that for a free electron multiplied by $\Omega S(q)$.

In conclusion, it remains for us to relate these results to the dielectric constant which we studied in the preceding section. Let us compare (2–38) and (2–47): we see that

$$\mathrm{Im}(1/\varepsilon) = \frac{2\pi i e^2}{q^2}[S(q,-\omega) - S(q,\omega)] \qquad \textit{(2-61)}$$

We give in Appendix A some supplementary details of the mathematical aspects of this relation. For the moment, we shall discuss only its physical significance: this theorem relates S, characteristic of the *correlations* of the system, to Im $(1/\varepsilon)$, characteristic of the *dissipation* of energy of an external source of frequency ω. Furthermore, $S(q,\omega)$ is just the mean square value of the density *fluctuation* of wave vector q and frequency ω. We thus discover the intimate relationship between three very different aspects of the problem: fluctuation, correlation, dissipation.

At nonzero temperatures, these relations are more complicated, but they persist. In classical physics, moreover, they have been known for a long time: as early as 1928 Nyquist demonstrated the connection between fluctuation and dissipation phenomena in his study of the thermal noise of a resistance. These few comments show the generality of the methods.

Chapter 3

General Properties of Green's Functions

We shall now set up a formalism based on the use of the "Green's functions" of the system, describing the propagation of one or more specified particles. This technique is frequently used in field theory; it is particularly fruitful, as it allows for a compact and coherent treatment of a large number of properties. Its application to the many-body problem, although relatively recent, has been the subject of a large number of articles; the reader will find a short Bibliography at the end of the book. The presentation which we shall adopt is similar to that of Galitskii and Migdal.

These methods have the advantage of allowing for a natural generalization to the case of nonzero temperature, thus establishing the basis of a new method of quantum-statistical mechanics. We shall limit ourselves to the study of the ground state and elementary excitations of the system.

The propagators in which we shall be interested are essentially the functions $f(r_i t_i, r_i' t_i')$, describing the probability amplitude for finding particles at the points $(r_i' t_i')$ when they have been introduced at the points $(r_i t_i)$. In a simple mathematical form, these functions pick out the physical information required to treat all problems involving only a small number of particles at one time. The objectives of this study are as follows:

a. Precise formulation of the propagation of a bare particle.

b. The setting up of approximate eigenstates corresponding to "clothed" particles or quasi particles, generally of finite lifetime. These quasi particles constitute the directly observable elementary excitations. The first excited states generally consist of "configurations" of quasi particles, more or less complex, whose components are to first approximation independent.

c. The demonstration of the existence of bound (collective) states, which do not have an anlog in an ideal gas, and which must be treated independently of the continuum mentioned in the preceding paragraph.

With the help of a certain number of hypotheses, we can thus justify the Landau model discussed in Chap. 1. Furthermore we can relate Green's functions to the correlation functions studied in Chap. 2, thus bringing the necessary unity to this review.

In the first section we discuss the general properties of the one-particle Green's function. The second section is devoted to analogous study of the two-particle Green's function, necessarily less detailed, because of the greater complexity of the problem. The conclusions of this chapter are general. Chapter 5 will be restricted to the study of Green's functions within the restricted framework of a perturbation expansion.

1. The Single-Particle Green's Function

a. Generalities

From now on we shall use the formalism of second quantization. The reader will find a brief outline of this method in Appendix B. Furthermore, except when noted to the contrary, we shall use the Heisenberg representation to describe the dynamic evolution of the system. (We shall denote the latter by the use of boldface letters for operators and wave vectors.)

The single-particle Green's function $G(rt,r't')$ is defined by the relation

$$G(rt, r' t') = i < \varphi_0 \mid T \{ \psi(r, t)\, \psi^*(r't') \} \mid \varphi_0 > \qquad \textit{(3–1)}$$

where $\boldsymbol{\psi}$, $\boldsymbol{\psi}^*$ are particle-creation and -destruction operators, and $|\varphi_0\rangle$ the normalized *actual* ground-state vector of the system (that is to say, taking account of all interactions). T is a "chronological" operator, which has the effect of ordering the factors within the brackets so that the time variable increases from *right to left*. Furthermore, in the case of a fermion system, the operation is accompanied by a change in sign each time that two operators are permuted with respect to the reference order indicated in the brackets. In short

$$T \{ \psi(r, t), \psi^*(r' t') \} = \begin{cases} \psi(r, t)\, \psi^*(r't') & \text{if } t' < t \\ \pm\, \psi^*(r't')\, \psi(r, t) & \text{if } t' > t \end{cases} \quad \begin{cases} + : \text{boson} \\ - : \text{fermion} \end{cases} \qquad \textit{(3–2)}$$

In the following we shall use the condensed notation $G(x,x')$, where x takes account of the four variables r,t. For obvious reasons of translational invariance, $G(x,x')$ is a function only of the difference $x - x'$ (this would no longer be true for a system with periodic structure). It is an intensive quantity, independent of volume.

For a Fermi gas we must introduce into (3–1) the spin of each particle. We thus set

$$G(rt\sigma, r't'\sigma') = i < \varphi_0 \mid T\{ \psi_\sigma(r, t)\, \psi^*_{\sigma'}(r't') \} \mid \varphi_0 > \qquad (3\text{–}1a)$$

In order to shorten the equations we introduce the symbol $\mathbf{r}$, representing the two quantities (r,σ), and also $\mathbf{x}$, corresponding to (r,t,σ). G is thus a function of $\mathbf{x}$ and $\mathbf{x}'$.

For an isolated system the total spin is a good quantum number. The Green's function is nonzero only if $\sigma = \sigma'$. If, furthermore, the system is isotropic (in particular, in the absence of a magnetic field), G is spin-independent. We shall consider only this single case, which often allows us to drop the spin indices. The transposition of our results to the more general case is easy (we just have to consider G as a 2×2 matrix). We shall come back to this question in the second section of Chap. 4.

We shall consider only a system of fermions. Most of the results which we shall state are easily transposed to the case of a system of bosons, except for those which are concerned with the structure of the excitation spectrum.

Physical interpretation of the Green's function. Let us assume $t > t'$. Setting aside the factor i, we find that $G(rt,r't')$ is then the scalar product of two states

$$\psi^*_{\sigma'}(r't') \mid \varphi_0 > \qquad \text{and} \qquad \psi^*_\sigma(r, t) \mid \varphi_0 >$$

The physical meaning of G is clear. G is the probability amplitude, when a particle has been added to the ground state at $(\mathbf{r}'t')$, for finding it at $(\mathbf{r}t)$, *without any other modification of the ground state* $|\varphi_0\rangle$. Let us note that the above states are not normalized. This is not at all surprising; their norm just gives the probability of *being able* to create a particle at $(\mathbf{r}t)$ or $(\mathbf{r}'t')$ in addition to the ground state (this is not always possible, because of the exclusion principle). As a consequence, the probability amplitude G can, if necessary, be decomposed into three factors:

(a) Probability of existence of $\psi^*(r't')\,|\varphi_0\rangle$
(b) Probability of existence of $\psi^*(rt)\,|\varphi_0\rangle$
(c) Probability that the first of these states evolve to the second

We see that, for $t > t'$, G describes the "propagation" of an extra particle, whence its name "propagator." Starting with the ground state supplemented by a particle in $(\mathbf{r}'t')$, we can follow the history of this "elementary deformation" of the wave function—in particular, the irreversible evolution toward incoherence: the additional particle is "dissolved" into the system. The main advantage of the Green's function G is that it isolates the information relative to the additional particle, without being involved with useless details concerning the structure of the ground state.

Let us now assume $t < t'$. Up to a factor $(-i)$, G is the scalar product of the states

$$\psi_\sigma(r, t) \mid \varphi_0 > \qquad \text{and} \qquad \psi_{\sigma'}(r', t') \mid \varphi_0 >$$

These states contain $N - 1$ particles (instead of $N + 1$ in the case $t > t'$); we are dealing with the propagation of a *hole* from $(\mathbf{r}t)$ to $(\mathbf{r}'t')$. Let us note that, owing to the chronological operator, propagation always takes place from past to future (the propagation of a hole being equivalent to the propagation of a particle backward in time).

General properties of $\mathbf{G}(x,0) = \mathbf{G}(x)$. $G(rt)$ is discontinuous at $t = 0$. We have

$$G(r, +0) - G(r, -0) = \mathrm{i} < \varphi_0 \mid [\psi_\sigma(r), \psi_\sigma^*(0)]_+ \mid \varphi_0 > = \mathrm{i}\delta(r) \qquad (3\text{–}3)$$

The discontinuity of G is thus highly singular.

Moreover, G is a complex function. From (3–1), we deduce that

$$G^*(\mathbf{r}t, \mathbf{r}'t') = +\mathrm{i} < \varphi_0 \mid \widetilde{T} \{ \psi_\sigma^*(r, t)\, \psi_{\sigma'}(r't') \} \mid \varphi_0 > \qquad (3\text{–}4)$$

where $\widetilde{T}$ is an "antichronological" operator, ordering the times in an increasing sequence from left to right. If we define an auxiliary function

$$\widetilde{G}(r, t) = \mathrm{i} < \varphi_0 \mid \widetilde{T} \{ \psi_\sigma(r, t)\, \psi_\sigma^*(0, 0) \} \mid \varphi_0 > \qquad (3\text{–}5)$$

By analogy with (3–1), we see that

$$G^*(r, t) = -\, \widetilde{G}(-r, -t)$$

Let us now turn to the study of the various Fourier transforms of G.

Let us first study the Fourier transform with respect to the space variables r and r'. We can write

$$\left\{\begin{aligned} G(\mathbf{r}t, \mathbf{r}'t') &= \frac{1}{\Omega}\sum_{kk'} G(\mathbf{k}t, \mathbf{k}'t')\exp[i(kr - k'r')] \\ G(\mathbf{k}t, \mathbf{k}'t') &= \frac{1}{\Omega}\iint d^3r\, d^3r'\, G(\mathbf{r}t, \mathbf{r}'t')\exp[-i(kr - k'r')] \end{aligned}\right. \tag{3-6}$$

(as in Chap. 1, $\mathbf{k}$ is an abridged notation for $(k\sigma)$). The second of these formulas follows from the "inversion" relations of the Fourier transform

$$\left\{\begin{aligned} &\sum_k e^{ikr} = \Omega\,\delta(r) \\ &\int d^3r\, e^{-ikr} = \Omega\,\delta_{k,0} \end{aligned}\right. \tag{3-7}$$

A glance at Eq. (B–7) shows that $G(\mathbf{k}t,\mathbf{k}'t')$ is given by

$$G(\mathbf{k}t, \mathbf{k}'t') = i < \varphi_0 \mid T\left\{ \mathbf{a}_\mathbf{k}(t)\, \mathbf{a}^*_{\mathbf{k}'}(t') \right\} \mid \varphi_0 > \tag{3-8}$$

For a uniform system $G(\mathbf{k}t,\mathbf{k}'t')$ is different from zero only if $\mathbf{k}' = \mathbf{k}$. The function

$$G(k, t) = G(\mathbf{k}t, \mathbf{k}0) = i < \varphi_0 \mid T\left\{ \mathbf{a}_\mathbf{k}(t)\, \mathbf{a}^*_\mathbf{k}(0) \right\} \mid \varphi_0 > \tag{3-9}$$

is just the Fourier transform of the function $G(r,t)$,

$$\begin{aligned} G(k, t) &= \int d^3r\, G(r, t)\, e^{-ikr} \\ G(r, t) &= \frac{1}{\Omega}\sum_k e^{ikr}\, G(k, t) \end{aligned} \tag{3-10}$$

In what follows we shall encounter each of the two forms of G, functions of one or two variables k. We have stressed this quite trivial point in order to indicate the factors of Ω entering into the equations. Note that $G(k,t)$ is independent of Ω.

Like $G(r,t)$, the function $G(k,t)$ is complex and discontinuous for $t = 0$. The magnitude of the discontinuity is given by

$$\begin{cases} G(k, +0) = i(1 - m_k) \\ G(k, -0) = -im_k \\ G(k, +0) - G(k, -0) = i \end{cases} \tag{3-11}$$

with

$$m_k = \langle \varphi_0 | a_{\mathbf{k}}^* a_{\mathbf{k}} | \varphi_0 \rangle$$

m_k measures the average occupation of the plane wave $\mathbf{k}$ in the ground state $|\varphi_0\rangle$. It gives the distribution of *bare* particles, not to be confused with the distribution of quasi particles $n(\mathbf{k})$ defined in Chap. 1.

For $t > 0$, $G(k,t)$ describes the propagation of a particle of wave vector $\mathbf{k}$, "emitted" at time zero and "detected" at time t. On the other hand, for $t < 0$, G describes the propagation of a hole of wave vector $\mathbf{k}$, emitted at time t and detected at the origin. It should be emphasized that except for the case of an ideal gas without interactions, neither $a^*_{\mathbf{k}}|\varphi_0\rangle$ nor $a_{\mathbf{k}}|\varphi_0\rangle$ is an *eigenstate* of the system. Both correspond to complicated combinations which are eventually damped [which is translated mathematically by the requirement $G(k,t) \to 0$ when $t \to \pm\infty$]. Moreover they have a norm < 1, equal to the probability $(1 - m_k)$ or m_k of being able to create or destroy a particle of wave vector $\mathbf{k}$ in the ground state.

We can give a more explicit representation of $G(k,t)$ by using the Heisenberg representation of the operators $a_{\mathbf{k}}$ and $a_{\mathbf{k}}^*$,

$$\mathbf{a}_{\mathbf{k}}(t) = e^{iHt} a_{\mathbf{k}} e^{-iHt} \tag{3-12}$$

We then have

$$G(k, t) = \begin{cases} i \langle \varphi_0 | e^{iHt} a_{\mathbf{k}} e^{-iHt} a_{\mathbf{k}}^* | \varphi_0 \rangle & \text{if } t > 0 \\ -i \langle \varphi_0 | a_{\mathbf{k}}^* e^{iHt} a_{\mathbf{k}} e^{-iHt} | \varphi_0 \rangle & \text{if } t < 0 \end{cases} \tag{3-13}$$

These relations serve as a starting point for the Lehmann expansion studied in the following section.

Note: We set the constant $\hbar$ equal to unity.

Let us turn now to the Fourier transform with respect to the time variables. We define the function $G(\mathbf{k}'\omega,\mathbf{k}'\omega')$ by

$$G(\mathbf{k}\omega, \mathbf{k}'\omega') = \frac{1}{2\pi}\iint dt\, dt'\, G(\mathbf{k}t, \mathbf{k}'t')\exp[+\,i(\omega t - \omega' t')]$$

$$G(\mathbf{k}t, \mathbf{k}'t') = \frac{1}{2\pi}\iint d\omega\, d\omega'\, G(\mathbf{k}\omega, \mathbf{k}'\omega')\exp[-\,i(\omega t - \omega' t')] \tag{3-14}$$

These relations simplify if it is noted that $G(\mathbf{k}t,\mathbf{k}'t')$ depends only on $t - t'$. Under these conditions $G(\mathbf{k}\omega,\mathbf{k}'\omega')$ takes the form

$$G(\mathbf{k}\omega, \mathbf{k}'\omega') = G(k\omega)\, \delta_{\mathbf{k},\mathbf{k}'}\, \delta(\omega - \omega') \tag{3-15}$$

the reduced functions $G(k,\omega)$ satisfying the pair of equations

$$\begin{cases} G(k, t) = \dfrac{1}{2\pi}\displaystyle\int d\omega\, G(k, \omega)\, e^{-i\omega t} \\ G(k, \omega) = \displaystyle\int dt\, G(k, t)\, e^{+i\omega t} \end{cases} \tag{3-16}$$

The second equation (3–16) can sometimes involve a certain amount of ambiguity. In case of doubt, it is the first equation (3–16) which must be consulted to determine $G(k,\omega)$.

We shall see later in this chapter that the Green's function $G(k,\omega)$ contains a great deal of information. It is much easier to manipulate than its transform $G(r,t)$; it is with $G(k,\omega)$ that we shall construct approximate eigenstates of the system, or "quasi particles." In order to get a preview of these results, let us calculate $G(k,\omega)$ for an ideal gas. The ground state is extremely simple in this case: the states of wave vector $|k| < k_F$ are filled; those for which $|k| > k_F$ are empty (k_F is the Fermi wave vector). It is furthermore easily verified that

$$\begin{cases} \mathbf{a}_{\mathbf{k}}(t) = \exp(-\,i\overset{0}{\varepsilon}_k t)\, a_{\mathbf{k}} \\ \mathbf{a}^*_{\mathbf{k}}(t) = \exp(i\overset{0}{\varepsilon}_k t)\, a^*_k \end{cases}$$

where $\overset{\circ}{\varepsilon}_k = k^2/2m$ is the energy of a particle of wave vector $\mathbf{k}$. From this we deduce

$$\begin{cases} |k| < k_F \qquad G_0(k, t) = \begin{cases} -\,i\exp(-\,i\overset{0}{\varepsilon}_k t) & \text{if } t < 0 \\ 0 & \text{if } t > 0 \end{cases} \\ |k| > k_F \qquad G_0(k, t) = \begin{cases} 0 & \text{if } t < 0 \\ i\exp(-\,i\overset{0}{\varepsilon}_k t) & \text{if } t > 0 \end{cases} \end{cases} \tag{3-17}$$

Let us return to the first equation (3–16): we obtain for $G(k,t)$ the above values by choosing for $G(k,\omega)$ the expression

$$\left\{\begin{aligned} G_0(k,\omega) &= \frac{1}{\overset{0}{\varepsilon}_k - \omega - i\eta} \\ \eta &= \begin{cases} +0 & \text{for } |k| > k_F \\ -0 & \text{for } |k| < k_F \end{cases} \end{aligned}\right. \tag{3-18}$$

We note the existence of a pole for $\omega = \overset{\circ}{\varepsilon}_k - i\eta$: we shall generalize this result in what follows, by showing that the poles of G give the energies of the quasi particles, that is to say, the elementary excitation spectrum.

b. Lehmann representation

Hitherto we have limited ourselves to a formal study of the properties of G. We now intend to analyze in more detail the analytic structure of $G(k,\omega)$. We shall use for this purpose an expansion method proposed many years ago by Lehmann, in relation to general problems of field theory.

Let us introduce the set of eigenstates of our system (without restriction on the total number of particles). This set is complete. Let $|\varphi_n\rangle$ be one of these states, of energy E_n: we shall define $\omega_{n0} = E_n - E_0$. Let us stress the fact that we are concerned with the *exact* eigenstates of the complete system, with interactions, and not with any kind of approximate states. Equations (3–19) can then be written

$$G(k,t) = \begin{cases} i\sum_n \left| \langle \varphi_n | \overset{*}{a}_k | \varphi_0 \rangle \right|^2 \exp(-i\omega_{n0}t) & \text{if } t > 0 \\ -i\sum_m \left| \langle \varphi_m | a_k | \varphi_0 \rangle \right|^2 \exp(+i\omega_{m0}t) & \text{if } t < 0 \end{cases} \tag{3-19}$$

If $t > 0$, the intermediate states $|\varphi_n\rangle$ contain $N + 1$ particles and have total momentum $\mathbf{k}$. On the other hand, if $t < 0$, these states correspond to $N - 1$ particles, with momentum $-\mathbf{k}$.

Let us limit ourselves to the case where the ground-state energy E_0 increases monotonically with the number of particles, and let us write

$$E_0(N+1) - E_0(N) = E_0(N) - E_0(N-1) = \mu \tag{3-20}$$

The quantity μ plays the role of a chemical potential. It is independent of N to order $1/N$. In the case of states of $N + 1$ particles, for example, the excitation energy ω_{n0} can then be decomposed in the form

$$\omega_{n0} = \mu + \xi_{n0}$$

where ξ_{n0} represents the *excitation* energy with respect to the ground state having the same number of particles, namely, $N + 1$. By definition, ξ_{n0} is positive. Similarly, for states of $N - 1$ particles we can write

$$\omega_{m0} = -\mu + \xi_{m0}$$

where ξ_{m0} is again positive.

Note: In this chapter we consider only isolated systems. Let us assume on the contrary that the system is in contact with a particle reservoir of infinite capacity: it is no longer the number of particles which is fixed, but the chemical potential μ. The interesting quantity is no longer the Hamiltonian H, but the operator $F = H - \mu N$, the minimization of which determines the ground state. It is easily seen that

$$\xi_{n0} = F_n - F_0$$

which clarifies the physical significance of ξ_{n0}.

In certain cases one can introduce, instead of G, a new function $\overline{G}$, defined by equations analogous to (3–13), H simply being replaced by F. It is easily verified that

$$\overline{G}(k, t) = e^{-i\mu t}\, G(k, t)$$

The Hamiltonian then disappears completely from the problem: we are concerned only with the operator F and the energy differences ξ_{n0}. Such a presentation is sometimes more compact and more elegant.

We now regroup the different states $|\varphi_n\rangle$ by defining two functions

$$\begin{cases} A_+(k, \omega) = \sum_n |< \varphi_n | a^*_{\mathbf{k}} | \varphi_0 >|^2\, \delta(\omega - \xi_{n0}) \\ A_-(k, \omega) = \sum_m |< \varphi_m | a_{\mathbf{k}} | \varphi_0 >|^2\, \delta(\omega - \xi_{m0}) \end{cases} \qquad (3\text{–}21)$$

A_+ and A_- are real positive functions, whose physical interpretation is simple. For example, let us make a spectral analysis of the state $a^*_{\mathbf{k}}|\varphi_0\rangle$; the total norm of the state is distributed among the different frequencies, the "density of norm" at the frequency $(\omega + \mu)$ being given by $A_+(k,\omega)$.

Similarly, $A_-(k,\omega)$ is the density of norm at the frequency $(\omega-\mu)$ resulting from the spectral analysis of the state $a_k|\varphi_0\rangle$. Let us remark that if $a_{\mathbf{k}}^*|\boldsymbol{\varphi}_0\rangle$ were an exact eigenstate of the system, of excitation energy ξ_k, the function $A_+(k,\omega)$ would reduce to the Dirac δ function $\delta(\omega-\xi_k)$. Actually, $a_{\mathbf{k}}^*|\boldsymbol{\varphi}_0\rangle$ is not an eigenstate; the δ function spreads out and gives rise to a finite density $A_+(k,\omega)$. $A_+(k,\omega)$ can, however, exhibit a more or less sharp peak: we shall see further on that this peak corresponds to an approximate elementary excitation.

By using (3–21) we can put equations (3–19) into the form

$$G(k,t)=\begin{cases} \mathrm{i}\displaystyle\int_0^\infty A_+(k,\omega)\exp\left[-\mathrm{i}(\mu+\omega)\,t\right]\mathrm{d}\omega & \text{for } t>0 \\ -\mathrm{i}\displaystyle\int_0^\infty A_-(k,\omega)\exp\left[-\mathrm{i}(\mu-\omega)\,t\right]\mathrm{d}\omega & \text{for } t<0 \end{cases} \qquad (3\text{–}22)$$

If we go back to the definition of $G(k,\omega)$, we see that

$$G(k,\omega)=\int_0^\infty \mathrm{d}\omega'\left\{\frac{A_+(k,\omega')}{\omega'-\omega+\mu-\mathrm{i}\eta}-\frac{A_-(k,\omega')}{\omega'+\omega-\mu-\mathrm{i}\eta}\right\} \qquad (3\text{–}23)$$

where η is a real positive infinitesimal quantity. (3–23) constitutes the "Lehmann representation" of the Green's function. This relation allows us to establish numerous analytic properties of G, discussed in detail in Appendix C. It should be pointed out that the complex function $G(k,\omega)$ actually depends only on the two real functions A_+ and A_-, defined in the interval $(0,+\infty)$; G has a very special structure, which is reminiscent of the correlation functions studied in Chap. 2.

c. Relation between G and the properties of the ground state

Let us first calculate the population m_k of the bare state of wave vector $\mathbf{k}$. From (3–11) we have

$$m_k=\mathrm{i}\left[G(k,t)\right]_{t=-0}=\frac{\mathrm{i}}{2\pi}\int_C \mathrm{d}\omega\, G(k,\omega) \qquad (3\text{–}24)$$

where C is a contour composed of the real axis and a semicircle at infinity in the upper half-plane. The total number of particles is trivially deduced from (3–24),

$$N=\sum_{\mathbf{k}} m_k=\frac{\mathrm{i}}{2\pi}\sum_{\mathbf{k}}\int_C \mathrm{d}\omega\, G(k,\omega) \qquad (3\text{–}25)$$

In order to calculate the ground-state energy E_0, we use a device due to Galitskii and Migdal. Let us consider the quantity

$$\sum_{\mathbf{k}} \overset{*}{a}_{\mathbf{k}} [T, a_{\mathbf{k}}]$$

where T is the total kinetic energy operator

$$T = \sum_{\mathbf{k}} \frac{k^2}{2m} \overset{*}{a}_{\mathbf{k}} a_{\mathbf{k}}$$

It is easily verified that

$$\sum_{\mathbf{k}} \overset{*}{a}_{\mathbf{k}} [T, a_{\mathbf{k}}] = -T \tag{3-26}$$

Let us replace T by the potential energy operator V, which for a binary law of interaction is of the type

$$V = \sum_{\mathbf{k}_1, \mathbf{k}_2, \mathbf{k}_3, \mathbf{k}_4} V(\mathbf{k}_1, \ldots, \mathbf{k}_4)\, \overset{*}{a}_{\mathbf{k}_1} \overset{*}{a}_{\mathbf{k}_2} a_{\mathbf{k}_3} a_{\mathbf{k}_4}$$

In the same way we see that

$$\sum_{\mathbf{k}} \overset{*}{a}_{\mathbf{k}} [V, a_{\mathbf{k}}] = -2V \tag{3-27}$$

As a consequence

$$\sum_{\mathbf{k}} < \varphi_0 \mid \overset{*}{a}_{\mathbf{k}} [H, a_{\mathbf{k}}] \mid \varphi_0 > = -< \varphi_0 \mid T + 2V \mid \varphi_0 > \tag{3-28}$$

Since, on the other hand, we can explicitly calculate the average value of the kinetic energy

$$< \varphi_0 \mid T \mid \varphi_0 > = \sum_{\mathbf{k}} \frac{k^2}{2m} m_k$$

we can deduce from it the average value of the ground-state energy E_0

$$E_0 = \langle \varphi_0 | T + V | \varphi_0 \rangle = \tfrac{1}{2} \sum_{\mathbf{k}} \left\{ \frac{k^2}{2m} m_k - \langle \varphi_0 | a^*_{\mathbf{k}} [H, a_{\mathbf{k}}] | \varphi_0 \rangle \right\} \tag{3-29}$$

The different quantities entering into (3–29) are simply related to the functions A_+ and A_-. In particular, from (3–22)

$$\langle \varphi_0 | a^*_{\mathbf{k}} [H, a_{\mathbf{k}}] | \varphi_0 \rangle = \int_0^\infty A_-(k, \omega) [\omega - \mu] \, d\omega = \left[\frac{dG(k, t)}{dt} \right]_{t=-0} \tag{3-30}$$

Returning to the definition of the Fourier transform, we finally obtain

$$\langle \varphi_0 | a^*_k [H, a_{\mathbf{k}}] | \varphi_0 \rangle = \frac{1}{2\pi i} \int_C G(k, \omega) \, \omega \, d\omega \tag{3-31}$$

where C is the contour defined above. Let us put this result back into (3–29) and use (3–24): we find

$$E_0 = \frac{i}{4\pi} \sum_{\mathbf{k}} \int_C d\omega \, G(k, \omega) [k^2/2m + \omega] \tag{3-32}$$

E_0 is thus completely determined when $G(k, \omega)$ is known. Later we shall sometimes need the interaction energy

$$E_{\text{int}} = \langle \varphi_0 | V | \varphi_0 \rangle$$

The latter is easily obtained from (3–28) and is given by

$$E_{\text{int}} = \frac{i}{4\pi} \sum_{\mathbf{k}} \int_C d\omega \, G(k\omega) \left[\omega - \frac{k^2}{2m} \right] \tag{3-33}$$

Relation (3–32) is very useful. It allows us to calculate various other properties, such as the pressure. The derivation given here is general. We shall reobtain this result in Chap. 5 by a method which is based on a perturbation expansion, and which, because of this, is questionable.

Note: We can write

$$-\langle \varphi_0 | a^*_{\mathbf{k}} [T + V, a_{\mathbf{k}}] | \varphi_0 \rangle = m_k (T_k + V_k)$$

m_k is the average number of particles in the state $\mathbf{k}$, T_k their kinetic energy $k^2/2m$, and V_k a kind of "average" interaction energy of these particles. The quantity $T_k + V_k$ is different from the Hartree-Fock energy (see Appendix C), since it takes account of the correlation energy. The ground-state energy E_0 is then written

$$E_0 = \sum_{\mathbf{k}} m_k(T_k + \tfrac{1}{2} V_k)$$

which is intuitively quite obvious.

d. Relation between G and the elementary excitations

Let us consider the state $a^*_{\mathbf{k}}|\varphi_0\rangle$: this is not an eigenstate, but a complicated combination of a large number of eigenstates. Its norm $1 - m_k$ is distributed among many frequencies. The density of norm at the frequency $(\omega + \mu)$ is given by $A_+(k,\omega)$. If $A_+(k,\omega)$ exhibits a more or less sharp maximum, in a band of width $2\Gamma_k$, centered at ξ_k, the components of $a^*_{\mathbf{k}}|\varphi_0\rangle$ corresponding to this peak oscillate at a frequency ξ_k and their phases remain coherent over a time $\sim 1/\Gamma_k$. We can therefore hope that, if we "filter" the state $a^*_{\mathbf{k}}|\varphi_0\rangle$ in such a way as to retain only the peak of $A_+(k,\omega)$, we shall obtain an approximate eigenstate, of energy $\xi_k + \mu$, of lifetime $1/\Gamma_k$.

In order to clarify this point, let us study the "propagation" of $a^*_{\mathbf{k}}|\varphi_0\rangle$, that is to say, the quantity (defined for $t > 0$)

$$< \varphi_0 \mid \mathbf{a}_{\mathbf{k}}(t)\, \mathbf{a}^*_{\mathbf{k}} \mid \varphi_0 > = -\mathrm{i}\, G(k, t) \tag{3-34}$$

By referring to (3–22) we see that

$$-\mathrm{i}\, G(k, t) = \mathrm{e}^{-\mathrm{i}\mu t} \int_0^\infty A_+(k, \omega)\, \mathrm{e}^{-\mathrm{i}\omega t}\, d\omega \tag{3-35}$$

We can deform the contour of integration (see Fig. 2), choosing $\alpha \gg 1/t$, so that $e^{-\alpha t}$ is negligible. We then have

$$-\mathrm{i}G(k, t)\, \mathrm{e}^{\mathrm{i}\mu t} \approx \int_0^{-\mathrm{i}\alpha} A_+(k, \omega)\, \mathrm{e}^{-\mathrm{i}\omega t}\, d\omega - 2\pi\mathrm{i} \sum_j \zeta_j\, \mathrm{e}^{-\mathrm{i}\omega_j t} \tag{3-36}$$

where the ξ_j are the poles of $A_+(k,\omega)$ in the shaded region and ζ_j their residues. One of these poles, of frequency $\omega = \xi_k - i\Gamma_k$, is located very near the real axis (the heavily shaded region in Fig. 2): it is this one which gives rise to the peak of A_+ indicated in Fig. 2.

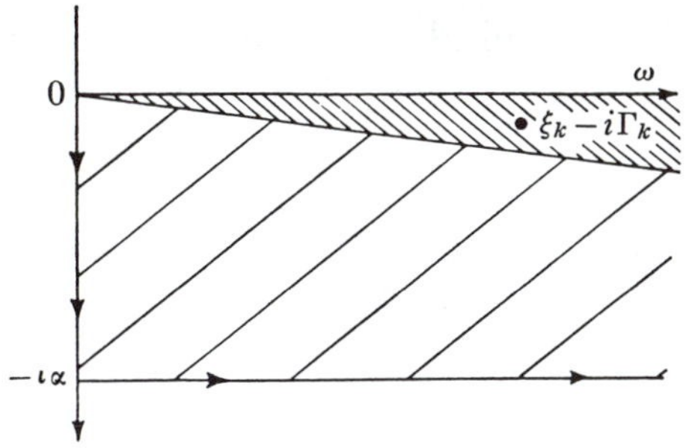

Figure 1

We are thus led to distinguish three distinct cases:

(a) t is small; α must be large; the first term of (3–36) is predominant. The variation of $G(t)$ is complicated, as all the components of $\overset{*}{a}_{\mathbf{k}}|\boldsymbol{\varphi}_0\rangle$ are important, whatever their frequency.

(b) t is very large, $\gg \Gamma_k^{-1}$. The contour of integration no longer encloses a pole. Only the first term of (3–36) remains, and its contribution is moreover extremely small, as the phases of the different components of $\overset{*}{a}_{\mathbf{k}}|\boldsymbol{\varphi}_0\rangle$ are completely incoherent. The variation of $G(t)$ has no periodicity.

(c) t is large, but $\lesssim \Gamma_k^{-1}$. The principal contribution arises from the pole $\varepsilon_k - i\Gamma_k$; we obtain

$$-i\, G(k, t) = -2\,\pi i\, \zeta_k \exp[(-i\,\xi_k - i\mu - \Gamma_k)\, t] \tag{3-37}$$

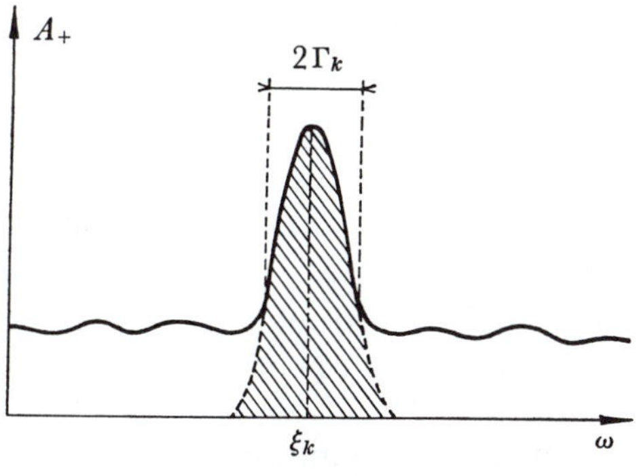

Figure 2

This periodic variation is typical of an excited state of energy $\xi_k + \mu$, of lifetime Γ_k^{-1}: this state contains $N + 1$ particles, with total wave vector **k**.

Equation (3–37) can be exact only if:

(a) The phases of the continuous background of $a_{\mathbf{k}}|\varphi_0\rangle$ are incoherent [first term of (3–36) negligible].

(b) The phases of the components of $\overset{*}{a}_{\mathbf{k}}|\varphi_0\rangle$ corresponding to the peak of A_+ are weakly dispersed; in other words, $\Gamma_k t \lesssim 1$.

These conditions can be realized only if Γ_k is sufficiently small. Physically this amounts to saying that only excitations whose lifetime is much longer than a certain "average period of excitation," characteristic of local fluctuations, can be observed.

The peak of $A_+(k,\omega)$ encloses an area

$$z_k = -2\pi i\, \zeta_k \tag{3–38}$$

Consequently, $\overset{*}{a}_{\mathbf{k}}|\varphi_0\rangle$ is divided into two parts:

(a) A coherent part, of norm z_k, oscillating with the frequency $\xi_k + \mu$ over times $< 1/\Gamma_k$, which describes a quasi particle.

(b) A uniform continuous incoherent background, of norm $1 - m_k - z_k$.

This continuous background is always present, but it quickly disappears from $G(t)$ because of the incoherence of the phases.

All the preceding discussion has been concerned with the state $\overset{*}{a}_{\mathbf{k}}|\varphi_0\rangle$, which has led us to the notion of quasi particles. We can repeat this study point by point by applying the method to the state $a_{\mathbf{k}}|\varphi_0\rangle$: in this way we shall arrive at the concept of quasi holes. Let us summarize the principal results. If $A_-(k,\omega)$ possesses a pole near the real axis, of energy $\xi_k - i\Gamma_k$, of residue $\zeta_k = -z_k/2\pi i$, we can decompose the state $a_{\mathbf{k}}|\varphi_0\rangle$ into two parts:

(a) A continuous background, of norm $m_k - z_k$.

(b) An elementary excitation or quasi hole, of energy $\mu - \xi_k$, of lifetime Γ_k^{-1}, of norm z_k. As before, this excitation actually exists only if Γ_k is sufficiently small.

This pole is manifested by a strong maximum of $A_-(k,\omega)$ on the real axis, the area under this maximum being equal to z_k.

We shall conclude this discussion with the help of the results of Appendix C. The elementary excitations are determined by the poles of the continuation of G obtained by a clockwise rotation of the real axis.

The poles located in the fourth quadrant arise from A_+ and correspond to quasi particles ($N + 1$ particles, total momentum $\mathbf{k}$); those located in the second quadrant arise from A_- and correspond to quasi holes ($N - 1$ particles, total momentum $-\mathbf{k}$). The position of the pole gives both the frequency and lifetime of the excitation. The residue of G, equal to z_k, gives the fraction of a quasi particle in the composite state $a^*_{\mathbf{k}}|\varphi_0\rangle$ (or $a_{\mathbf{k}}|\varphi_0\rangle$). Because of their finite lifetime, these elementary excitations are only approximate. (This is fundamentally different from the case of quantum electrodynamics.)

For the moment the above results remain very formal. In order to be able to obtain practical results, we need information concerning the structure of the elementary excitation spectrum, as a function of k. At this stage, we can say nothing general, and we must consider each particular case. This will be the object of the following chapters.

The study of G having been discontinued, it would be well to summarize our results. In practice, G contains all information concerning *one* particle or *one* hole. In certain cases $G(k,t)$ has, within a limited period of time, a periodic behavior, characteristic of an elementary excitation of *quasi particle*. We can think of these excitations as *bare* particles, carrying along a cloud of neighboring particles. Within a short interval of time, G departs from its periodic form: this represents the *inertia* of the cloud clothing the bare particles. Furthermore, when t increases, we find that the particle is damped and is "dissolved" into the ground state. All this shows that the notion of quasi particle is of limited validity and contains much less information than the complete Green's function $G(k,t)$.

Of course, the single-particle Green's function is not sufficient to treat all problems. In particular, G contains no information concerning the interaction between quasi particles. It is in order to treat such problems that we introduce the *two*-particle Green's function.

2. The Two-Particle Green's Function

a. General properties

By analogy with G we define the two-particle Green's function, $K(\mathbf{x}_1 \ldots \mathbf{x}_4)$ by the relation

$$K(\mathbf{x}_1, \mathbf{x}_2, \mathbf{x}_3, \mathbf{x}_4) = \langle \varphi_0 | T \left\{ \psi_{\sigma_1}(x_1)\, \psi_{\sigma_2}(x_2)\, \psi^*_{\sigma_3}(x_3)\, \psi^*_{\sigma_4}(x_4) \right\} | \varphi_0 \rangle \quad (3\text{–}39)$$

$K(\mathbf{x}_i)$ is an intensive function, independent of Ω. It is a complex function, invariant with respect to simultaneous translation of the four coordinates

$\mathbf{x}_i$. From the definition of the chronological operator T we have

$$K(1\ 2\ 3\ 4) = -K(2\ 1\ 3\ 4) = -K(1\ 2\ 4\ 3) = K(2\ 1\ 4\ 3) \quad \textit{(3–40)}$$

As a consequence it is sufficient for the analysis of $K(\mathbf{x}_i)$ to consider, for example, the three cases

$$\begin{cases} t_1, t_2 > t_3, t_4 \\ t_1, t_2 < t_3, t_4 \\ t_1, t_3 > t_2, t_4 \end{cases}$$

all other cases being obtainable by permutation.

The two-particle Green's function depends on the spin indices σ_i. For an isolated system the total spin is conserved: we must have

$$\sigma_1 + \sigma_2 = \sigma_3 + \sigma_4$$

If, furthermore, the ground state is isotropic, it is invariant under a reversal of all spins. We then have only two independent components of K, corresponding to $\sigma_1 = \pm \sigma_2$.

$K(\mathbf{x}_i)$ is discontinuous when t_1 or t_2 is equal to t_3 or t_4. Consider, for instance, $t_1 = t_4$:

$$K(\mathbf{r}_1(t+0), \mathbf{x}_2, \mathbf{x}_3, \mathbf{r}_4\, t) - K(\mathbf{r}_1(t-0), \mathbf{x}_2, \mathbf{x}_3, \mathbf{r}_4\, t) = -\mathrm{i}\, G(\mathbf{x}_2, \mathbf{x}_3)\, \delta(\mathbf{r}_1 - \mathbf{r}_4) \quad \textit{(3–41)}$$

where we have denoted by $\delta(\mathbf{r}_1 - \mathbf{r}_4)$ the product

$$\delta(r_1 - r_4)\, \delta_{\sigma_1, \sigma_4} \quad \textit{(3–41a)}$$

The discontinuities of K thus immediately give the single-particle Green's function G.

The physical meaning of $K(\mathbf{x}_i)$ depends on the order of the time variables.

(a) If t_1, $t_2 > t_3$, t_4, K describes the propagation of a pair of additional particles and gives information about the energy states of the $(N + 2)$-particle system.

(b) Similarly the case t_1, $t_2 < t_3$, t_4 corresponds to the propagation of a pair of holes [$(N - 2)$-particle system.] We shall see that these two terms of $K(\mathbf{x}_i)$ are particularly useful in the study of superconductivity.

(c) Finally, if t_1, $t_3 > t_2$, t_4, $K(\mathbf{x}_i)$ describes the propagation of a particle hole pair and gives us information about the excited states of the N-particle system—in particular, about the bound states and the collective excitations (such as phonons and plasmons).

The function $K(\mathbf{x}_i)$ describes the propagation of any pair of excitations. We shall write it in the form

$$K(1, 2, 3, 4) = G(1, 3)\, G(2, 4) - G(1, 4)\, G(2, 3) + \delta K(1, 2, 3, 4) \qquad (3\text{–}42)$$

The first two terms of (3–42) correspond to the propagation of two excitations totally independent of one another, the indistinguishability of the particles being taken into account. This is the "free" part of K. On the other hand, $\delta K(1,2,3,4)$ describes the *interaction* of these two excitations and contains all the new information in $K(\mathbf{x}_i)$. We shall call it the "bound" part of K. By referring to (3–41) we see that the discontinuities of K are absorbed into the free part. The bound part δK is thus continuous. We shall define

$$\delta K(\mathbf{x}_i) = \iiiint d^4\mathbf{x}'_i\, G(\mathbf{x}_1, \mathbf{x}'_1)\, G(\mathbf{x}_2, \mathbf{x}'_2)\, \gamma(\mathbf{x}'_i)\, G(\mathbf{x}'_3, \mathbf{x}_3)\, G(\mathbf{x}'_4, \mathbf{x}_4) \qquad (3\text{–}43)$$

where we have used the condensed notation

$$\int d^4\mathbf{x} = \int d^3r\, dt \sum_{\sigma} \qquad (3\text{–}43a)$$

We thus separate the interaction process into three steps:

(a) Propagation from $\mathbf{x}_1, \mathbf{x}_2$ to $\mathbf{x}'_1, \mathbf{x}'_2$

(b) The so-called act of interaction, extending from $\mathbf{x}'_1, \mathbf{x}'_2$ to $\mathbf{x}'_3, \mathbf{x}'_4$

(c) Propagation from $\mathbf{x}'_3, \mathbf{x}'_4$ to $\mathbf{x}_3, \mathbf{x}_4$

The function $\gamma(\mathbf{x}_i)$ describes the interaction of the two elementary excitations. We have extracted from it all propagation factors which would complicate it. Like δK, it is a continuous function of the $\mathbf{x}_i$.

We can ask what the range in space and time of the function $K(\mathbf{x}_i)$ is. The free part has an infinite range. Consider, for instance, the term $G(1,3)\, G(2,4)$. It does not depend on the distance between the groups (1,3) and (2,4), which may be arbitrarily large. Within each of these groups the Fourier transform $G(k, t_1 - t_3)$ remains sensible over an interval of time of order Γ_k^{-1}; the range, though finite, is limited only by real transitions. The situation is quite different for the bound part δK, the range of which depends in an essential way on the system under consideration. For certain systems, which we shall call "normal," the interaction is localized; there exist a finite range and characteristic relaxation time, beyond which $\gamma(\mathbf{x}_i)$ becomes negligible. On the other hand, for "superfluid" systems, the quantity $\gamma(\mathbf{x}_1,\mathbf{x}_2,\mathbf{x}_3,\mathbf{x}_4)$ is "localized" as far as the variables $x_1 - x_2$ and $x_3 - x_4$ are concerned, whereas it exhibits a long range in the variable

$$\frac{x_1 + x_2}{2} - \frac{x_3 + x_4}{2}$$

We shall return to this question in the following chapters.

We can easily calculate the Fourier transform of $K(\mathbf{x}_i)$ with respect to the space variables. By analogy with (3–6), we put

$$\begin{cases} K(\mathbf{k}_i, t_i) = \dfrac{1}{\Omega^2} \displaystyle\int \exp[-\mathrm{i}(k_1 r_1 + k_2 r_2 - k_3 r_3 - k_4 r_4)]\, K(\mathbf{r}_i t_i)\, \mathrm{d}^3 r_i \\ K(\mathbf{r}_i, t_i) = \dfrac{1}{\Omega^2} \displaystyle\sum_{k_i} \exp[\mathrm{i}(k_1 r_1 + \cdots - k_4 r_4)]\, K(\mathbf{k}_i, t_i) \end{cases} \tag{3–44}$$

Analogous relations lead to the definition of $\delta K(\mathbf{k}_i,t_i)$ and $\gamma(\mathbf{k}_i,t_i)$. Owing to translation invariance, all these quantities contain a factor

$$\delta_{\mathbf{k}_1+\mathbf{k}_2,\mathbf{k}_3+\mathbf{k}_4}$$

A glance at Eqs. (B–7) shows that

$$K(\mathbf{k}_i, t_i) = \langle \varphi_0 \,|\, T\{\, \mathbf{a}_{\mathbf{k}_1}(t_1)\, \mathbf{a}_{\mathbf{k}_2}(t_2)\, \mathbf{a}^*_{\mathbf{k}_3}(t_3)\, \mathbf{a}^*_{\mathbf{k}_4}(t_4) \,\}\,|\, \varphi_0 \rangle \tag{3–45}$$

As with $K(\mathbf{r}_i,t_i)$, we can separate a free part from a bound part. The free part is given by

$$G(\mathbf{k}_1 t_1, \mathbf{k}_3 t_3)\, G(\mathbf{k}_2 t_2, \mathbf{k}_4 t_4) - G(\mathbf{k}_1 t_1, \mathbf{k}_4 t_4)\, G(\mathbf{k}_2 t_2, \mathbf{k}_3 t_3)$$

It is different from 0 only if

$$\begin{cases} \mathbf{k}_1 = \mathbf{k}_3 \\ \mathbf{k}_2 = \mathbf{k}_4 \end{cases} \qquad \text{or} \qquad \begin{cases} \mathbf{k}_1 = \mathbf{k}_4 \\ \mathbf{k}_2 = \mathbf{k}_3 \end{cases}$$

In this case it gives a contribution independent of the volume Ω, which depends only on the two independent $\mathbf{k}$ variables.

The bound part of $K(\mathbf{k}_i,t_i)$ is more complex. In a normal system, where γ is of finite range, δK is of order $1/\Omega$ [see (3–44) and (3–46)]. On the other hand, there is only a single relation involving momentum conservation: $\mathbf{k}_1 + \mathbf{k}_2 = \mathbf{k}_3 + \mathbf{k}_4$. δK thus depends on three independent variables, which offsets the above factor $1/\Omega$. In a superfluid system, the preceding conclusions remain valid, unless $\mathbf{k}_1 + \mathbf{k}_2 = 0$ (which implies

$\mathbf{k}_3 + \mathbf{k}_4 = 0$); in this case, δK is independent of Ω—a consequence of the long range with respect to

$$\frac{r_1 + r_2 - r_3 - r_4}{2}$$

Of course, these conclusions apply also to $\gamma(\mathbf{k}_i, t_i)$.

The relation (3–43) which defines the interaction operator γ is easily rewritten and becomes, after Fourier transformation

$$\delta K(\mathbf{k}_i, t_i) = \int dt_i' G(\mathbf{k}_1, t_1 - t_1')\, G(\mathbf{k}_2, t_2 - t_2')\, \gamma(\mathbf{k}_i, t_i')\, G(\mathbf{k}_3, t_3' - t_3)\, G(\mathbf{k}_4, t_4' - t_4) \tag{3–46}$$

Let us now turn to the Fourier transformation in the time variables, and let us define $K(\mathbf{k}_i, \omega_i)$ by the relation

$$K(\mathbf{k}_i t_i) = \frac{1}{(2\pi)^2} \int K(\mathbf{k}_i, \omega_i) \exp[-i(\omega_1 t_1 + \omega_2 t_2 - \omega_3 t_3 - \omega_4 t_4)]\, d\omega_i \tag{3–47}$$

This definition is not ambiguous, although the inverse relation

$$K(\mathbf{k}_i, \omega_i) = \frac{1}{(2\pi)^2} \int K(\mathbf{k}_i, t_i) \exp[i(\omega_1 t_1 + \cdots - \omega_4 t_4)]\, dt_i \tag{3–48}$$

sometimes poses problems of interpretation. The study of the analytic properties of $K(\mathbf{k}_i, \omega_i)$ is difficult: this question is touched on in Appendix D. The definitions (3–47) and (3–48) extend naturally to δK and γ. Because of translational invariance, $\delta K(\mathbf{k}_i, \omega_i)$ contains a factor $\delta(\omega_1 + \omega_2 - \omega_3 - \omega_4)$, assuring conservation of energy. As far as the free part of $K(\mathbf{k}_i, \omega_i)$ is concerned, it involves *two* δ functions, separately assuring conservation of energy in each of the factors G.

In this new representation, Eq. (3–46) takes a simple form,

$$\delta K(\mathbf{p}_i) = G(\mathbf{p}_1)\, G(\mathbf{p}_2)\, G(\mathbf{p}_3)\, G(\mathbf{p}_4)\, \gamma(\mathbf{p}_i) \tag{3–49}$$

where the notation $\mathbf{p}$ is here used to designate all the four variables $(\mathbf{k}, \omega)$. The function $\gamma(\mathbf{p}_i)$ is highly singular. This difficulty is easily avoided by defining

$$\gamma(\mathbf{p}_i) = \Gamma(\mathbf{p}_i)\, \delta(\omega_1 + \omega_2 - \omega_3 - \omega_4)\, \delta_{\mathbf{k}_1 + \mathbf{k}_2 . \mathbf{k}_3 + \mathbf{k}_4} \tag{3–50}$$

The operator $\Gamma(\mathbf{p}_i)$ is called the "interaction operator." It is a regular function, whose analytic properties we can eventually study.

b. Direct analysis of K

Let us consider the particular case $t_1 t_4 > t_2 t_3$, and let us introduce the complete set of eigenstates $|\boldsymbol{\varphi}_s\rangle$ of the real system. We can write

$$K(\mathbf{x}_i) = \sum_s \tilde{\chi}_s(1,4)\, \chi_s(2,3) \qquad \textit{(3–51)}$$

the functions χ_s and $\tilde{\chi}_s$ being defined by

$$\begin{cases} \chi_s(\mathbf{x}, \mathbf{x}') = \langle \varphi_s \mid T\{\psi_\sigma(x)\, \psi^*_{\sigma'}(x')\} \mid \varphi_0 \rangle \\ \tilde{\chi}_s(\mathbf{x}, \mathbf{x}') = \langle \varphi_0 \mid T\{\psi_\sigma(x)\, \psi^*_{\sigma'}(x')\} \mid \varphi_s \rangle \end{cases} \qquad \textit{(3–52)}$$

$\tilde{\chi}_s\ (\mathbf{x},\mathbf{x}')$ is simply the "transposed" function of $\chi_s(\mathbf{x},\mathbf{x}')$. Similarly we have

$$\begin{cases} K(\mathbf{x}_i) = \sum_s \tilde{\eta}_s(1, 2)\, \xi_s(3, 4) & \text{if } t_1, t_2 > t_3, t_4 \\ K(\mathbf{x}_i) = \sum_s \eta_s(1, 2)\, \tilde{\xi}_s(3, 4) & \text{if } t_1, t_2 < t_3, t_4 \end{cases} \qquad \textit{(3–53)}$$

where the functions η_s, ξ_s and their transposes $\tilde{\eta}_s$, $\tilde{\xi}_s$ are defined by

$$\begin{cases} \eta_s(\mathbf{x}, \mathbf{x}') = \langle \varphi_s \mid T\{\psi_\sigma(x)\, \psi_{\sigma'}(x')\} \mid \varphi_0 \rangle \\ \xi_s(\mathbf{x}, \mathbf{x}') = \langle \varphi_s \mid T\{\psi^*_\sigma(x)\, \psi^*_{\sigma'}(x')\} \mid \varphi_0 \rangle \end{cases} \qquad \textit{(3–54)}$$

The functions χ_s, η_s, ξ_s are analogous to the propagator G, except for the fact that they are nondiagonal. (We can apply to each of them the methods developed in the preceding section.) They are divided into a number of categories, according to the nature of the state $|\boldsymbol{\varphi}_s\rangle$.

(a) $|\boldsymbol{\varphi}_s\rangle = |\boldsymbol{\varphi}_0\rangle$. χ_s is just the propagator G; the corresponding term of (3–51) is simply half the free part of $K(\mathbf{x}_i)$.

(b) $|\boldsymbol{\varphi}_s\rangle$ corresponds to two more or less independent elementary excitations, taken in the continuum of excited states. The corresponding terms χ_s serve on the one hand to complete the free part of $K(\mathbf{x}_i)$, and on the other hand to describe the scattering of two elementary excitations

(c) $|\boldsymbol{\varphi}_s\rangle$ corresponds to a bound state of two elementary excitations. This state might be a particle-hole pair (case of χ_s), or it might be a hole-hole or particle-particle pair (cases of η_s and ξ_s). The bound states of the first type are encountered in normal systems, whereas the others are characteristic of superfluid systems.

We consider only functions χ_s corresponding to particle-hole bound states. These give us new information which the simple study of G could not furnish. Their physical meaning is simple. They give the probability amplitude that, if we add to $|\boldsymbol{\varphi}_0\rangle$ a hole at $\mathbf{x}$ and a particle at $\mathbf{x}'$, we obtain the excited state $|\boldsymbol{\varphi}_s\rangle$: χ_s is thus the *wave function* of the particle and hole in the excited state $|\boldsymbol{\varphi}_s\rangle$. We can put it in the form

$$\chi_s(\mathbf{x}, \mathbf{x}') = \exp\left\{ i \left[\omega_{s0} \frac{(t + t')}{2} - q_s \frac{(r + r')}{2} \right] \right\} f_s(t - t', r - r', \sigma, \sigma') \quad \textit{(3–55)}$$

where ω_{s0} is the excitation energy of the state $|\boldsymbol{\varphi}_s\rangle$ and q_s is its total momentum. The function $f_s(r - r', t - t', \sigma, \sigma')$ describes the relative motion of the particle and hole. For $t = t'$, f_s is just the internal wave function of the bound state, in the usual sense of the word. Let us remark that in general this wave function $f_s(r - r', 0, \sigma, \sigma')$ is not sufficient to describe the phenomenon completely: because of polarization, the particle hole interaction is retarded, and it is then essential to know the correlations between the two excitations at different instants, which involves the calculation of $f_s(t,r)$ for $t \neq 0$.

Let us put expression (3–55) into (3–51) and take the complete Fourier transform, with respect to the four variables $\mathbf{x}_i$. It can then be verified that $K(\mathbf{k}_i, \omega_i)$ is singular when

$$\omega_2 - \omega_3 = -\omega_1 + \omega_4 = \omega_{s0} - i\eta \quad \textit{(3–56)}$$

where η is a positive infinitesimal (see Appendix D). Thus particle-hole bound states appear in $K(\mathbf{k}_i, \omega_i)$ in the form of poles with respect to the variable $\omega_2 - \omega_3$ (assuming the $\mathbf{k}_i$ to be given). It can likewise be verified that the bound states of two particles or of two holes appear in the form of poles with respect to the variable $\omega_1 + \omega_2 = -(\omega_3 + \omega_4)$. We thus have here a method to characterize and differentiate the bound states. By permutation of the variables we generate other singularities of $K(\mathbf{k}_i, \omega_i)$. Note that these singularities exist whatever the individual values of ω_i are, provided that, for example, (3–56) is satisfied.

The preceding discussion applies in principle only to discrete bound states, well separated from the continuum. In practice, there exist "collective" states resulting from the interaction between quasi particles, which are found immersed in the continuum (for example, phonons and plasmons). These states can then decay into independent excitations and have a finite lifetime. They are always given by the poles of $K(\mathbf{k}_i, \omega_i)$, the latter being found now in the complex plane. For example, the singularity (3–56) will become

$$\omega_2 - \omega_3 = \omega_{s0} - i\,\Gamma_s$$

where Γ_s^{-1} is the lifetime of the state. In conclusion, whereas the poles of G furnish the elementary excitations of $N \pm 1$ particles, the poles of K [that is to say, of $\Gamma(\mathbf{k}_i, \omega_i)$] give us information about the "bound" elementary excitations of N and $N \pm 2$ particles.

c. Correlation functions

The manipulation of functions of four complex variables is "tricky," and it is difficult to draw any practical conclusions from it. Thus, we are going to limit our goals and consider a reduced version of the Green's function $K(\mathbf{x}_i)$, much less valuable, but much easier to discuss. Let us define

$$L(\mathbf{x}, \mathbf{x}') = L(rt\sigma, r't'\sigma') = K(rt\sigma, r't'\sigma', r't'_+\sigma', rt_+\sigma) \qquad (3\text{–}57)$$

Let $\rho(r,\sigma)$ be the density of particles with spin σ at the point r, given by

$$\rho(r, \sigma) = \psi^*_\sigma(r)\, \psi_\sigma(r)$$

L can be written in the form

$$L(\mathbf{x}, \mathbf{x}') = < \varphi_0 \mid P\{\, \rho(rt\sigma)\ \rho(r't'\sigma')\,\} \mid \varphi_0 > \qquad (3\text{–}58)$$

where P is a chronological operator involving no change of sign. L thus measures the correlation between the density of spin σ at the point r and that of spin σ' at the point r'.

For an isotropic system L depends only on the relative orientation of the spins σ and σ'. Thus we can write

$$L(\mathbf{x}, \mathbf{x}') = L_s(x, x') \pm L_a(x, x') \qquad (3\text{–}59)$$

the signs $\pm$ corresponding to $\sigma = \pm\sigma'$. The antisymmetric term L_a measures the exchange correlations of the spins.

In many instances we do not care about spin but are interested only in the function

$$\bar{S}(x, x') = \sum_{\sigma,\sigma'} L(rt\sigma, r't'\sigma') = 4L_s(x, x') \qquad (3\text{–}60)$$

$\bar{S}$ measures the correlation of densities, with no reference to spin. We have already introduced this quantity, in Appendix A [see Eqs. (A–3), (A–1), and (2–54)]; we know that $\bar{S}$ is closely related to the usual correlation function $S(r,t)$.

We shall now study $\bar{S}$ in the light of the present chapter. Most of the following results are also valid for the antisymmetric part L_a. Owing to translational invariance $\bar{S}$ depends only on $x - x'$. Using (A–3) and (2–54), we see that

$$\begin{cases} \bar{S}(r, t) = S(r, |t|) \\ [\bar{S}(r, t)]^* = S(r, -|t|) \end{cases} \tag{3–61}$$

(We thereby assume that $|\varphi_0\rangle$ is reflection-invariant.) As a consequence, knowledge of $\bar{S}$ implies that of S. Let us remark that $\bar{S}$ is an even function of t, continuous at $t = 0$.

The Fourier transform of $\bar{S}(r,t)$ with respect to r is given by

$$\bar{S}(q, t) = \int \bar{S}(r, t)\, e^{-iqr}\, d^3r = \frac{1}{\Omega} < \varphi_0 \,|\, P\{\, \rho_{-q}(t)\, \rho_q(0)\,\} \,|\, \varphi_0 > \tag{3–62}$$

where the density fluctuation ρ_q is defined by (B–19). It is easily verified that $\bar{S}(q,t)$ is an intensive quantity independent of the volume Ω. The inverse relation is written

$$\bar{S}(r, t) = \frac{1}{\Omega} \sum_q e^{iqr}\, \bar{S}(q, t) \tag{3–63}$$

The Fourier transform with respect to t has already been calculated in Appendix A. It is given by (A–11),

$$\bar{S}(q, \omega) = \frac{1}{2i\pi} \int_{-\infty}^{+\infty} d\omega'\, S(q, \omega') \left\{ \frac{1}{\omega' - i\eta + \omega} + \frac{1}{\omega' - i\eta - \omega} \right\} \tag{3–64}$$

We see that $\bar{S}(q,\omega)$ is an even function. We know that $S(q,\omega)$ is real, given by (2–47),

$$S(q, \omega) = \frac{2\pi}{\Omega} \sum_s \left| < \varphi_s \,|\, \rho_q \,|\, \varphi_0 > \right|^2 \delta(\omega_{s0} - \omega) \tag{3–65}$$

(3–64) is just the Lehmann representation of $\bar{S}(q,\omega)$.

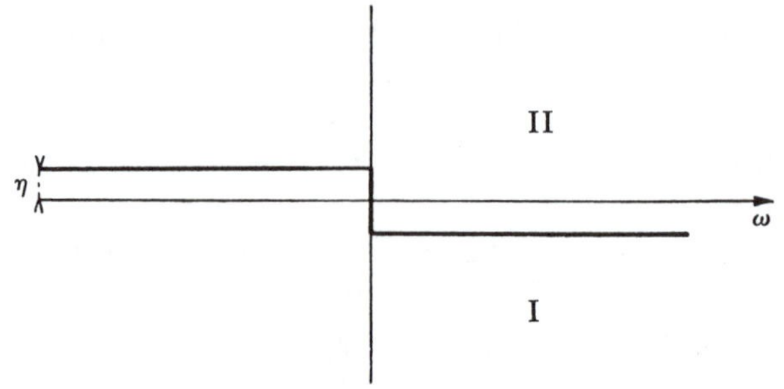

Figure 3

The intermediate states $|\varphi_s\rangle$ contain N particles, with total wave vector q and total spin 0. Most of these states are made up of independent particle-hole pairs, whose energy varies continuously over the positive real axis. Consequently, $\bar{S}$ exhibits two branch cuts, indicated in Fig. 3. $\bar{S}$ is analytic in each of the regions **I** and **II**. The values of the function in **I** and **II** are, moreover, related by the equation

$$-\bar{S}(q, \omega^*) = [\bar{S}(q, \omega)]^* \tag{3-66}$$

valid everywhere except on the real axis. $\bar{S}$ is discontinuous across the cut. The discontinuity is given by

$$\bar{S}_{II} - \bar{S}_I = \begin{cases} 2 \operatorname{Re} \bar{S} = S(q, \omega) & \text{if } \omega > 0 \\ -2 \operatorname{Re} \bar{S} = -S(q, -\omega) & \text{if } \omega < 0 \end{cases} \tag{3-67}$$

When $\omega \to \infty$, $\bar{S}$ is of order $1/\omega^2$, with a coefficient which is deduced from the sum rule studied in Chap. 2. The latter can be written [see (2–39), (2–40), and (2–47)] as

$$\int_0^\infty S(q, \omega)\, \omega\, d\omega = \frac{\pi}{\Omega} \frac{Nq^2}{m} \tag{3-68}$$

As a consequence

$$\bar{S}(q, \omega) \underset{\omega \to \infty}{\to} \frac{iNq^2}{m\Omega} \frac{1}{\omega^2} \tag{3-69}$$

[Let us recall that (3–68) and (3–69) are valid only if the binary law of interaction is velocity-independent.] By combining (3–69) with the analytic properties of $\bar{S}$ we can, as with G, obtain a number of interesting relations. We leave to the reader the task of establishing them.

If we cross the cuts, we find another set of values for $\bar{S}$, located on the next Riemann sheet. This new function has the same poles as $S(q,|\omega|)$, occurring in pairs of opposite values. We are interested only in poles close to the real axis, giving rise to a sharp peak in the spectral density $S(q,\omega)$. These poles correspond to *bound states* of a particle-hole pair, with total wave vector q and total spin 0. These bound states may be either of the "exciton" type (discrete and stable states) or of the "collective-mode" type (damped oscillations). The poles appear in the analytic continuation of S into regions II and IV (see Fig. 3). The value of the pole yields the energy and damping of the bound state.

$\bar{S}$ describes only the collective modes involving a density fluctuation —that is, the longitudinal, spin-independent modes. In order to examine more complicated collective modes, we must study other correlation functions. For instance, longitudinal spin waves will show up as poles of $L_a(q,\omega)$; transverse modes will appear in the current correlation functions (see Chap. 6). Quite generally, any correlation function

$$< \varphi_0 \mid P\{\,\mathbf{A}(r,t,\sigma)\,\mathbf{B}(r',t',\sigma')\,\} \mid \varphi_0 > \qquad (3\text{–}70)$$

where A and B are one-electron operators, can be expressed in terms of the two-particle Green's function $K(k_i,t_i,\sigma_i)$, in which we set

$$\begin{cases} \sigma_1 = \sigma_4 & t_4 = t_1 + 0 \\ \sigma_2 = \sigma_3 & t_3 = t_2 + 0 \end{cases}$$

The discussion given above for $\bar{S}$ applies to all these correlation functions.

For the purpose of characterizing the bound states, it is sufficient to study the correlation functions, which are much simpler than the complete Green's function $K(\mathbf{x}_i)$. This advantage, however, is balanced by severe limitations: $\bar{S}$ can describe only processes in which a particle-hole pair is created at time 0, annihilated at time t (or the inverse). In practice, the creation and annihilation of a pair are of finite duration, the appearance of the particle and that of the hole not being simultaneous. $\bar{S}$ then becomes inadequate, and it is necessary to resort to the complete Green's function K. Once the problem is solved, it is possible to go back from K to $\bar{S}$ to calculate the energy of the elementary excitations.

d. Dynamical equations relating the Green's functions

Until now we have introduced no dynamical considerations. To fill this gap, let us consider the equation of motion

$$\mathrm{i}\frac{\partial}{\partial t}\psi_\sigma(x) = [\psi_\sigma(x), H] \tag{3-71}$$

The Hamiltonian is given to us by (B–12) and (B–14). We thus obtain

$$\begin{aligned}\mathrm{i}\frac{\partial}{\partial t}\psi_\sigma(r,t)\\ = -\frac{1}{2m}\nabla^2\psi_\sigma(r,t) + \sum_{\sigma''}\int d^3r''\, V(r''-r)\,\psi^*_{\sigma''}(r'',t)\,\psi_{\sigma''}(r'',t)\,\psi_\sigma(r,t)\end{aligned} \tag{3-72}$$

Let us now take the chronological product of (3–72) and $\psi^*_{\sigma'}(r',t')$, and let us take the average value of the equation thus obtained in the ground state $|\varphi_0\rangle$. We then find

$$\begin{aligned}\mathrm{i}<\varphi_0|\,T\left\{\frac{\partial\psi_\sigma(r,t)}{\partial t}\,\psi^*_{\sigma'}(r't')\right\}|\,\varphi_0>\\ = -<\varphi_0|\,T\left\{\frac{\nabla^2}{2m}\psi_\sigma(r,t)\,\psi^*_{\sigma'}(r',t')\right\}|\,\varphi_0>\\ +\sum_{\sigma''}\int d^3r''\,V(r-r'')<\varphi_0|\,T\{\psi^*_{\sigma''}(r'',t)\,\psi_{\sigma''}(r'',t)\,\psi_\sigma(r,t)\,\psi^*_{\sigma'}(r',t')\}\,|\varphi_0>\end{aligned} \tag{3-73}$$

Note that, because of the discontinuity of the chronological product,

$$\begin{aligned}\frac{\partial}{\partial t}T\left\{\psi_\sigma(r,t)\,\psi^*_{\sigma'}(r',t')\right\}\\ = T\left\{\frac{\partial\psi_\sigma(r,t)}{\partial t}\,\psi^*_{\sigma'}(r',t')\right\} + \delta(t-t')\,\delta_{\sigma,\sigma'}\,\delta(r-r')\end{aligned} \tag{3-74}$$

By using (3–74), we can put (3–73) into the form

$$\begin{aligned}\left[\frac{\partial}{\partial t}-\frac{\mathrm{i}\nabla^2}{2m}\right]G(\mathbf{r}t,\mathbf{r}'t')\\ -\int d^3\mathbf{r}''\,V(r-r'')\,K(\mathbf{r}''t,\mathbf{r}t,\mathbf{r}''t_+,\mathbf{r}'t') = \mathrm{i}\,\delta(t-t')\,\delta(\mathbf{r}-\mathbf{r}')\end{aligned} \tag{3-75}$$

[where we have used the notations (3–41a) and (3–43a)]. (3–75) is the *dynamical equation* relating G and K. By the same method we can establish a whole sequence of analogous equations, relating G and K to the three-, four-, etc. particle Green's functions. In principle this infinite set of equations determines the solution of the problem. Quite obviously, we are obliged to limit the number of these equations by some sort of approximation: this is the essence of the method of Martin and Schwinger.

The same method can be applied to the calculation of the ground-state energy,

$$E_0 = < \varphi_0 \mid H \mid \varphi_0 > \tag{3-76}$$

Let us first consider the kinetic energy, given by

$$T = -\frac{i}{2m} \int d^3\mathbf{r} \left[\nabla^2 G(\mathbf{r}t, \mathbf{r}'t_+)\right]_{r'=r} \tag{3-77}$$

(3–77) is easily transformed, thanks to (3–10), and becomes:

$$T = +i \sum_{\mathbf{k}} \frac{k^2}{2m} G(k, -0) = \sum_{\mathbf{k}} m_k \frac{k^2}{2m}$$

We find again the result obtained previously. The interaction energy, on the other hand, is equal to

$$E_{\text{int}} = \tfrac{1}{2} \iint d^3\mathbf{r}\, d^3\mathbf{r}'\, V(r - r')\, K(\mathbf{r}'t, \mathbf{r}t, \mathbf{r}t_+, \mathbf{r}'t_+) \tag{3-78}$$

(3–78) can be transformed by using the dynamical equation (3–75). We then recover the expression (3–33) for E_{int}, obtained by a completely different method.

The Green's function entering (3–78) is very close to $\bar{S}(r - r', 0)$—yet it is different, because of the operator ordering. Using the commutation rules, we see that

$$\sum_{\sigma,\sigma'} K(\mathbf{r}'t, \mathbf{r}t, \mathbf{r}t_+, \mathbf{r}'t_+) = S(r - r') - \frac{N}{\Omega}\, \delta(r - r')$$

(3–78) can then be written as

$$E_{\text{int}} = \tfrac{1}{2} \iint d^3r\, d^3r'\, V(r - r') \left\{ S(r - r') - \frac{N}{\Omega}\, \delta(r - r' \right\} \tag{3-79}$$

Let us introduce the Fourier transforms of V[(B–15)] and of $S(r,0)$ [(2–52)]. (3–79) then becomes

$$E_{\text{int}} = \tfrac{1}{2} \sum_{q} V_q \{ S(q) - N/\Omega \} \tag{3-80}$$

where $S(k)$ is the "instantaneous" correlation function. Upon comparing (3–80) with (3–33), we are tempted to identify the terms of the two expressions for each value of k. This would be a gross error: although the two sums are equal, the same is not true for each term taken individually [to satisfy oneself of this, it is sufficient to expand (3–33) with the help of the dynamical equation].

3. Conclusion

This chapter may seem somewhat disconnected. However, it must be realized that this is the effect of the formalism used, which is easily adapted to the most diverse problems. Among the advantages of using Green's functions, the following are particularly interesting:

a. G and K select information relative to one or two particles. In most cases, this is all that we need. We thus leave unexplored the useless details relative to the structure of the ground state.

b. G and K naturally introduce the damping of the elementary excitations, which would not be done by a formalism based on the use of canonical transformations.

c. G and K enable us to take account of the deformation of quasi particles due to the inertia of the surrounding cloud of particles. We can follow the transition between an "instantaneous" response, involving only one bare particle in a static Hartree-Fock potential, and an "adiabatic" response, involving the dressed quasi particle.

Chapter 4

The Structure of the Elementary Excitation Spectrum

1. Normal Systems

a. Definition

A "normal" system is characterized by an interaction operator $\gamma(\mathbf{x}_i)$ which is *localized* in space and in time. $\gamma(\mathbf{x}_i)$ is negligible unless *the four* $\mathbf{x}_i$ are grouped in a volume of finite dimensions (l,τ), where l is the effective range of the interaction, and τ is a relaxation time. The Fourier transform $\gamma(\mathbf{k}_i,t_i)$ is thus a regular function of the k_i, of order $1/\Omega$. We note that $\gamma(\mathbf{k}_i,t_i)$ can have a long temporal range, arising from the poles of $\gamma(\mathbf{k}_i,\omega_i)$, that is to say, from the bound states; this range will be limited by dissipative effects. Aesthetics would have us deduce the structure of the elementary excitation spectrum from this single definition. This is probably not possible. We shall need further assumptions, which, however, can be shown to be quite natural.

Our first objective is the determination of the poles of G. Let us consider the dynamical equation (3–75):

$$\left(\frac{\partial}{\partial t}-\frac{i\nabla^2}{2m}\right) G(\mathbf{x},\mathbf{x}') - \int d^3\mathbf{r}''\, V(r-r'')\, K(\mathbf{r}''t,\mathbf{x},\mathbf{r}''t_+,\mathbf{x}') = i\,\delta(\mathbf{x}-\mathbf{x}') \qquad (4\text{–}1)$$

We shall split K in a free part K_L, given by (3–42), and a bound part δK; we write δK in the form (3–43),

$$\delta K(\mathbf{x}_i) = \iiiint d^4\mathbf{x}_i'\, G(\mathbf{x}_1,\mathbf{x}_1')\, G(\mathbf{x}_2,\mathbf{x}_2')\, \gamma(\mathbf{x}_i')\, G(\mathbf{x}_3',\mathbf{x}_3)\, G(\mathbf{x}_4',\mathbf{x}_4) \qquad (4\text{–}2)$$

Equation (4–1) can then be decomposed as follows:

$$\left(\frac{\partial}{\partial t}-\frac{i\nabla^2}{2m}\right) G(\mathbf{x}, \mathbf{x}')$$
$$-\int d^3\mathbf{r}''\, V(r-r'')\,\{G(\mathbf{r}''t, \mathbf{r}''t_+)\, G(\mathbf{x}, \mathbf{x}') - G(\mathbf{x}, \mathbf{r}''t_+)\, G(\mathbf{r}''t, \mathbf{x}')\} \tag{4–3}$$
$$-\int d^3\mathbf{r}'' \iiiint d^4\mathbf{x}_i'\, V(r-r'')\, G(\mathbf{r}''t, \mathbf{x}_1')\, G(\mathbf{x}, \mathbf{x}_2')\, \gamma(\mathbf{x}_i')\, G(\mathbf{x}_3', \mathbf{r}''t)\, G(\mathbf{x}_4', \mathbf{x}')$$
$$= i\delta(\mathbf{x}-\mathbf{x}')$$

This expression is long and cumbersome. We can make it more compact by writing

$$\left(\frac{\partial}{\partial t}-\frac{i\nabla^2}{2m}\right) G(\mathbf{x}, \mathbf{x}') - i\int d^4\mathbf{x}'''\, M(\mathbf{x}, \mathbf{x}''')\, G(\mathbf{x}''', \mathbf{x}') = i\delta(\mathbf{x}-\mathbf{x}') \tag{4–4}$$

where $M(\mathbf{x},\mathbf{x}''')$ is given by (4–3). As was true for K, it is convenient to split M into a free part and a bound part.

$$M = M_L + \delta M \tag{4–5}$$

Comparing (4–3) with (4–4), we see readily that

$$M_L(\mathbf{x}, \mathbf{x}''') = -i\int d^3\mathbf{r}''\, V(r-r'')\,\{G(\mathbf{r}''t, \mathbf{r}''t_+)\, \delta(\mathbf{x}-\mathbf{x}''') - G(\mathbf{x}, \mathbf{r}''t_+)\, \delta(t-t''')\, \delta(\mathbf{r}''-\mathbf{r}''')\} \tag{4–6}$$

$$\delta M(\mathbf{x}, \mathbf{x}''') = -i\int d^3\mathbf{r}'' \iiint d^4\mathbf{x}_1'\, d^4\mathbf{x}_2'\, d^4\mathbf{x}_3'\, V(r-r'') \times G(\mathbf{r}''t, \mathbf{x}_1')\, G(\mathbf{x}, \mathbf{x}_2')\, \gamma(\mathbf{x}_1', \mathbf{x}_2', \mathbf{x}_3', \mathbf{x}''')\, G(\mathbf{x}_3', \mathbf{r}''t) \tag{4–7}$$

Because of translational invariance in space and time, M depends only on the difference $\mathbf{x} - \mathbf{x}'''$. We can thus define its Fourier transform $M(\mathbf{k},\omega)$ exactly as we did for $G(\mathbf{k},\omega)$. After transforming equation (4–4), we get the very simple result

$$\left[\frac{k^2}{2m}-\omega-M(k, \omega)\right] G(k, \omega) = 1 \tag{4–8}$$

The poles of G will thus be the roots of the factor $[k^2/2m - \omega - M(\mathbf{p})]$. This important result will ultimately yield the quasi-particle energy. $M(\mathbf{p})$ is called the "self-energy," or "mass," operator. We shall meet it again in Chap. 5, in the framework of perturbation theory, where it has a very simple graphical interpretation.

If M is a nonreal function, we get G trivially from (4–8). However, if M is real, we need *additional* information in order to define the behavior of G near its real poles. In practice, we shall add to M a small term $\pm i\eta$ ($\eta > 0$, infinitesimal), which fixes the position of the pole with respect to the real axis. This choice depends on boundary conditions, supplementing (3–75), expressing the fact that $|\boldsymbol{\varphi}_0\rangle$ is the ground state. We shall not dwell on this point, which is fully analyzed by Martin and Schwinger.

In computing M, the simplest approximation is to neglect δM. Since $M_L(x,x''')$ contains a factor $\delta(t-t''')$, the Fourier transform $M_L(k,\omega)$ does not depend on ω. Using (4–6), we readily find

$$M_L(\mathbf{p}) = -\frac{i}{\Omega}\sum_{\mathbf{k}'} V_0 G(\mathbf{k}', 0_-) + \frac{i}{\Omega}\sum_{q} V_q G(\mathbf{k}-q, 0_-) \tag{4-9}$$

$-\ M_L(\mathbf{p})$ is just the average Hartree-Fock energy felt by the particle k. The only effect of the interaction is a correction to the kinetic energy $k^2/2m$.

Let us now consider $\delta M(p)$. In order to discuss solutions of (4–8), we must know, at least qualitatively, the behavior of the function $\delta M(\omega)$. For this purpose, we Fourier-transform (4–7) with respect to space variables; using the definition (B–15), we obtain

$$\begin{aligned}\delta M(\mathbf{k}t, \mathbf{k}t') = -\frac{i}{\Omega}\sum_{\mathbf{k}', q} V_q \int dt_1\, dt_2\, dt_3\, \gamma(\mathbf{k}'t_1, (\mathbf{k}-q)\, t_2, (\mathbf{k}'-q)\, t_3, \mathbf{k}t')\\ G(k', t-t_1)\, G(k-q, t-t_2)\, G(k'-q, t_3-t)\end{aligned} \tag{4-10}$$

δM thus appears as a convolution product with respect to time, involving one factor γ and three factors G. Each of these factors could be long ranged, limited only by dissipation. At first sight, it looks as if the convolution product could also be long ranged. Actually, this does not occur because of the summation over $\mathbf{k}'$ and q. For large volumes of t, the phases of the factors G and γ vary as $f(\mathbf{k}',q)$; they quickly become incoherent. When we sum over k' and q, the terms of (4–10) undergo destructive interference; the result goes to zero.

Consequently, $M(k, t - t')$ is a localized function of the time variable $t - t'$, which is negligible when $t - t'$ exceeds some typical relaxation time. As a result, $M(k,\omega)$ is a smooth function of ω. Therefore, we may reasonably hope that the equation

$$\frac{k^2}{2m} - \omega - M(k, \omega) = 0$$

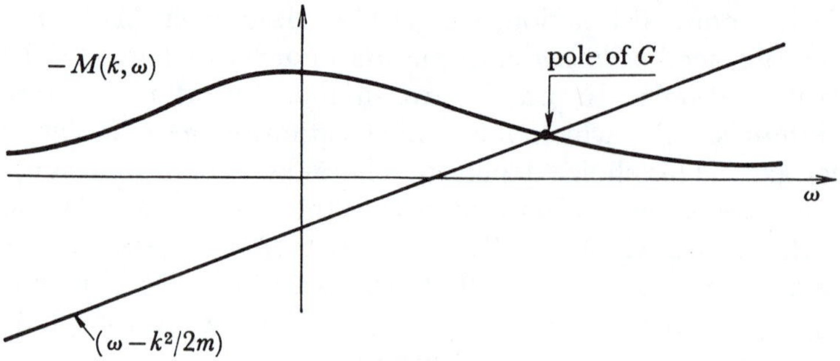

Figure 1

will have only a single solution for each value of k (Fig. 1). This is only an argument of reasonableness. We shall instead assume that the Green's function $G(k,\omega)$ of a normal system *has only one pole for each value of* k.

We showed in Chap. 3 that the poles of G are found in quadrants II and IV (indicated in Fig. 2). Quadrant IV corresponds to quasi particles, quadrant II to quasi holes. For large values of k, the pole of G will certainly be of the quasi-particle type, whereas for $k = 0$ we are dealing with a quasi hole. When k increases from 0 to ∞ in any given direction the pole crosses the real axis for a certain value of momentum k_F (for an isotropic gas, k_F is independent of direction). In this way we define a *surface of discontinuity*, or *Fermi surface*: inside we have quasi holes, outside quasi particles.

We shall assume that $\gamma(\mathbf{k}_i,t_i)$ is continuous when $k_i = k_F$ (this result is obvious in the case of a perturbation expansion). M is then continuous when k crosses k_F, which implies the continuity of the pole of $G(k,\omega)$.

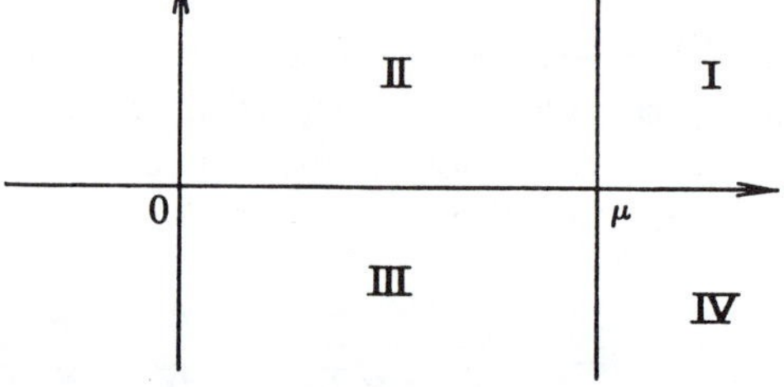

Figure 2

Let $\varepsilon_k = \mu + \xi_k - i\Gamma_k$ be the position of the pole. We know that

$$\xi_k,\ \Gamma_k \begin{cases} > 0 & \text{if} \quad k > k_F \\ < 0 & \text{if} \quad k < k_F \end{cases}$$

ξ_k and Γ_k must therefore tend to 0 when $k \to k_F$. The energy of quasi particles and quasi holes *at the Fermi surface* is thus equal to the chemical potential μ, their damping being zero. This important result constitutes the "Van Hove theorem"; we shall prove it rigorously in the following chapter, using perturbation theory.

According to (4–8), the residue $-z_k$ of G can be written as

$$-z_k = -\left.\frac{1}{1 + \partial M/\partial\omega}\right|_{\omega=\varepsilon_k} \qquad \textit{(4–11)}$$

z_k is thus continuous when k crosses k_F. Similarly, the continuity of $\nabla_k M$ implies that of the group velocity of the quasi particle, given by the usual kinematical relation:

$$v = \nabla_k \varepsilon_k \qquad \textit{(4–12)}$$

In summary, the only change at the Fermi surface is the passage of the pole across the real axis; all other characteristics of G are continuous.

b. Structure of the excitation spectrum

We can summarize the results of the preceding paragraph in Fig. 3, which gives us a clear idea of the elementary excitation spectrum as a function of the parameter k. These excitations exist only when they have a long lifetime, that is to say, in the immediate neighborhood of the Fermi surface. By combining them in more or less complex *configurations*, we can generate a whole continuum of excited states $|\varphi_n\rangle$, of small excitation energy (this does not exclude the existence of other excited states of a different type, such as phonons). Let us consider, for example, the $N + 1$ particle states, of total momentum $\mathbf{k}$ [involved in the definition (3–21) of $A_+(k,\omega)$]. If $(\omega_{n0} - \mu)$ is small, such a state consists of a configuration of p quasi holes of wave vector $\mathbf{k}_i$ and $p + 1$ quasi particles of wave vector $\mathbf{k}'_i$, all of them close to the Fermi surface, such that

$$\sum_i \mathbf{k}'_i - \sum_i \mathbf{k}_i = \mathbf{k} \qquad \textit{(4–13)}$$

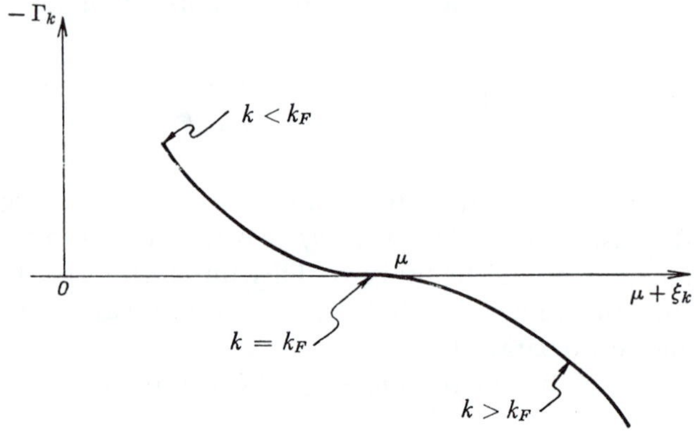

Figure 3

If $p \ll N$, we can neglect the interactions between the different elementary excitations. The energy is simply additive, given by

$$\omega_{n0} = \sum_{\mathbf{k}_i'} \xi_{k_i'} - \sum_{\mathbf{k}_i} \xi_{k_i} + \mu \,, \quad \text{or} \quad \begin{cases} \xi_{k_i'} > 0 \\ \xi_{k_i} < 0 \end{cases} \tag{4–14}$$

These configurations form a continuum. It is precisely this continuum which provides the damping of the individual excitations. The latter can in fact decay into more complex configurations, having the same total momentum and energy. The transition probability (as well as Γ_k) is proportional to the *density of final states* at the energy considered. Let us return to the above example, and let $(\mu + \xi)$ be the energy of the initial quasi particle. The density at the energy $(\mu + \xi)$ of configurations of p holes and $(p + 1)$ particles, satisfying (4–13), is of order ξ^{2p}. [In order to calculate the number of configurations between μ and $\mu + \xi$ we first choose $(2p - 1)$ elements in a shell of thickness ξ about the Fermi surface; then we determine the next to last **k** so that it and the last, given by (4–13), are both in the above shell: this gives a factor $\xi^{2p-1}\, \xi^2$, whose derivative gives the density of states.] When $\xi \to 0$, the configurations of two particles and one hole are thus preponderant. The density of states and Γ_k are of order ξ^2. This conclusion is naturally carried over to the case of $N - 1$ particle excited states, of wave vector $-\mathbf{k}$, which enter the definition of $A_-(k,\omega)$.

We should emphasize that the configurations whose density of states we have just calculated are not exact eigenstates of the system; they have a finite lifetime, due to scattering phenomena, elastic and inelastic. Nevertheless, the density of real states will be of the same order of magnitude as the density of approximate states. In fact, a scattering process can spread a discrete state over a range of frequency $d\xi$ only if, from the start, this range contains eigenstates into which the initial state can eventually decay. The density of real states in the neighborhood of $\xi = 0$ is thus also of order ξ^2.

The functions $A_{\pm}(\omega)$ are proportional to the density of states with an excitation energy $\xi = \omega$. A_+ and A_- are thus of order ω^2 when $\omega \to 0$. The Lehmann representation (3–33) then shows that the singularity of G at $\omega = \mu$ is relatively weak, of the type

$$(\omega - \mu)^2 \ln (\omega - \mu)$$

When $k \to k_F$, we can separate A_+ and A_- into two parts (see Fig. 4):

a. A regular slowly varying function, arising from configurations of *several* elementary excitations. This contribution is of order ω^2 when $\omega \to 0$. It gives rise to the incoherent part of G, which we shall designate by G_{inc}.

b. A peak corresponding to the *elementary* excitation k, centered about a frequency $\xi_k = v(k - k_F)$. The width Γ_k of this peak is of order $(k - k_F)^2$. This singularity appears in A_+ when $k > k_F$, in A_- when $k < k_F$. The area subtended by the peak is continuous, equal to z_k.

We can condense this information in the following form, valid in the neighborhood of $k = k_F$:

$$G(k, \omega) = G_{\text{inc}}(k, \omega) + \frac{z}{\mu + v(k - k_F) - \omega \pm i\, u(k - k_F)^2} \qquad \textit{(4–15)}$$

The minus sign corresponds to $k > k_F$, and vice versa. z, v, and u are

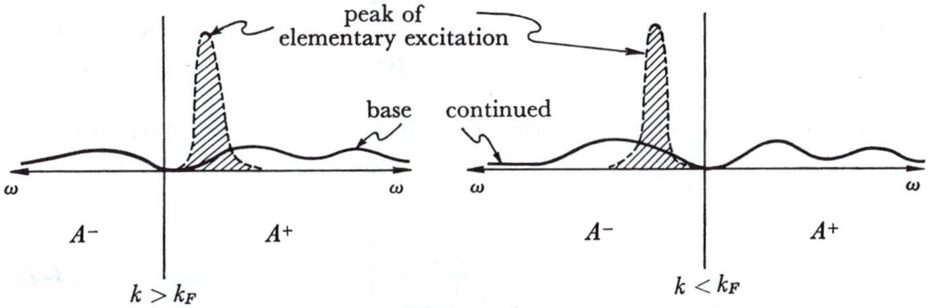

Figure 4

positive constants. Equation (4–15) will be extremely useful in what follows.

When we leave the Fermi surface, Γ_k increases rapidly and most of the preceding results become false. The notion of quasi particle is then illusory; it is necessary to resort to the complete study of G. Let us note that this prevents us from constructing a complete set of quasi-particle states.

c. *Wave functions of the elementary excitations*

Until now we have spoken frequently about elementary excitations, without ever specifying their nature. We propose to fill this gap as far as possible. We shall treat only the case of quasi particles, but all the following considerations are also valid for quasi holes.

We saw in Chap. 3 that we obtain quasi particles by "filtering" the state $\mathbf{a}_{\mathbf{k}}^*(t)\,|\boldsymbol{\varphi}_0\rangle$, in such a way as to conserve only the peak of the spectral density $A_+(k,\omega)$, corresponding to the pole $(\xi_k - i\Gamma_k)$. To attain this objective, we define

$$|\,\varphi_{\mathbf{k}}> = C\int_{-\infty}^{+\infty} dt'\,\exp[-\,i(\xi_k + \mu)\,t']\,f(t')\,\mathbf{a}_{\mathbf{k}}^*(t')\,|\,\varphi_0> \qquad (4\text{–}16)$$

C is a normalization constant and $f(t')$ a positive function which is negligible outside an interval of width $1/\alpha$. The choice of $f(t')$ is of no importance. For calculational convenience we put

$$f(t')\begin{cases} = e^{\alpha t} & \text{if } t < 0 \\ = 0 & \text{if } t > 0 \end{cases} \qquad (4\text{–}17)$$

Physically, we make particles created within a time $1/\alpha$ interfere with carefully chosen phase relations. The components of

$$\mathbf{a}_{\mathbf{k}}^*(t)\,|\,\varphi_0>$$

whose frequencies are found in a band of width $1/\alpha$ about $(\mu + \xi_k)$ experience constructive interference and give an important contribution. Conversely, the components outside of this passband vanish by destructive interference.

In order to demonstrate this interference effect, we carry out the spectral analysis of $|\boldsymbol{\varphi}_{\mathbf{k}}\rangle$. The density of norm $N_k(\omega)$ at the frequency $(\omega + \mu)$ is given by

$$N_k(\omega) = C^2\,\frac{A_+(k,\,\omega)}{(\omega - \xi_k)^2 + \alpha^2} \qquad (4\text{–}18)$$

The impedance of our filter is proportional to

$$\frac{\alpha}{(\omega-\xi_k)^2+\alpha^2}$$

The width of the passband is just equal to α. [In particular, if $\alpha \to 0$, the impedance is of order $\delta(\omega - \xi_k)$.] Now we must fix the value of α. We are obviously interested in making it small, so as to obtain as narrow a band of filtration as possible. However, α must remain larger than the width Γ_k of the peak of A_+; if we want to transmit the components of $a_{\mathbf{k}}^*|\varphi_0\rangle$ representing the quasi particle without deformation, the impedance must be constant over the whole range of corresponding frequencies. Under these conditions, the state $|\varphi_{\mathbf{k}}\rangle$ decays in a time Γ_k^{-1}. (It contains a factor $e^{-\Gamma_k t}$.) If we filter too much by taking $\alpha \ll \Gamma_k$, we can verify that $|\varphi_{\mathbf{k}}\rangle$ decays as $e^{-\alpha t}$; we have replaced the free oscillation of the quasi particle, of time constant Γ_k, by a forced oscillation, purely formal, of time constant α.

We are thus placed in a dilemma: α must in any case be greater than Γ_k. Nevertheless, we would like to choose it small, so as to eliminate as much as possible the continuous spectrum of $a_{\mathbf{k}}^*|\varphi_0\rangle$. This clearly shows that the notion of quasi particle makes sense only if Γ_k is small, that is to say, for k near the Fermi surface.

Let us assume this condition realized. We can then neglect the incoherent part of $A_+(k,\omega)$ in (4–18) and write

$$A_+(k,\omega) \sim \frac{z_k}{\pi}\frac{\Gamma_k}{(\omega-\xi_k)^2+\Gamma_k^2} \tag{4–19}$$

Let us put this expression for A_+ back into (4–18) and normalize $|\varphi_{\mathbf{k}}\rangle$. We find (assuming $\alpha \gg \Gamma_k$)

$$|\varphi_{\mathbf{k}}\rangle = (\alpha/\sqrt{z_k})\int_{-\infty}^{0} dt' \exp(-i\varepsilon_k t')\exp(\alpha t')\,\mathbf{a}_{\mathbf{k}}^*(t')\,|\varphi_0\rangle \tag{4–20}$$

where $\varepsilon_{\mathbf{k}}$ is the complex quasi-particle energy.

Let us now consider the state

$$e^{-iHt}|\varphi_{\mathbf{k}}\rangle$$

It is normalized, like $|\varphi_{\mathbf{k}}\rangle$. On the other hand, we easily verify that, for values of $t \ll \Gamma_k^{-1}$,

$$\langle\varphi_{\mathbf{k}}|e^{-iHt}|\varphi_{\mathbf{k}}\rangle \approx \exp[-i(E_0+\varepsilon_k)\,t]$$

The modulus of this matrix element is equal to 1; it follows that

$$e^{-iHt} \mid \varphi_{\mathbf{k}} > \, = \exp[-i(E_0 + \varepsilon_k)\, t] \mid \varphi_{\mathbf{k}} > \tag{4–21}$$

Thus, in a sufficiently short time interval, $|\varphi_{\mathbf{k}}\rangle$ exhibits the properties of an eigenstate of the real system, of energy $(E_0 + \varepsilon_k)$: it is just the *wave function of our elementary excitation.*

In fact, (4–21) is rigorous only if $\Gamma_k = 0$; $|\varphi_{\mathbf{k}}\rangle$ then describes an *exact* elementary excitation. If Γ_k is small (4–21) is approximately true when

$$-\Gamma_k^{-1} \ll t \ll \Gamma_k^{-1} \tag{4–22}$$

the error in this interval being of order Γ_k. The state vector $|\varphi_{\mathbf{k}}\rangle$ is only an approximate eigenstate. Outside the interval (4–22), the phases of the different components of $|\varphi_{\mathbf{k}}\rangle$, which coincide by construction at the origin, spread out again; $|\varphi_{\mathbf{k}}\rangle$ becomes incoherent and no longer resembles an eigenstate; the elementary excitation has decayed. When Γ_k is large, all the preceding considerations become void; we can no longer clearly define the quasi particle, much less its wave function.

Therefore let us remain in the neighborhood of the Fermi surface, and let us ignore the lack of precision which marks the definition of $|\varphi_{\mathbf{k}}\rangle$. What are the properties of the excited system containing the quasi particle **k** in addition to $|\varphi_0\rangle$? In order to answer this question, we shall calculate the Green's function in this new state,

$$\begin{cases} G_{\mathbf{k}}(\mathbf{k}'t) = i < \varphi_{\mathbf{k}} \mid T\{\, a_{\mathbf{k}'}(t)\, a^*_{\mathbf{k}'}(0)\,\} \mid \varphi_{\mathbf{k}} > \\ \qquad\qquad = G(\mathbf{k}'t) + \delta_{\mathbf{k}} G(\mathbf{k}'t) \end{cases} \tag{4–23}$$

If we put (4–16) into (4–23), we see that $G_{\mathbf{k}}(\mathbf{k}',t)$ resembles a two-particle Green's function K. For these two expressions to be identical it is necessary that the operators $a^*_{\mathbf{k}}(t')$ entering into $|\varphi_{\mathbf{k}}\rangle$ all correspond to t' earlier than the times $(0,t)$, the operators $a_{\mathbf{k}}(t'')$ contained in $\langle\varphi_{\mathbf{k}}|$ corresponding to t'' later than these times. This leads us to define $|\varphi_{\mathbf{k}}\rangle$ and $\langle\varphi_{\mathbf{k}}|$ differently. We shall put

$$\begin{aligned} \mid \varphi_{\mathbf{k}} > &= (\alpha/\sqrt{z_k}) \int_{-\infty}^{0} dt' \exp[-i\varepsilon_k(t' + \tau')] \exp(\alpha t')\, a^*_{\mathbf{k}}(t' + \tau') \mid \varphi_0 > \\ < \varphi_{\mathbf{k}} \mid &= (\alpha/\sqrt{z_k}) \int_{0}^{\infty} dt'' \exp[+i\varepsilon_k(t'' + \tau'')] \exp(-\alpha t'') < \varphi_0 \mid a_{\mathbf{k}}(t'' + \tau'') \end{aligned} \tag{4–24}$$

where τ' and τ'' are chosen so as to bracket the interval $(0,t)$:

$$\tau' \ll (0, t) \ll \tau'' \tag{4–25}$$

It is easily verified that the definitions (4–24) are both equivalent to (4–16). Let us put (4–24) into the definition of $G_{\mathbf{k}}(\mathbf{k}',t)$; we find

$$G_{\mathbf{k}}(\mathbf{k}', t) = \frac{i\alpha^2}{z_k} \int_{-\infty}^{0} dt' \int_{0}^{\infty} dt'' \exp[i\varepsilon_k(t'' + \tau'' - t' - \tau')] \exp[\alpha(t' - t'')] \times \langle \varphi_0 | T\{ \mathbf{a}_{\mathbf{k}}(t'' + \tau'')\, \mathbf{a}_{\mathbf{k}'}(t)\, \mathbf{a}^*_{\mathbf{k}'}(0)\, \mathbf{a}^*_{\mathbf{k}}(t' + \tau') \} | \varphi_0 \rangle \tag{4–26}$$

The last factor in (4–26) is just the two-particle Green's function $K(\mathbf{k}_i,t_i)$.

Let us first consider the case $\mathbf{k}' \neq \mathbf{k}$. The free part of the factor K reduces to

$$-\, G(\mathbf{k}, t'' + \tau'' - t' - \tau')\, G(\mathbf{k}', t)$$

The corresponding part of (4–26) is just

$$\langle \varphi_{\mathbf{k}} | \varphi_{\mathbf{k}} \rangle\, G(\mathbf{k}', t) = G(\mathbf{k}', t)$$

The correction to $G(k',t)$ arising from the quasi particle $\mathbf{k}$ is thus given by

$$\delta_{\mathbf{k}} G(\mathbf{k}', t) = \frac{i\alpha^2}{z_k} \int_{-\infty}^{0} dt' \int_{0}^{\infty} dt'' \exp[i\varepsilon_k(t'' + \tau'' - t' - \tau'] \exp[\alpha(t' - t'')] \times \delta K \{ \mathbf{k}(t'' + \tau''), \mathbf{k}'t, \mathbf{k}'\, 0, \mathbf{k}(t' + \tau') \} \tag{4–27}$$

$\delta_{\mathbf{k}}G$ is of order $1/\Omega$, like δK. It is negligible with respect to $G(k',t)$, which is practically unmodified by the introduction of a quasi particle **k**. In particular, the poles of $G(k',t)$ are unchanged; *the elementary excitations are independent of each other.* [However, there are cases where $\delta_{\mathbf{k}}G$ plays an important role; for example, if we want to calculate the energy of the state $|\varphi_{\mathbf{k}}\rangle$ using (3–32), the set of $\delta_{\mathbf{k}}G(k',t)$ for all values of k′ will give a contribution independent of Ω, which we cannot neglect in the calculation of $\varepsilon_{\mathbf{k}}$.] Note that these conclusions are no longer valid in a superfluid system, for which δK can be independent of Ω.

Let us now turn to the case $\mathbf{k}' = \mathbf{k}$. We neglect the bound part of (4–26), of order $1/\Omega$, and we keep only the free part,

$$G(\mathbf{k}, t'' + \tau'')\, G(\mathbf{k}, t - \tau' - t') - G(\mathbf{k}, t'' + \tau'' - t' - \tau')\, G(\mathbf{k}, t) \tag{4–28}$$

Let us put this expression into (4–26) and integrate over t' and t''. Because of the filtering from the start, only the coherent part of G contributes [see (4–15)],

$$G(\mathbf{k}, t) - G_{\text{inc}}(\mathbf{k}, t) = \begin{cases} iz_k \exp(-i\varepsilon_k t) & \text{if } t > 0 \\ 0 & \text{if } t < 0 \end{cases} \tag{4–29}$$

We thus obtain

$$G_{\mathbf{k}}(\mathbf{k}, t) = G(\mathbf{k}, t) - iz_k \exp(-i\varepsilon_k t) \tag{4–30}$$

The coherent part of (4–30) is equal [see (4–29)] to

$$G_{\mathbf{k}}(\mathbf{k}, t) - G_{\text{inc}}(\mathbf{k}, t) = \begin{cases} 0 & \text{if } t > 0 \\ -iz_k \exp(-i\varepsilon_k t) & \text{if } t < 0 \end{cases} \tag{4–31}$$

In other words, the introduction of the quasi particle **k** causes the pole of $G(k,\omega)$ to move from the lower half-plane to the upper. The elementary excitation of wave vector **k**, which was a quasi particle, becomes a quasi hole (which brings us back to the ground state). Therefore, *the quasi particles are fermions.*

The distribution function for bare particles in the state $|\varphi_{\mathbf{k}}\rangle$ is given by

$$m_{\mathbf{k}'} + \delta_{\mathbf{k}} m_{\mathbf{k}'} = iG_{\mathbf{k}}(\mathbf{k}', -0) \tag{4–32}$$

When $\mathbf{k}' = \mathbf{k}$, we see from (4–30) that

$$\delta_{\mathbf{k}} m_{\mathbf{k}} = z_k \tag{4–33}$$

z_k thus represents the *fraction of bare particle k contained in the quasi particle.* When $\mathbf{k}' \neq \mathbf{k}$, $\delta_{\mathbf{k}} m_{\mathbf{k}'}$ is of order $1/\Omega$. As the state contains $N + 1$ particles, we must have

$$\sum_{\mathbf{k}' \neq \mathbf{k}} \delta_{\mathbf{k}} m_{\mathbf{k}'} = 1 - z_k \tag{4–34}$$

All these components $\delta_{\mathbf{k}} m_{\mathbf{k}'}$ correspond to the "clothing" which dresses the bare particle **k** to form a quasi particle, that is to say, to the cloud of "daughter" particles carried along by the "mother" particle. Let us note that (4–33) and (4–34) imply

$$0 \leqslant z_k \leqslant 1 \tag{4–35}$$

When k crosses k_F, the excitations change from quasi particle to quasi hole. From (4–33), m_k must have a discontinuity z_{k_F}; this is easily verified by referring to (3–24). For $k < k_F$ the pole of G is the contour of integration, whereas it is no longer so for $k > k_F$ hence the predicted discontinuity z_{k_F} (Fig. 5). It is important not to confuse m_k

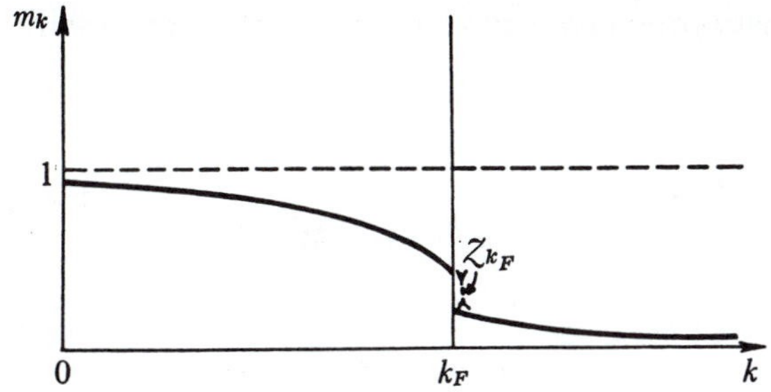

Figure 5

with the distribution function of quasi particles n_k, used by Landau, whose nature is indicated in Fig. 6.

d. Quasi-particle creation and destruction operators

We can make the formalism more compact by writing (for $k > k_F$)

$$| \varphi_{\mathbf{k}} > = A^*_{\mathbf{k}} | \varphi_0 >$$

$$A^*_{\mathbf{k}} = \frac{\alpha}{\sqrt{z_k}} \int_{-\infty}^{0} \exp(- i\varepsilon_k t') \exp(\alpha t') \, \mathbf{a}^*_{\mathbf{k}}(t') \, dt' \qquad \textit{(4–36)}$$

Similarly it can be shown that the state containing a quasi hole of wave vector $k < k_F$ is given by

$$| \varphi_{\mathbf{k}} > = A_{\mathbf{k}} | \varphi_0 >$$

where $A_{\mathbf{k}}$ is the operator conjugate to (4–36). Let us recall that $\varepsilon_k > \mu$ for $k > k_F$ and $< \mu$ for $k < k_F$. The matrix elements of $A^*_{\mathbf{k}}$ and $A_{\mathbf{k}}$

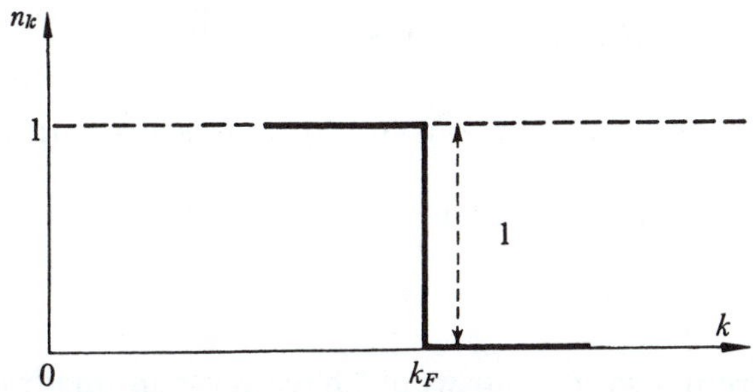

Figure 6

between two rigorous eigenstates of the system are easily deduced from (4–36),

$$\begin{cases} <\varphi_n \mid A_{\mathbf{k}}^* \mid \varphi_m> = \dfrac{\alpha}{\sqrt{z_k}} \dfrac{1}{\mathrm{i}(\xi_{nm} - \xi_k) + \alpha} <\varphi_n \mid a_{\mathbf{k}}^* \mid \varphi_m> \\ <\varphi_n \mid A_{\mathbf{k}} \mid \varphi_m> = \dfrac{\alpha}{\sqrt{z_k}} \dfrac{1}{\mathrm{i}(\xi_{nm} + \xi_k) + \alpha} <\varphi_n \mid a_{\mathbf{k}} \mid \varphi_m> \end{cases} \tag{4–37}$$

The operators $A_{\mathbf{k}}$ and $A_{\mathbf{k}}^*$ make sense only if they are applied to the ground state or to weakly excited states (otherwise the structure of the quasi particle changes, which leads to a new definition of $A_{\mathbf{k}}^*$). We have seen that quasi particles of different **k** are independent of one another to order $1/\Omega$: we can thus write, to within an error of the same order,

$$\begin{cases} [A_{\mathbf{k}}, A_{\mathbf{k}'}]_+ \mid \varphi_n> = 0 \\ [A_{\mathbf{k}}^*, A_{\mathbf{k}'}^*]_+ \mid \varphi_n> = 0 \qquad \text{if } \mathbf{k} \neq \mathbf{k}' \\ [A_{\mathbf{k}}, A_{\mathbf{k}'}^*]_+ \mid \varphi_n> = 0 \end{cases} \tag{4–38}$$

Furthermore, the quasi particles are fermions. We cannot create more than one of them for each value of **k**; as a consequence

$$A_{\mathbf{k}}^* A_{\mathbf{k}}^* \mid \varphi_n> = A_{\mathbf{k}} A_{\mathbf{k}} \mid \varphi_n> = 0 \tag{4–39}$$

Let us now consider the state $A_{\mathbf{k}}^*|\varphi_0\rangle$ $(k > k_F)$. If we destroy the quasi particle **k**, we return to the state $|\varphi_0\rangle$. [For example, we can verify that the propagator $G(k,t)$ takes on the same value as in the ground state.] We can express this conclusion by the equations

$$\begin{cases} A_{\mathbf{k}} A_{\mathbf{k}}^* \mid \varphi_0> = \mid \varphi_0> \qquad \text{for } k > k_F \\ A_{\mathbf{k}}^* A_{\mathbf{k}} \mid \varphi_0> = \mid \varphi_0> \qquad \text{for } k < k_F \end{cases} \tag{4–40}$$

Moreover, the following relations are automatically satisfied:

$$\begin{cases} A_{\mathbf{k}} \mid \varphi_0> = 0 \qquad \text{if } k > k_F \\ A_{\mathbf{k}}^* \mid \varphi_0> = 0 \qquad \text{if } k < k_F \end{cases} \tag{4–41}$$

[If (4–41) were not so, $G(k,\omega)$ would have poles in quadrants I and III; this is incompatible with its analytic structure.] By combining (4–40) and (4–41), and with the help of (4–38), we finally see that

$$[A_{\mathbf{k}}, A^*_{\mathbf{k}}]_+ \mid \varphi_n > = \mid \varphi_n > \qquad (4\text{–}42)$$

$A_{\mathbf{k}}$, $A^*_{\mathbf{k}}$ thus have all the properties of *quasi-particle creation and destruction operators*, in so far as they are applied only to weakly excited states.

The operators, $A_{\mathbf{k}}$, $A^*_{\mathbf{k}}$ are the analogs for dressed particles of the operators $a_{\mathbf{k}}$, $a^*_{\mathbf{k}}$, which correspond to bare particles. As much as possible, it is obviously desirable to replace $a^*_{\mathbf{k}}$ by $A^*_{\mathbf{k}}$ everywhere. This is the essence of renormalization methods, which we shall discuss at the end of this chapter. Let us again stress the fact that all the above relations are subject to an error of order Γ_k; they are valid only in the neighborhood of the Fermi surface. Furthermore, the notion of an elementary excitation becomes unclear if one is interested in very rapid phenomena—in fact, one no longer has time to filter the composite state $a^*_{\mathbf{k}}|\varphi_0\rangle$ to extract from it a quasi particle; mathematically, this leads to a deviation of $G(k,t)$ from the asymptotic form (4–29), the physical origin of the phenomenon being the inertia of the cloud which surrounds the bare particle. The range of validity of the operators $A_{\mathbf{k}}$, $A^*_{\mathbf{k}}$ is thus narrowly limited.

e. Evolution of the ground state as a function of the number of particles

Let us add to the ground state $|\varphi_0(N)\rangle$ the δN quasi particles nearest to the Fermi surface. If $\delta N/N$ is infinitesimal, these quasi particles have an infinite lifetime; the state obtained is a rigorous eigenstate of the $(N + \delta N)$-particle system. Among the states composed only of quasi holes and quasi particles, this one has the minimum energy. If we exclude the possibility of more complex bound states (for example, molecules), we obtain in this way the ground state of $(N + \delta N)$ particles. Let $\mathbf{k}_1 \dots \mathbf{k}_N$ be the corresponding wave vectors,

$$\mid \varphi_0(N + \delta N) > = A^*_{\mathbf{k}_1} \dots A^*_{\mathbf{k}_{\delta N}} \mid \varphi_0(N) > \qquad (4\text{–}43)$$

What are the elementary excitations of this new $(N+ \delta N)$-particle system? According to section **1c** of this chapter, they are identical to those of an N-particle system except for those of wave vector $\mathbf{k}_i$, which become quasi holes: the Fermi surface S_F is simply displaced, so as to enclose δN more points ($\delta N/2$ if we take spin into account). The volume V_F interior to S_F changes by

$$\delta V_F = \frac{4\pi^3}{\Omega} \delta N \qquad (4\text{–}44)$$

For an isotropic system the Fermi surface is spherical; its radius k_F changes by

$$\delta k_F = \pi^2 \frac{\delta N}{k_F^2 \Omega} \tag{4–45}$$

It is tempting to integrate the differential relation (4–44) and to write

$$V_F = 4\pi^3 N/\Omega \tag{4–46}$$

Actually, (4–46) is exact only if the ground state $|\varphi_0(N)\rangle$ is a continuous function of N from $N \sim 0$ up to its actual value. [When N is small, the density N/Ω is so small that the interaction between particles can be neglected; we have an ideal gas, for which (4–46) is exact.] Relation (4–46) thus is applicable only to a *limited* class of normal systems. One can imagine pathological cases for which (4–44) is satisfied, although (4–46) is not. We ignore these difficulties, and we limit ourselves to systems all of whose properties vary continuously with N.

The volume of the Fermi surface does not depend on the interaction; it is a *geometric* factor, independent of the dynamical properties of the system. For an isotropic gas, S_F is a sphere whose radius

$$k_F = [3\pi^2 N/\Omega]^{1/3} \tag{4–47}$$

is independent of the nature of the interaction. Conversely, for an anisotropic system, the *shape* of S_F can change, its volume always being given by (4–46). We shall see in Chap. 5 that this change of shape complicates the formulation of a perturbation treatment.

Let us now consider the ground-state energy E_0. We know that

$$\mathrm{d}E_0/\mathrm{d}N = \mu(N) \tag{4–48}$$

where the chemical potential $\mu(N)$ is equal to the quasi-particle energy at the Fermi surface. In order to integrate (4–48), we assume that S_F expands in a monotonic way when the number of particles increases from 0 to N. To each wave vector k inside $S_F(N)$ there now corresponds a value $N(k)$ for which S_F passes through the point k. Let us set

$$\mu(k) = \mu[N(k)]$$

We can then integrate (4–48), which becomes

$$E_0 = 2 \sum_{\substack{k \text{ in} \\ S_F}} \mu(k) = \frac{\Omega}{4\pi^3} \int_{V_F} d^3k \; \mu(k) \tag{4-49}$$

(the factor 2 accounts for spin). (4–49) simply means that, in order to construct the state $|\varphi_0(N)\rangle$, we stack up the N particles, each of them providing an energy equal to the chemical potential at the corresponding density. This decomposition of E_0 is most natural from the physical point of view; Klein has used it to improve the usual perturbation methods.

f. Renormalization of the propagators

Until now we have considered the propagation of bare particles only. We now generalize this result to the physical case of dressed particles. We shall assume implicitly that the notion of quasi particle makes sense, which imposes two restrictions:

(a) The wave vectors must be very near the Fermi surface.

(b) The time intervals considered must be shorter than the lifetime Γ_k^{-1}. Keeping this in mind, we can neglect the inaccuracy inherent in the definition of the quasi particles.

The single-quasi-particle Green's function $\mathcal{G}(k,t)$ is defined by

$$\mathcal{G}(k, t) = i < \varphi_0 \,|\, T \left\{ \mathbf{A}_\mathbf{k}(t) \, \mathbf{A}^*_\mathbf{k}(0) \right\} |\, \varphi_0 > \tag{4-50}$$

In order to characterize its properties, we calculate the spectral densities of $\mathcal{G}$ for positive and negative t [equivalent to the functions $A_+(k,\omega)$ and $A_-(k,\omega)$ defined in Chap. 3]. We obtain, using (4–37),

$$\begin{cases} \mathcal{A}_+(k, \omega) = A_+(k, \omega) \cdot \dfrac{\alpha^2}{z_k[(\omega - \xi_k)^2 + \alpha^2]} \\ \mathcal{A}_-(k, \omega) = A_-(k, \omega) \cdot \dfrac{\alpha^2}{z_k[(\omega + \xi_k)^2 + \alpha^2]} \end{cases} \tag{4-51}$$

For simplicity let us assume that $k > k_F$; ξ_k is then positive. Going back to Fig. 4 we see that A_- is completely incoherent, whereas A_+ contains an incoherent part supplemented by a coherent peak of width $\Gamma_k < \alpha$. If we neglect the width of this peak, we can write

$$[A_+(k, \omega)]_{\text{coh}} = z_k \delta(\omega - \xi_k) \tag{4-52}$$

Let us return to (4–51); in the limit $\alpha \to 0$ (which implies $\Gamma_k \to 0$), we have

$$\begin{cases} \mathcal{A}_+(k, \omega) = \delta(\omega - \xi_k) \\ \mathcal{A}_-(k, \omega) = 0 \end{cases} \qquad k > k_F \tag{4–53}$$

For $k < k_F$ ($\xi_k < 0$), we would similarly have

$$\begin{cases} \mathcal{A}_+(k, \omega) = 0 \\ \mathcal{A}_-(k, \omega) = \delta(\omega + \xi_k) \end{cases} \qquad k < k_F \tag{4–54}$$

The spectral densities $\mathscr{A}_+$ and $\mathscr{A}_-$ thus have the same structure as for a free particle in an ideal gas. Only the energy has been modified, going from $k^2/2m$ to ε_k; we have *renormalized G.*

We can condense this information by writing the Fourier transform $\mathscr{G}(k,\omega)$,

$$\mathcal{G}(k, \omega) = \frac{1}{\varepsilon_k - \omega} \tag{4–55}$$

$$\text{Im}\,(\varepsilon_k) \cdot (k - k_F) < 0 \tag{4–56}$$

From (4–55) one can easily find $\mathscr{G}(k,t)$.

Let us now turn to the two-particle Green's function. In a similar way we define

$$\mathcal{K}(\mathbf{k}_i, t_i) = < \varphi_0 \mid T \left\{ A_{\mathbf{k}_1}(t_1)\, A_{\mathbf{k}_2}(t_2)\, A^*_{\mathbf{k}_3}(t_3)\, A^*_{\mathbf{k}_4}(t_4) \right\} \mid \varphi_0 > \tag{4–57}$$

Again $\mathscr{K}(\mathbf{k}_i,t_i)$ breaks up into a free part and a bound part. The free part is evidently given by

$$\mathcal{G}(\mathbf{k}_1 t_1, \mathbf{k}_3 t_3)\, \mathcal{G}(\mathbf{k}_2 t_2, \mathbf{k}_4 t_4) - \mathcal{G}(\mathbf{k}_1 t_1, \mathbf{k}_4 t_4)\, \mathcal{G}(\mathbf{k}_2 t_2, \mathbf{k}_3 t_3) \tag{4–58}$$

It is of no great interest. The best way to evaluate the bound part is to return to the definition (3–46) of the interaction operator,

$$\delta K(\mathbf{k}_i, t_i) = \int dt'_i\, G(k_1, t_1 - t'_1) \cdots G(k_4, t'_4 - t_4)\, \gamma(\mathbf{k}_i, t'_i) \tag{4–59}$$

To obtain $\delta\mathscr{K}$ it is sufficient to replace the external $a_{\mathbf{k}}$ by $A_{\mathbf{k}}$. Hence

$$\delta\mathcal{K}(\mathbf{k}_i, t_i) = \int dt'_i\, g(k_1, t_1 - t'_1)\, g(k_2, t_2 - t'_2)\, \tilde{g}(k_3, t'_3 - t_3)\, \tilde{g}(k_4, t'_4 - t_4)\, \gamma(\mathbf{k}_i, t'_i) \tag{4–60}$$

where the functions g and $\tilde{g}$ are defined by

$$\begin{cases} g(k, t-t') = \mathrm{i} < \varphi_0 \mid T \{ \mathbf{A_k}(t)\, \mathbf{a^*_k}(t') \} \mid \varphi_0 > \\ \tilde{g}(k, t-t') = \mathrm{i} < \varphi_0 \mid T \{ \mathbf{a_k}(t)\, \mathbf{A^*_k}(t') \} \mid \varphi_0 > \end{cases} \tag{4-61}$$

In order to study g and $\tilde{g}$, we calculate the corresponding spectral densities $a_\pm(k,\omega)$ and $\tilde{a}_\pm(k,\omega)$. By returning to (4–37) we see that

$$\begin{cases} a_+(k, \omega) = [\tilde{a}_+(k, \omega)]^* = \dfrac{\alpha}{\sqrt{z_k}} \dfrac{A_+(k, \omega)}{\mathrm{i}(\xi_k - \omega) + \alpha} \\ a_-(k, \omega) = [\tilde{a}_-(k, \omega)]^* = \dfrac{\alpha}{\sqrt{z_k}} \dfrac{A_-(k, \omega)}{\mathrm{i}(\xi_k + \omega) + \alpha} \end{cases} \tag{4-62}$$

Again we can neglect the incoherent parts of $A_+(k,\omega)$ and $A_-(k,\omega)$, *to within an error of the order of* α. By comparing (4–51) and (4–62) we see that

$$\tilde{g}(k, t) = g(k, t) = \sqrt{z_k}\, \mathcal{G}(k, t) \tag{4-63}$$

This allows us to rewrite (4–60) in a "renormalized" form,

$$\delta\mathcal{K}(\mathbf{k}_i, t_i) = \int dt'_i \mathcal{G}(k_1, t_1 - t'_1) \ldots \mathcal{G}(k_4, t'_4 - t_4)\, \gamma^R(\mathbf{k}_i, t'_i) \tag{4-64}$$

where the renormalized interaction γ^R is simply defined by

$$\gamma^R(\mathbf{k}_i, t'_i) = \sqrt{z_1 z_2 z_3 z_4}\, \gamma(\mathbf{k}_i, t'_i) \tag{4-65}$$

γ^R describes the interaction between quasi particles. Let us emphasize again that all these results are valid only if the quasi particles considered have very long lifetimes.

g. Scattering of the elementary excitations

Let us consider the function $\delta\mathcal{K}(\mathbf{k}_i, t_i)$, and let us assume $t_1, t_4 \ll t_2, t_3$; we know that $\delta\mathcal{K}$ then describes the propagation of a quasi-particle–quasi-hole pair. Let us increase the interval between (t_1,t_4) and (t_2,t_3); in general, $\delta\mathcal{K}$ oscillates. However, for certain values of $\mathbf{k}_i$ such that

$$\varepsilon_{k_1} - \varepsilon_{k_4} = \varepsilon_{k_3} - \varepsilon_{k_2} \tag{4-66}$$

$\delta\mathcal{K}$ increases uniformly with time; we have *real* scattering of the two excitations. In order to study real scattering processes it is thus necessary to know the asymptotic form of $\mathscr{K}(\mathbf{k}_i,t_i)$ when the t_i spread apart from one another.

The scattering processes that $\mathscr{K}$ can describe are of several different types. Let us designate a quasi particle by P, a quasi hole by T. We can have

$$\begin{cases} P_1 + P_1' \rightarrow P_2 + P_2' \\ T_1 + T_1' \rightarrow T_2 + T_2' \\ T_1 + P_1 \rightarrow T_2 + P_2 \\ P_1 \rightleftarrows P_2 + P_2' + T_1 \\ T_1 \rightleftarrows T_2 + T_2' + P_1 \end{cases} \tag{4-67}$$

The first three processes are elastic, the last two inelastic. We shall see that knowledge of $\mathscr{K}$ directly furnishes the corresponding amplitudes and cross sections. In addition we can describe through $\mathscr{K}$, although less clearly, scattering processes involving the bound states (PP), (PT), and (TT).

$$\begin{cases} P_1 + P_1' \rightleftarrows (PP) \\ T_1 + T_1' \rightleftarrows (TT) \\ P_1 + T_1 \rightleftarrows (PT) \end{cases} \tag{4-68}$$

The last of these reactions corresponds, for example, to the annihilation of a plasmon with the emission of an independent electron-hole pair.

For simplicity, we consider only the scattering $P_1 + P_1' \rightarrow P_2 + P_2'$. The other cases are treated in the same way. We shall just point out in passing the terms which give rise to the processes (4–68).

Let us consider the quantity $\mathscr{K}(\mathbf{k}_1t', \mathbf{k}_2t', \mathbf{k}_3t, \mathbf{k}_4t)$, and let us assume $\mathbf{k}_1$ to be different from $\mathbf{k}_3$ and $\mathbf{k}_4$. From (4–64) we have

$$\mathcal{K}(\ldots) = \int dt_i' \mathcal{G}(k_1, t' - t_1')\,\mathcal{G}(k_2, t' - t_2')\,\mathcal{G}(k_3, t_3' - t)\,\mathcal{G}(k_4, t_4' - t)\,\gamma^R(\mathbf{k}_i, t_i') \tag{4-69}$$

Let us first assume that $\gamma^R(\mathbf{k}_i,t_i)$ is localized in time; the four t_i' are grouped about their center of gravity, τ. τ can in general vary from t to t'. The integration over τ involves a factor

$$\exp(-i\{\varepsilon_{k_1} + \varepsilon_{k_2} - \varepsilon_{k_3} - \varepsilon_{k_4}\}\tau)$$

arising from the four factors $\mathscr{G}$; if the quasi-particle energy is not conserved, $\mathscr{K}$ is a periodic function of $(t'-t)$; the scattering is virtual. On the other hand, if

$$\delta\varepsilon_k = \varepsilon_{k_1} + \varepsilon_{k_2} - \varepsilon_{k_3} - \varepsilon_{k_4} \tag{4–70}$$

is zero, $\mathscr{K}$ is proportional to $(t'-t)$; we have real scattering of the two quasi particles. Actually, $\gamma^R(\mathbf{k}_i,t_i)$ contains terms which are long-range in time. These terms arise from the bound states of two particles $|\boldsymbol{\varphi}_s\rangle$ and have the structure [see (3–55)]

$$\exp\left[+\frac{i}{2}\,\omega_{s0}(t_3' + t_4' - t_1' - t_2')\right] g_s(t_3' - t_4')\, g_s'(t_1' - t_2') \tag{4–71}$$

It can be similarly shown that the corresponding contribution to $\mathscr{K}$ is proportional to $(t'-t)$ if one of the following conditions is realized:

$$\begin{cases} \varepsilon_{k_1} + \varepsilon_{k_2} = \omega_{s0} \\ \varepsilon_{k_3} + \varepsilon_{k_4} = \omega_{s0} \end{cases} \tag{4–72}$$

In this way we display the real scattering $P_1 + P_1' \leftrightarrows (P,P)$.

The above considerations indicate the origin of real scattering; the wavelets coming from the scattering at different instants are all in phase only if energy is conserved.

Let us now turn to a more quantitative study of the problem. The scattering amplitude of the quasi particles is given by

$$\lambda = \langle \varphi_0 \mid \mathbf{A}_{\mathbf{k}_1}(t')\, \mathbf{A}_{\mathbf{k}_2}(t')\, \mathbf{A}^*_{\mathbf{k}_3}(t)\, \mathbf{A}^*_{\mathbf{k}_4}(t) \mid \varphi_0 \rangle \tag{4–73}$$

In order to calculate λ, it is convenient to use the Fourier transform $\mathscr{K}(\mathbf{k}_i,\omega_i)$. By going back to (3–50) and (4–65), we see that

$$\lambda = \frac{1}{(2\pi)^2}\sqrt{z_1 z_2 z_3 z_4}\int d\omega_i \exp\{-i[(\omega_1+\omega_2)\,t' - (\omega_3+\omega_4)\,t]\}\,\mathcal{G}(k_1\omega_1) \ldots \mathcal{G}(k_4\omega_4)\,\Gamma(\mathbf{k}_i\omega_i)\,\delta(\omega_1+\omega_2-\omega_3-\omega_4) \tag{4–74}$$

(we implicitly assume $\mathbf{k}_1 + \mathbf{k}_2 = \mathbf{k}_3 + \mathbf{k}_4$). The δ function is inconvenient; we eliminate it by writing

$$\delta(x) = \lim_{\alpha\to 0}\frac{1}{2\pi}\int_{-\infty}^{+\infty} du\, \exp(iux - \alpha\mid u\mid) \tag{4–75}$$

The integration over $\omega_1 \dots \omega_4$ gives three types of contribution, coming from:

(a) The poles of $\mathscr{G}$.

(b) The poles of Γ.

(c) Integration around the cuts of Γ.

A priori, there is no reason to drop one of these terms. However, for certain values of the $\mathbf{k}_i$, one of them can give a result proportional to $(t' - t)$; the two others are then negligible (this is *false* if energy is not conserved). The elastic scattering of two elementary excitations arises from the poles of the four factors $\mathscr{G}$, whereas the scattering $P_1 + P_1' \to (PP)$ arises from the poles of Γ.

We limit ourselves to the scattering $P_1 + P_1' \to P_2 + P_2'$. Only the poles of the four factors $\mathscr{G}$, $\omega_i = \varepsilon_{k_i}$, contribute. All four of these poles are below the real axis. (4–74) becomes

$$\lambda \simeq 2\pi \sqrt{z_1 z_2 z_3 z_4}\, \Gamma(\mathbf{k}_i, \varepsilon_{k_i}) \times \int_t^{t'} du \exp\{-i[(\varepsilon_{k_1} + \varepsilon_{k_2})(t' - u) - (\varepsilon_{k_3} + \varepsilon_{k_4})(t - u)]\} \tag{4–76}$$

The integration over u is now trivial and gives

$$\lambda = 2\pi \sqrt{z_1 z_2 z_3 z_4}\, \Gamma(\mathbf{k}_i, \varepsilon_{k_i}) \times \exp\{-i[(\varepsilon_{k_1} + \varepsilon_{k_2})t' - (\varepsilon_{k_3} + \varepsilon_{k_4})t]\} \left\{ \frac{e^{i\delta\varepsilon_k t'} - e^{i\delta\varepsilon_k t}}{i\,\delta\varepsilon_k} \right\} \tag{4–77}$$

Actually, we are interested not in the amplitude λ, but in the *transition probability*,

$$W = \frac{|\lambda|^2}{t' - t} \tag{4–78}$$

which, according to (4–77), can be written as

$$W = z_1 z_2 z_3 z_4 \,|\, 2\pi\Gamma(\mathbf{k}_i, \varepsilon_{k_i}) \,|^2 \cdot \frac{4 \sin^2(\frac{1}{2}\, \delta\varepsilon_k[t' - t])}{(t' - t)\,(\delta\varepsilon_k)^2} \tag{4–79}$$

If $(t' - t)$ tends to infinity, (4–79) reduces to

$$W = 2\pi\, z_1 z_2 z_3 z_4 \,|\, 2\pi\Gamma(\mathbf{k}_i, \varepsilon_{k_i}) \,|^2\, \delta(\varepsilon_{k_1} + \varepsilon_{k_2} - \varepsilon_{k_3} - \varepsilon_{k_4}) \tag{4–80}$$

The fundamental result (4–80) resembles the classical result obtained in the Born approximation. The "effective" matrix element characterizing the scattering is here equal to

$$2\pi\Gamma(\mathbf{k}_i, \varepsilon_{k_i}) \sqrt{z_1 z_2 z_3 z_4}$$

In this case, complete knowledge of the function $\Gamma(\mathbf{k}_i, \omega_i)$ is unnecessary. This simplification arises from the fact that we are studying a real transition, spread out over a very large interval of time $(t' - t)$. This is no longer the case, unfortunately, when one is interested in periods of time of the order of the relaxation time of $\gamma(\mathbf{x}_i)$.

Let us recall that all the preceding considerations make sense only if the lifetime Γ_k^{-1} of the four states $\mathbf{k}_i$ is much larger than $(t' - t)$. (We cannot speak of "scattering" unless the incident and emerging elements are well defined.) (4–80) thus applies only in the neighborhood of the Fermi surface.

h. The Landau model

The *normal* systems whose properties we have just studied possess all the characteristics of the Landau model studied in Chap. 1; the distribution of elementary excitations is the same in both cases. Only the states near the Fermi level can be interpreted in terms of quasi particles; this limits us to low temperatures, which conforms to the hypothesis of Chap. 1.

Thus, we can consider this chapter as a microscopic justification of Landau's phenomenological model. It remains only to calculate the "interaction energy" of two quasi particles $f(\mathbf{k},\mathbf{k}')$, that is, the variation of $\varepsilon_{k'}$ when a quasi particle $\mathbf{k}$ is introduced. To characterize this variation we consider the Green's function $G(\mathbf{k}',t)$, successively for the ground state $|\varphi_0\rangle$ and for $A^*_{\mathbf{k}}|\varphi_0\rangle$. When t is large, we can write

$$\begin{cases} G(\mathbf{k}', t) \rightarrow \mathrm{i} z_{k'} \exp(-\mathrm{i}\varepsilon_{k'} t) \\ G_{\mathbf{k}}(\mathbf{k}', t) \rightarrow \mathrm{i}(z_{k'} + \delta_{\mathbf{k}} z_{\mathbf{k}'}) \exp\{-\mathrm{i}[\varepsilon_{k'} + f(\mathbf{k}, \mathbf{k}')]\, t\} \end{cases} \qquad (4\text{–}81)$$

The quantities $\delta_k z_k$ and $f(\mathbf{k},\mathbf{k}')$ are small (of order $1/\Omega$). Thus we can expand (4–81),

$$\delta_{\mathbf{k}} G(\mathbf{k}', t) \rightarrow [\mathrm{i}\, \delta_{\mathbf{k}} z_{\mathbf{k}'} + z_{k'} f(\mathbf{k}, \mathbf{k}')\, t] \exp(-\mathrm{i}\varepsilon_{k'} t) \qquad (4\text{–}82)$$

If t is large enough, the first term is negligible; we immediately obtain $f(\mathbf{k},\mathbf{k}')$. By returning to (4–27) we finally arrive at the relation

$$f(\mathbf{k}, \mathbf{k}') = \frac{i\alpha^2}{z_k z_{k'} t} \int_{-\infty}^{0} dt' \int_{0}^{\infty} dt'' \exp\left[i\varepsilon_k(t'' + \tau'' - t' - \tau') + i\varepsilon_{k'} t\right] \exp\left[\alpha(t' - t'')\right]$$
$$\times \delta K\left[\mathbf{k}(t'' + \tau''), \mathbf{k}'t, \mathbf{k}'0, \mathbf{k}(t' + \tau')\right] \qquad (4\text{–}83)$$

The factor δK appearing in (4–83) can be calculated just as the scattering amplitude λ was in the preceding section. By transposing (4–76) we obtain

$$\delta K\left[\mathbf{k}(t'' + \tau''), \mathbf{k}'t, \mathbf{k}'0, \mathbf{k}(t' + \tau')\right] = \Gamma(\mathbf{k}\varepsilon_k, \mathbf{k}'\varepsilon_{k'}, \mathbf{k}'\varepsilon_{k'}, \mathbf{k}\varepsilon_k)$$
$$\times 2\pi z_k^2 z_{k'}^2 \int_0^t du \exp\left[-i\varepsilon_k(t'' + \tau'' - t' - \tau') - i\varepsilon_{k'} t\right] \qquad (4\text{–}84)$$

Conservation of energy is assured here by construction. We finally find

$$f(\mathbf{k}, \mathbf{k}') = 2\pi\, i z_k z_{k'} \Gamma(\mathbf{k}\varepsilon_k, \mathbf{k}'\varepsilon_{k'}, \mathbf{k}'\varepsilon_{k'}, \mathbf{k}\varepsilon_k) \qquad (4\text{–}85)$$

(4–85) is identical to the result obtained by Landau using an entirely different method.

Unfortunately, we have ignored a major difficulty: $\Gamma(\mathbf{p},\mathbf{p}',\mathbf{p}',\mathbf{p})$ is defined in an ambiguous way. In fact, we shall show later (by a perturbation method) that the limit of $\Gamma(\mathbf{p}, \mathbf{p}', \mathbf{p}' + \delta p, \mathbf{p} - \delta p)$ for $\delta p \to 0$ depends on the way in which δp is made to tend to 0. If we choose to let first $\delta k \to 0$, then $\delta\omega \to 0$, we obtain a limit $\Gamma^0(\mathbf{p},\mathbf{p}',\mathbf{p}',\mathbf{p})$, different from the limit $\Gamma^\infty(\mathbf{p},\mathbf{p}',\mathbf{p}',\mathbf{p})$ obtained by first taking $\delta\omega \to 0$, then $\delta k \to 0$. The factor Γ appearing in (4–85) corresponds to the limit Γ^0; indeed, δk is then zero by construction, whereas $\delta\omega$ is only approximately zero, with an error of the order of the natural width of the states $\mathbf{k}$ and $\mathbf{k}'$. On the other hand, let us consider the forward-scattering amplitude, corresponding to the process

$$P_{\mathbf{k}} + P_{\mathbf{k}'} \to P_{\mathbf{k}} + P_{\mathbf{k}'}$$

This amplitude is given by (4–77) and can be written as

$$\lambda = 2\pi z_k z_{k'} \Gamma\left[\mathbf{k}'\varepsilon_{k'}, \mathbf{k}\varepsilon_k, \mathbf{k}\varepsilon_k, \mathbf{k}'\varepsilon_{k'}\right] \exp\left[i(\varepsilon_k + \varepsilon_{k'})(t - t')\right] . (t' - t) \qquad (4\text{–}86)$$

If $\mathbf{k}$ and $\mathbf{k}'$ lie on the Fermi surface, the energy transfer in any collision is necessarily zero (because of the exclusion principle), whereas the momentum transfer measures the scattering angle. The limit of forward scattering thus corresponds to $\delta\omega = 0$, $\delta k \to 0$; in (4–86), we must use the limit Γ^∞. This would be wrong if $\mathbf{k}$ and $\mathbf{k}'$ were not on the Fermi surface.

This is a delicate question; one must handle $\Gamma(\mathbf{p},\mathbf{p}',\mathbf{p}',\mathbf{p})$ with great caution.

2. The Superfluid State

a. Generalities

Definition. We shall denote as "superfluid" all systems possessing the following two characteristics:

(a) There exists a *bound* state connecting two particles or two holes; the pairs thus formed obey Bose–Einstein statistics.

(b) In the ground state $|\varphi_0\rangle$, these pairs form a *condensed* phase, analogous to that in superfluid helium.

In order to clarify this definition, let us return to the decomposition of $K(\mathbf{x}_i)$ studied in Chap. 3, Sec. 2. When $t_1, t_4 > t_2, t_3$, we can write K in the form (3–51).

$$K(\mathbf{x}_i) = \sum_s \tilde{\chi}_s(1, 4)\, \chi_s(2, 3)$$

where the matrix element χ_s is defined by

$$\chi_s(\mathbf{x}, \mathbf{x}') = < \varphi_s \mid T\{\psi_\sigma(x)\, \psi^*_{\sigma'}(x')\} \mid \varphi_0 >$$

As we have already mentioned, the state $|\varphi_s\rangle$ can belong to three distinct categories:

(a) $|\varphi_s\rangle = |\varphi_0\rangle$: χ_s reduces to the Green's function G, which is independent of the volume of the system Ω.

(b) $|\varphi_s\rangle$ corresponds to a state containing a quasi particle of wave vector $\mathbf{k}$ and a quasi hole of wave vector $\mathbf{k}'$, independent of one another. χ_s gives the probability amplitude for finding these two components at $\mathbf{x}'$ and $\mathbf{x}$ respectively. The probability $|\chi_s|^2$ is of order $1/\Omega^2$. χ_s is thus of order $1/\Omega$. The number of states $|\varphi_s\rangle$ of this type is the number of pairs $(\mathbf{k},\mathbf{k}')$; it is of order Ω^2: the total contribution of these elements to K thus remains independent of Ω. This contribution gives a long-range term which, added to the term (a), yields the free part of K and a short-range term which describes the interaction of the hole and the particle.

(c) Finally, $|\varphi_s\rangle$ can be a bound state of the exciton type, of total wave vector q. This state is spread out over the whole volume Ω, which makes χ_s of order $1/\sqrt{\Omega}$. On the other hand, the number of these bound states, that is to say, the number of possible values of q, is of order Ω. The corresponding contribution to K is thus of the same order as cases

(a) and (b); however, it is short-ranged because of the destructive interference of the phase factors when we sum over q.

Let us now compare these results with the expansion (3–53) of $K(\mathbf{x}_i)$, valid when $t_1, t_2 > t_3, t_4$,

$$K(\mathbf{x}_i) = \sum_s \tilde{\eta}_s(1, 2)\, \xi_s(3, 4)$$

where the $\tilde{\eta}_s$ and ξ_s are defined by (3–54). There, again, the type of the states $|\varphi_s\rangle$ can vary. For a superfluid system, three cases are possible:

(b′) $|\varphi_s\rangle$ contains two independent quasi particles. The discussion is the same as for case (b) above. We thus generate the usual free part of K (product of two factors G), modified at short distances by terms describing the interaction of the two particles.

(c′) $|\varphi_s\rangle$ corresponds to a bound state of two particles, of variable total wave vector q; the discussion of case (c) can be carried over just as it is.

(a′) $|\varphi_s\rangle$ corresponds to a *condensed* bound state; the total wave vector q is then unique (q is zero for a symmetric system); the pairs of particles pile up in this state just, as in the condensation of a Bose–Einstein gas—the *population* of this bound state is of order Ω. The function ξ_s, which more or less plays the role of a matrix element of the "creation-of-a-bound-pair" operator, is proportional to the square root of the population. The factor $\sqrt{\Omega}$ thus obtained compensates the factor $1/\sqrt{\Omega}$ characteristic of the bound states (c). The contribution of this bound state K is thus essential.

Very often the state $|\varphi_s\rangle$ obtained by adding a condensed pair to $|\varphi_0\rangle$ is just the $(N + 2)$-particle ground state $|\varphi_0(N + 2)\rangle$. As an intermediate state, it plays a role for superfluid systems equivalent to that of the state $|\varphi_0(N)\rangle$ appearing in case (a). The corresponding contributions to K are both long-ranged and are of the same order; it is necessary to treat them on an equal footing. This leads us to modify the decomposition of the function K and to choose as a new "free part" the quantity

$$G(1, 3)\, G(2, 4) - G(1, 4)\, G(2, 3) + \tilde{\eta}_0(1, 2)\, \xi_0(3, 4)$$

where the index 0 corresponds to

$$|\varphi_s\rangle = |\varphi_0(N + 2)\rangle$$

The quantities $\tilde{\eta}_0$ and ξ_0, play a role analogous to G. We shall study them in detail in the following section.

This qualitative discussion indicates the new feature introduced for a superfluid—the existence of a *unique*, *condensed*, *bound* state of $(N+2)$ particles, which contributes to the long-ranged free part of K in the same way as the ground state. This requires a complete revision of the formalism set up for a normal system.

Generalized single-particle Green's functions. We must consider on an equal footing the quantities

$$\begin{aligned} P &= \langle \varphi_0(N) \mid T\{ \psi_\sigma(x)\, \psi^*_{\sigma'}(x') \} \mid \varphi_0(N) \rangle \\ Q &= \langle \varphi_0(N+2) \mid T\{ \psi^*_\sigma(x)\, \psi^*_{\sigma'}(x') \} \mid \varphi_0(N) \rangle \\ R &= \langle \varphi_0(N-2) \mid T\{ \psi_\sigma(x)\, \psi_{\sigma'}(x') \} \mid \varphi_0(N) \rangle \end{aligned} \tag{4-87}$$

P, Q, R are continuous functions of the number of particles N; their variation when N goes to $(N+2)$ is of order $1/\Omega$ and thus completely negligible.

The quantity P is simply related to the usual Green's function,

$$P = -\mathrm{i}G(x\sigma, x'\sigma') \tag{4-88}$$

P depends only on the difference $(x - x')$. The structure of Q and R can be deduced from the general relation (3–55). The state $|\varphi_0(N+2)\rangle$ has a zero total wave vector and an energy

$$E_0(N \pm 2) = E_0(N) \pm 2\,\mu \tag{4-89}$$

where μ is the chemical potential. We can thus put Q and R into the form

$$\begin{cases} Q = -\mathrm{i}\exp[\mathrm{i}\mu(t+t')]\,\check{G}(x\sigma, x'\sigma') \\ R = -\mathrm{i}\exp[-\mathrm{i}\mu(t+t')]\,\hat{G}(x\sigma, x'\sigma') \end{cases} \tag{4-90}$$

where the quantities $\hat{G}$ and $\check{G}$ depend only on the difference $(x - x')$. For reasons of symmetry we shall slightly modify the definition of G, by introducing a new propagator

$$\begin{cases} \bar{G}(x\sigma, x'\sigma') = \exp[\mathrm{i}\mu(t-t')]\,G(x\sigma, x'\sigma') \\ P = -\mathrm{i}\exp[-\mathrm{i}\mu(t-t')]\,\bar{G}(x\sigma, x'\sigma') \end{cases} \tag{4-91}$$

What is the physical significance of the functions $\hat{G}$, $\bar{G}$, $\check{G}$? We already know that $\bar{G}$ describes the propagation of an additional particle or hole. By analogy, $\check{G}(\hat{G})$ describes the *coherent* creation (annihilation) of a pair of bound particles. For example, $\check{G}$ gives the probability amplitude for two additional particles $(x\sigma)$ and $(x'\sigma')$ to "collapse" into the condensed phase while losing all their individuality. In other words, if we add a single particle to $|\varphi_0(N)\rangle$, it can either:

(a) Propagate: this is a process described by $\bar{G}$.

(b) Be exchanged with a particle in the condensed phase: this is a process described by $\check{G}$.

Similarly, the propagation of an additional hole is governed either by $\bar{G}$ or by $\hat{G}$.

Before studying the properties of these functions, we shall introduce an abbreviated notation, involving only the difference $(x - x')$,

$$\check{\bar{\hat{G}}}(x\sigma, x'\sigma') = \check{\bar{\hat{G}}}_{\sigma\sigma'}(x - x') \qquad (4\text{–}92)$$

$\hat{G}$, $\bar{G}$, and $\check{G}$ are 2×2 matrices labeled by the indices σ and σ'.

The symmetry properties of the propagators. A priori, the only obvious symmetry relation arises from the definition of the T-product and is written as

$$\check{\bar{\hat{G}}}_{\sigma\sigma'}(x) = -\check{\bar{\hat{G}}}_{\sigma'\sigma}(-x) \qquad (4\text{–}93)$$

To be able to go further we must make some hypotheses about the nature of the ground state $|\varphi_0(N)\rangle$ and about the law of interaction. We shall *postulate* that

(a) The interaction forces are independent of spin. The total spin of the system is a good quantum number. All the properties are invariant under spin reversal.

(b) The states $|\varphi_0(N)\rangle$ and $|\varphi_0(N+2)\rangle$ are *nondegenerate*, invariant under rotation in ordinary space, reflection, and time reversal.

These assumptions are met in the usual superconductors, in which the bound pairs are formed in S states. If, however, the pairs have a nonzero orbital angular momentum, the resulting wave function shows rotational degeneracy; the above assumptions are violated. Such an anisotropic situation might occur in liquid ^{3}He. The generalization of this section to such cases is straightforward.

Remark: The cubic boundary conditions which we have adopted are not compatible with rotation invariance. We avoid this difficulty by provisionally enclosing the system in a sphere instead of putting it into a cube: the functions $\hat{G}$, $\bar{G}$, and $\check{G}$ will not be modified.

The nondegeneracy implies zero total spin and orbital angular momentum (which assumes N to be even). The parity of the states $|\varphi_0\rangle$

is therefore positive. Furthermore, we shall assume the phases of $|\varphi_0(N)\rangle$ and $|\varphi_0(N\pm 2)\rangle$ to be adjusted so that the corresponding state vectors are invariant with respect to time reversal.

Since all ground states have zero total spin, we have

$$\begin{cases} \overline{G}_{+-} = \overline{G}_{-+} = 0 \\ \check{\hat{G}}_{++} = \check{\hat{G}}_{--} = 0 \end{cases} \tag{4–94}$$

Let us make a rotation of the spins only, denoted by S. The operators ψ and ψ^* are transformed according to the same law,

$$S\psi_\pm(r,t)\,S^{-1} = \pm\,\psi_\mp(r,t) \tag{4–95}$$

The result is that

$$\begin{cases} \overline{G}_{++} = \overline{G}_{--} \\ \check{\hat{G}}_{+-} = -\,\check{\hat{G}}_{-+} \end{cases} \tag{4–96}$$

The reflection invariance imposes that all the functions $\hat{G}$, $\bar{G}$, $\check{G}$, be even in r. By comparing the results (4–93) and (4–96), we see that $\hat{G}_{\sigma,\sigma'}$ and $\check{G}_{\sigma,\sigma'}$ are even functions of t.

It remains to exploit the time-reversal invariance. The corresponding transformation is antiunitary (see, for example, A. Messiah's "Quantum Mechanics"). The operators $\psi_\pm(r,t)$ and $\psi^*_\pm(r,t)$ are both transformed according to the law

$$\psi_\pm(r,t) \to \pm\,\psi_\mp(r,-t) \tag{4–97}$$

Let us first apply this transformation to $G(x\sigma,x'\sigma')$. We find

$$\overline{G}(rt\sigma, r't'\sigma') = \\ i\exp[i\mu(t-t')]\,[<\varphi_0(N)\mid \tilde{T}\{\psi_{-\sigma}(r,-t)\,\psi^*_{-\sigma'}(r',-t')\}\mid\varphi_0(N)>]^*$$

where $\tilde{T}$ is the "antichronological" operator. Using the definition of the Hermitian conjugate operator, we put this relation into the form

$$\begin{aligned} \overline{G}(rt\sigma, r't'\sigma') &= \\ &= i\exp[i\mu(t-t')] < \varphi_0(N)\mid T\{\psi_{-\sigma'}(r',-t')\,\psi^*_{-\sigma}(r,-t)\mid\varphi_0(N)> \\ &= \overline{G}(r',-t',-\sigma';\,r,-t,-\sigma) \end{aligned}$$

This result may be obtained directly from spin rotation and reflection invariance; in this case, time reversal introduces nothing new. On the other hand, if we repeat the same reasoning for $\hat{G}(rt\sigma, r't'\sigma')$, we obtain a new relation,

$$\widehat{G}(rt\sigma, r't'\sigma') = -\check{G}(r', -t', -\sigma'; r, -t, -\sigma) = -\check{G}(rt\sigma, r't'\sigma') \quad \textit{(4–98)}$$

which relates the matrices $\hat{G}$ and $\check{G}$.

Let us collect the preceding results. We see that the three matrices $\hat{G}$, $\bar{G}$, and $\check{G}$ depend only on two independent functions $F_1(x)$ and $F_2(x)$,

$$\begin{cases} \bar{G}_{++} = \bar{G}_{--} = F_1(x) \\ \widehat{G}_{+-} = -\widehat{G}_{-+} = -\check{G}_{+-} = \check{G}_{-+} = F_2(x) \end{cases} \quad \textit{(4–99)}$$

F_1 and F_2 are even in r; furthermore, F_2 is even in t. We thus exhaust all the information furnished by simple symmetry considerations.

The properties of $\mathbf{F_1(x)}$ and $\mathbf{F_2(x)}$. The function $F_1(x)$ is related to the usual Green's function $G(x)$,

$$F_1(x) = e^{i\mu t} G(x) \quad \textit{(4–100)}$$

After Fourier transformation, (4–100) reads:

$$F_1(k, \omega) = G(k, \mu + \omega) \quad \textit{(4–101)}$$

The properties of F_1 are thus obtained by simple transposition of the results of Chap. 3. We shall not repeat them. Let us simply remark that the branch point of F_1 is now located at the origin, $\omega = 0$.

Let us turn now to the study of F_2, which is characteristic of the superfluid state (for a normal gas, $F_2 = 0$). We shall write it in the form

$$F_2(r, t) = i e^{i\mu t} < \varphi_0(N) \mid T\{\psi_+(r, t)\, \psi_-(0, 0)\} \mid \varphi_0(N + 2) > \quad \textit{(4–102)}$$

F_2 is continuous at $t = 0$ [since $\psi_+(r,0)$ and $\psi_-(0,0)$ anticommute]. On the other hand, its derivatives at the origin are generally discontinuous. Its Fourier transform with respect to r is

$$F_2(k, t) = i e^{i\mu t} < \varphi_0(N) \mid T\{\mathbf{a}_{k,+}(t)\, \mathbf{a}_{-k,-}(0)\} \mid \varphi_0(N + 2) > \quad \textit{(4–103)}$$

$F_2(k,t)$ is an even function of k and of t.

Just as for F_1, we can express $F_2(k,t)$ in the form of a Lehmann expansion. For this purpose we introduce the spectral density $A_0(k,\omega)$,

$$A_0(k, \omega) = \sum_n < \varphi_0(N) \mid a_{k,+} \mid \varphi_n > < \varphi_n \mid a_{-k,-} \mid \varphi_0(N+2) > \delta(\omega - \xi_{n0}) \tag{4–104}$$

$$= -\sum_n < \varphi_0(N) \mid a_{-k,-} \mid \varphi_n > < \varphi_n \mid a_{k,+} \mid \varphi_0(N+2) > \delta(\omega - \xi_{n0})$$

where the positive excitation energy ξ_{n0} is given by

$$E_n - E_0(N) = \mu + \xi_{n0} \tag{4–105}$$

The expression (4–103) can now be put into the form

$$F_2(k, t) = \mathrm{i} \int_0^\infty A_0(k, \omega) \exp(-\mathrm{i}\omega \mid t \mid) \, d\omega \tag{4–106}$$

(4–106) clearly shows that F_2 is an even function of t.

Let us transform A_0 by time reversal; we find

$$A_0(k, \omega) = -\left\{ \sum_{n'} < \varphi_0(N) \mid a_{-k,-} \mid \varphi_{n'} > < \varphi_{n'} \mid a_{k,+} \mid \varphi_0(N+2) > \delta(\omega - \xi_{n0}) \right\}^*$$

where the state $|\varphi_{n'}\rangle$ is the transform of $|\varphi_n\rangle$; these two states have the same energy. By comparison with (4–104) we see that

$$A_0(k, \omega) = \{ A_0(k, \omega) \}^*$$

A_0 is therefore a *real* function, which confirms its analogy with the functions A_+ and A_- introduced in Chap. 3.

Let us now turn to the Fourier transform of F_2 with respect to t.

$$F_2(k, \omega) = \int_{-\infty}^{+\infty} dt \, F_2(k, t) \, e^{\mathrm{i}\omega t} \tag{4–107}$$

This is an even function of ω. Using (4–106), we find

$$F_2(k, \omega) = \int_0^\infty A_0(k, \omega') \left[\frac{1}{\omega' - \omega - \mathrm{i}\eta} + \frac{1}{\omega' + \omega - \mathrm{i}\eta} \right] \tag{4–108}$$

(4–108) is the Lehmann representation of F_2. F_1 and F_2 thus have the same analytic structure: cuts above the real negative axis and below the real positive axis, branch point at the origin, poles on the neighboring Riemann sheets. The poles nearest the physical sheet are located in the second and fourth quadrants; as F_2 is even in ω, its poles appear automatically in pairs of opposite values and residues. In contrast to F_1, F_2 is of order $1/\omega^2$ when $\omega \to \infty$.

This brief outline of the properties of F_2 is sufficient for our needs; we shall not explore the question further. Let us only remark that, in order to know all single-particle propagators, it is enough to calculate the three real functions A_+, A_-, and A_0, defined in the interval $0 < \omega < +\infty$.

b. Two-particle Green's function

Condensed notation. The existence of propagators $\hat{G}$ and $\check{G}$ leads us to modify the definition of the free part of K and the interaction operator γ. The notation used until now proves to be quite inconvenient for carrying out this new decomposition. To simplify the discussion, we introduce the following condensed notation:

$$\psi_\sigma^\alpha(x) = \begin{cases} \psi_\sigma(x) & \text{if} \quad \alpha = +1 \\ \psi_\sigma^*(x) & \text{if} \quad \alpha = -1 \end{cases} \tag{4–109}$$

The most general single-particle Green's function can then be written as

$$\begin{aligned} &G(x\sigma\alpha, x'\sigma'\alpha') = \\ &\quad \mathrm{i}\exp[\mathrm{i}\mu(\alpha t + \alpha' t')] < \varphi_0(N - \alpha - \alpha') \mid T\{\psi_\sigma^\alpha(x)\,\psi_{\sigma'}^{\alpha'}(x')\} \mid \varphi_0(N) > \end{aligned} \tag{4–110}$$

Since these functions remain constant if N varies by a few units, we can, in the definition (4–10), avoid specifying the number of particles contained in each of the states $|\varphi_0\rangle$; it is understood once for all that this number is close to N and that the difference between the states on the right and left is adjusted so as to assure conservation of the total number of particles. We shall thus write

$$G(x\sigma\alpha, x'\sigma'\alpha') = \mathrm{i}\exp[\mathrm{i}\mu(\alpha t + \alpha' t')] < \varphi_0 \mid T\{\psi_\sigma^\alpha(x)\,\psi_{\sigma'}^{\alpha'}(x')\} \mid \varphi_0 > \tag{4–111}$$

The different elements of G are given by

$$\begin{cases} G(x\sigma_+, x'\sigma'_+) = \hat{G}(x\sigma, x'\sigma') \\ G(x\sigma_-, x'\sigma'_-) = \check{G}(x\sigma, x'\sigma') \\ G(x\sigma_+, x'\sigma'_-) = \bar{G}(x\sigma, x'\sigma') \\ G(x\sigma_-, x'\sigma'_+) = -\bar{G}(x'\sigma', x\sigma) \end{cases} \tag{4–112}$$

Conservation of total spin imposes the condition

$$\alpha\sigma + \alpha'\sigma' = 0 \tag{4–113}$$

The elements of G which do not satisfy this condition are zero. Let us recall that $G(x\sigma\alpha, x'\sigma'\alpha')$ depends only on the difference $(x - x')$.

The Fourier transform with respect to the spatial variables is defined in the following way:

$$G(kt\sigma\alpha, k't'\sigma'\alpha') = (1/\Omega) \iint d^3r\, d^3r'\, G(rt\sigma\alpha, r't'\sigma'\alpha') \exp[-i(\alpha kr + \alpha' k' r')] \tag{4–114}$$

Because of translational invariance, this transform contains a factor $\delta_{\alpha k, -\alpha' k'}$, which expresses conservation of momentum. $G(kt\sigma\alpha, k't'\sigma'\alpha')$ can be written in the form

$$G(kt\sigma\alpha, k't'\sigma'\alpha') = i < \varphi_0 \mid T\left\{ \mathbf{a}^{\alpha}_{k\sigma}(t)\, \mathbf{a}^{\alpha'}_{k'\sigma'}(t') \right\} \mid \varphi_0 > \exp[i\mu(\alpha t + \alpha' t')] \tag{4–115}$$

where the operator $a^{\alpha}_{k\sigma}$ is defined by

$$a^{\alpha}_{k\sigma} = \begin{cases} a_{k\sigma} & \text{if } \alpha = +1 \\ a^{*}_{k\sigma} & \text{if } \alpha = -1 \end{cases} \tag{4–116}$$

This expression will be used in Chap. 7.

The Fourier transform with respect to time is defined as

$$G(k\omega\sigma\alpha, k'\omega'\sigma'\alpha') = \frac{1}{2\pi} \iint dt\, dt'\, G(kt\sigma\alpha, k't'\sigma'\alpha') \exp[i(\alpha\omega t + \alpha'\omega' t')] \tag{4–117}$$

Again, translational invariance introduces a factor $\delta(\alpha\omega + \alpha'\omega')$. In order to simplify the notation, which is becoming quite cumbersome, we introduce the symbol ξ to represent the set (k,ω,σ,α), with the following conventions:

$$\begin{cases} \xi \rightarrow (k, \omega, \sigma, \alpha) \\ \sum_{\xi} \rightarrow \sum_{k} \sum_{\sigma} \sum_{\alpha} \int d\omega \\ \delta(\xi, \xi') = \delta_{\alpha k, -\alpha' k'}\, \delta_{\alpha\sigma, -\alpha'\sigma'}\, \delta(\alpha\omega + \alpha'\omega') \end{cases} \tag{4–118}$$

Conservation of spin, momentum, and energy is expressed by the presence in $G(\xi,\xi')$ of a factor $\delta(\xi,\xi')$.

This definition of G is perfectly symmetric; we shall use it later. In certain cases, however, it is preferable to introduce another decomposed form. By comparing (4–99) and (4–112), we can write $G(x\sigma\alpha, x'\sigma'\alpha')$ in the form

$$G(x\sigma\alpha, x'\sigma'\alpha') = F_1(x - x')\,\delta_{\sigma,\sigma'}\,\delta_{\alpha,+}\,\delta_{\alpha',-} - F_1(x' - x)\,\delta_{\sigma,\sigma'}\,\delta_{\alpha,-}\,\delta_{\alpha',+} + \alpha\sigma F_2(x - x')\,\delta_{\sigma,-\sigma'}\,\delta_{\alpha,\alpha'} \tag{4–119}$$

Let us take the Fourier transform of (4–119), using the definitions (4–118); we find

$$\begin{aligned} G(\xi, \xi') &= G(p\sigma\alpha, p'\sigma'\alpha') \\ &= \delta(\xi, \xi')\,\{\alpha F_1(p)\,\delta_{\sigma\sigma'} + \alpha\sigma F_2(p)\,\delta_{\sigma,-\sigma'}\} \end{aligned} \tag{4–120}$$

Let us now introduce a new Green's function $\tilde{G}(p\sigma\alpha, p'\sigma'\alpha')$ defined by

$$\tilde{G}(p\sigma\alpha, p'\sigma'\alpha') = -\alpha' G(p\sigma\alpha, p'\sigma'\alpha') \tag{4–121}$$

(this function appears naturally in the perturbation expansion studied in Chap. 7). By comparing (4–120) and (4–121), we see that

$$\begin{cases} \tilde{G}(\xi, \xi') = \delta(\xi, \xi')\,\tilde{G}_{\sigma\sigma'}(p) \\ \tilde{G}_{\sigma\sigma'}(p) = F_1(p)\,\delta_{\sigma,\sigma'} - \sigma F_2(p)\,\delta_{\sigma,-\sigma'} \end{cases} \tag{4–122}$$

The index α has disappeared [it subsists only in the factor $\delta(\xi,\xi')$]. This is the great advantage of $\tilde{G}$ over G.

We shall represent $\tilde{G}_{\sigma\sigma'}(p)$ by a 2×2 matrix, denoted $\tilde{\mathbf{G}}(p)$, which is equal to

$$\tilde{\mathbf{G}}(p) = \begin{pmatrix} F_1 & -F_2 \\ F_2 & F_1 \end{pmatrix} \tag{4–123}$$

Let us introduce the matrices

$$\mathbf{1} = \begin{pmatrix} 1 & 0 \\ 0 & 1 \end{pmatrix} \qquad \boldsymbol{\beta}_1 = \begin{pmatrix} 0 & -1 \\ 1 & 0 \end{pmatrix} \qquad \boldsymbol{\beta}_2 = \begin{pmatrix} 1 & 0 \\ 0 & -1 \end{pmatrix} \tag{4–124}$$

With these definitions, $\tilde{G}$ is written as

$$\tilde{\mathbf{G}} = F_1\mathbf{1} + F_2\boldsymbol{\beta}_1 \tag{4–125}$$

Generalized two-particle Green's function. The Green's function K defined in Chap. 3 is written as

$$K(x_i, \sigma_i) = < \varphi_0 \mid T\{ \psi_{\sigma_1}(x_1) \dots \psi^*_{\sigma_4}(x_4) \} \mid \varphi_0 >$$

By analogy with $\bar{G}$, we first extract from K the exponential factor coming from the chemical potential μ, by setting

$$\overline{K}(x_i, \sigma_i) = K(x_i, \sigma_i) \exp[\mathrm{i}\mu(t_1 + t_2 - t_3 - t_4)] \tag{4-126}$$

After Fourier transformation, (4–126) becomes

$$\overline{K}(k_i, \omega_i, \sigma_i) = K(k_i, (\mu + \omega_i), \sigma_i) \tag{4-127}$$

We can in fact introduce a more general Green's function,

$$K(x_i, \sigma_i, \alpha_i) = < \varphi_0 \mid T\{ \psi^{\alpha_1}_{\sigma_1}(x_1) \dots \psi^{\alpha_4}_{\sigma_4}(x_4) \} \mid \varphi_0 > \exp\left(\mathrm{i}\mu \sum_i \alpha_i t_i\right) \tag{4-128}$$

The function K corresponds to the particular choice: $\alpha_1 = \alpha_2 = +1$, $\alpha_3 = \alpha_4 = -1$. Conservation of total spin implies

$$\sum_i \alpha_i \sigma_i = 0$$

For a normal system, only the terms such that

$$\sum_i \alpha_i = 0$$

are different from 0. This condition disappears when we turn to a superfluid gas; it is then necessary to take account of the 16 possible combinations of α_i.

K satisfies a number of symmetry relations, which one obtains by generalizing the discussion given for G. To begin with, K is antisymmetric with respect to permutation of any two of the variables $(x_i \sigma_i \alpha_i)$. Furthermore, invariance with respect to rotation of spins implies

$$K(x_i, \sigma_i, \alpha_i) = \sigma_1 \dots \sigma_4 K(x_i, -\sigma_i, \alpha_i) \tag{4-129a}$$

Time-reversal invariance leads to

$$K(x_i, \sigma_i, \alpha_i) = \sigma_1 \ldots \sigma_4 \, K(-x_i, -\sigma_i, -\alpha_i) \qquad \textit{(4–129b)}$$

(where we have also used reflection invariance). By combining these two identities, we obtain the relation

$$K(x_i, \sigma_i, \alpha_i) = K(-x_i, \sigma_i, -\alpha_i) \qquad \textit{(4–129c)}$$

Let us turn now to the decomposition of K. The free part is obtained by contracting the operators ψ in pairs in all possible ways. By comparing (4–111) and (4–128) we see that

$$[K(x_i, \sigma_i, \alpha_i)]_{\text{free}} = G(1,3)\,G(2,4) - G(1,4)\,G(2,3) - G(1,2)\,G(3,4) \qquad \textit{(4–130)}$$

where, for simplification, we have set

$$G(i,j) = G(x_i\sigma_i\alpha_i, x_j\sigma_j\alpha_j)$$

(4–130) is the generalization to superfluid systems of Eq. (3–42). Note the existence of an additional term, characteristic of the superfluid. By way of example, let us write down the free part of $\bar{K}(\mathbf{x}_i)$ in some cases:

$$\left\{\begin{array}{l} \left.\begin{array}{l}\sigma_1 = \sigma_4 = +1 \\ \sigma_2 = \sigma_3 = +1\end{array}\right\} (\bar{K}(\mathbf{x}_i))_{\text{free}} = \\ \qquad F_1(x_1 - x_3)\,F_1(x_2 - x_4) - F_1(x_1 - x_4)\,F_1(x_2 - x_3) \\ \left.\begin{array}{l}\sigma_1 = \sigma_4 = +1 \\ \sigma_2 = \sigma_3 = -1\end{array}\right\} (\bar{K}(\mathbf{x}_i))_{\text{free}} = \\ \qquad -F_1(x_1 - x_4)\,F_1(x_2 - x_3) - F_2(x_1 - x_2)\,F_2(x_3 - x_4) \end{array}\right. \qquad \textit{(4–131)}$$

Let us now turn to the *bound part*, which we shall call $\Delta K(x_i\sigma_i\alpha_i)$. Since we eliminated all long-range terms, ΔK is localized in space and time. This is no longer so if we carry out a Fourier transformation of the spatial variables. The function $\Delta K(\mathbf{k}_i,t_i)$ thus obtained describes an interaction process followed or preceded by independent propagation of the four quasi particles $\mathbf{k}_i$. This independent propagation can extend over an appreciable period of time. It is therefore important to isolate the "interaction" process, which is *localized in time* (unless additional bound states come into play). For this purpose we set

$$\Delta K(x_i\sigma_i\alpha_i) = \sum_{\sigma_i'\alpha_i'} \int dx_i' \, \gamma(x_i'\sigma_i'\alpha_i') \prod_{i=1}^{i=4} G(x_i\sigma_i\alpha_i, x_i'\sigma_i'\alpha_i') \qquad \textit{(4–132)}$$

γ is the interaction operator. (4–132) is just the transposition of (3–43), to which it reduces if $F_2 = 0$ (normal gas). The situation is complicated, as γ now has 16 components instead of 1, describing the different processes obtained by combining the scattering of a particle, and the annihilation and the creation of a condensed pair.

The complexity of (4–132) is characteristic of superfluidity. There exist new types of long-range propagation, which compel us to refine the definition of the *localized* interaction operator. Note that propagation no longer conserves the number of excited particles; two particles can collapse into a condensed pair, and vice versa. Of course, the total number of particles remains constant nevertheless.

Let us now define the Fourier transforms of K and γ. As far as K is concerned, we proceed by analogy with G [Eqs. (4–114) and (4–117)], and we put

$$\begin{cases} K(k_i t_i \sigma_i \alpha_i) = \dfrac{1}{\Omega^2} \displaystyle\int d^3 r_i K(r_i t_i \sigma_i \alpha_i) \exp\left(-\mathrm{i} \sum_i \alpha_i k_i r_i\right) \\ K(k_i \omega_i \sigma_i \alpha_i) = \dfrac{1}{(2\pi)^2} \displaystyle\int dt_i K(k_i t_i \sigma_i \alpha_i) \exp\left(+\mathrm{i} \sum_i \alpha_i \omega_i t_i\right) \end{cases} \qquad (4\text{–}133)$$

For γ, we note that, at the junction of γ and G, the interaction *stops* while propagation *begins*. This leads us to the following definition of γ:

$$\begin{cases} \gamma(k_i t_i \sigma_i \alpha_i) = \dfrac{1}{\Omega^2} \displaystyle\int d^3 r_i\, \gamma(r_i t_i \sigma_i \alpha_i) \exp\left(+\mathrm{i} \sum_i \alpha_i k_i r_i\right) \\ \gamma(k_i \omega_i \sigma_i \alpha_i) = \dfrac{1}{(2\pi)^2} \displaystyle\int dt_i\, \gamma(k_i t_i \sigma_i \alpha_i) \exp\left(-\mathrm{i} \sum_i \alpha_i \omega_i t_i\right) \end{cases} \qquad (4\text{–}134)$$

These definitions contain the more restrictive definitions (3–47) and (3–48).

In this new representation, definition (4–132) becomes

$$\Delta K(p_i \sigma_i \alpha_i) = \sum_{k'_i} \int d\omega'_i \sum_{\sigma'_i, \alpha'_i} \gamma(p'_i \sigma'_i \alpha'_i) \prod_{i=1}^{4} G(p_i \sigma_i \alpha_i, p'_i \sigma'_i \alpha'_i) \qquad (4\text{–}135)$$

We can condense the writing of (4–135) by using the symbols (4–118). We find

$$\Delta K(\xi_i) = \sum_{\xi_i'} \gamma(\xi_i') \prod_{i=1}^{4} G(\xi_i, \xi_i') \tag{4–136}$$

The summation over the ξ_i' is in fact trivial. Applying (4–121) and (4–122), we obtain

$$\Delta K(p_i\sigma_i\alpha_i) = \sum_{\sigma_i'} \left\{ \gamma\left(\frac{\sigma_i' p_i}{\sigma_i}, \sigma_i', -\frac{\sigma_i'\alpha_i}{\sigma_i}\right) \prod_{i=1}^{4} \frac{\sigma_i'\alpha_i}{\sigma_i} \tilde{G}_{\sigma_i,\sigma_i'}(p_i) \right\} \tag{4–137}$$

These relations generalize Eq. (3–49).

*c. **Dynamical equations relating the Green's functions***

We start from the Schrödinger equation (3–72),

$$\begin{aligned} \frac{i\partial}{\partial t}\psi_\sigma(r,t) = &-\frac{1}{2m}\nabla^2\psi_\sigma(r,t) \\ &+\int d^3r'' \sum_{\sigma''} V(r-r'')\,\psi^*_{\sigma''}(r'',t)\,\psi_{\sigma''}(r'',t)\,\psi_\sigma(r,t) \end{aligned} \tag{4–138}$$

$\psi^*_\sigma(r,t)$ satisfies the equation conjugate to (4–138). We can write these two as a single equation by using the condensed notation defined above:

$$\begin{aligned} i\alpha\frac{\partial}{\partial t}\psi^\alpha_\sigma(r,t) = &-\frac{1}{2m}\nabla^2\psi^\alpha_\sigma(r,t) \\ &-\int d^3r'' \sum_{\sigma''} V(r-r'')\,\psi^*_{\sigma''}(r'',t)\,\psi^\alpha_\sigma(r,t)\,\psi_{\sigma''}(r'',t) \end{aligned} \tag{4–139}$$

[To establish (4–139), we rely on the anticommutation relations of ψ and ψ^*.]

As in Chap. 3 [Eqs. (3–73) to (3–75)] we take the chronological product of (4–139) with $\psi^{\alpha'}_{\sigma'}(r',t')$. This leads us to the following dynamical equation,

$$\begin{aligned} &\left(\alpha\frac{\partial}{\partial t} - i\frac{\nabla^2}{2m} - i\mu\right) G(x\sigma\alpha, x'\sigma'\alpha') \\ &-\int d^3r'' \sum_{\sigma''} V(r-r'')\,K(r''t_-\sigma''+, x\sigma\alpha, r''t_+\sigma''-, x'\sigma'\alpha') = i\alpha\,\delta(x-x')\,\delta_{-\alpha,\alpha'}\,\delta_{\sigma,\sigma'} \end{aligned} \tag{4–140}$$

where $\delta_{-\alpha,\alpha'}$ and $\delta_{\sigma\sigma'}$ are the usual Kronecker δ symbols. (4–140) is the generalized version of (3–75). As for a normal system, we can write down an infinite set of dynamical equations, relating G and K to the higher-order Green's functions. These equations, supplemented by appropriate boundary conditions, in principle allow the solution of the problem.

Our ambitions are much more modest. As in the beginning of this chapter, we replace the Green's function K in (4–140) by its decomposed form. The contribution of the free part is easily evaluated. The bound part is analyzed in the same way as for a normal system [Eqs. (4–3) to (4–7)]. (4–140) can thus be written as

$$\left(\alpha\frac{\partial}{\partial t} - \mathrm{i}\frac{\nabla^2}{2m} - \mathrm{i}\mu\right) G(x\sigma\alpha, x'\sigma'\alpha')$$
$$- \mathrm{i}\int \mathrm{d}x'' \sum_{\alpha''\sigma''} M(x, \sigma, -\alpha\,; x''\sigma''\alpha'')\, G(x''\sigma''\alpha'', x'\sigma'\alpha') = \mathrm{i}\alpha\, \delta(x - x')\, \delta_{\alpha,-\alpha'}\, \delta_{\sigma,\sigma'} \tag{4–141}$$

(4–141) defines the self-energy operator M. Let us point out the index $-\alpha$ assigned to M; the reason for this choice will become evident within the framework of the perturbation methods developed in Chap. 7.

By analogy with (4–134), we define the Fourier transforms of M in the following way:

$$M(kt\sigma\alpha, k't'\sigma'\alpha') = \frac{1}{\Omega}\int \mathrm{d}^3r\, \mathrm{d}^3r'\, M(x\sigma\alpha, x'\sigma'\alpha') \exp[\mathrm{i}(\alpha kr + \alpha'k'r')]$$
$$M(k\omega\sigma\alpha, k'\omega'\sigma'\alpha') = \frac{1}{2\pi}\int \mathrm{d}t\, \mathrm{d}t'\, M(kt\sigma\alpha, k't'\sigma'\alpha') \exp[-\mathrm{i}(\alpha\omega t + \alpha'\omega't')] \tag{4–142}$$

Note the difference from (4–114) and (4–117).

As for a normal gas, $M(k,t)$ is a localized function of t yielding a smooth $M(k,\omega)$. This is obvious for the free part of M, which contains a factor $\delta(t)$. For the bound part, the reason for it is more subtle: this bound part is essentially the convolution of a factor γ and three factors G. Such an element, strictly speaking, can be long-ranged. But we must also sum over two wave vectors; the destructive interference of the phase factors causes M to be short-ranged.

Once these transformations are made, the dynamical equation takes a very simple form, especially if we use the symbols (4–118),

$$(k^2/2m - \omega - \mu)\, G(\xi, \xi') - \sum_{\xi''} M(\xi^-, \xi'')\, G(\xi'', \xi') = \alpha\delta_{\sigma,\sigma'}\, \delta(\xi, \xi') \tag{4–143}$$

[We have designated by ξ^- the set of indices $(k,\omega,\sigma,-\alpha)$, in contrast to ξ, which represents (k,ω,σ,α).]

Let us multiply both sides of (4–143) by $-\alpha'$. Owing to (4–121), we obtain

$$(k^2/2m - \omega - \mu)\,\tilde{G}(\xi, \xi') - \sum_{\xi''} M(\xi^-, \xi'')\,\tilde{G}(\xi''\ \xi') = \delta_{\sigma,\sigma'}\,\delta(\xi, \xi') \qquad \textit{(4–144)}$$

In order to simplify this equation, we first replace $\tilde{G}$ by its decomposed expression (4–122). Furthermore, relying on translational invariance, we write M in the form

$$M(\xi, \xi') = M_{\sigma\sigma'}(p\,;\,\alpha)\,\delta(\xi, \xi') \qquad \textit{(4–145)}$$

(4–145) expresses conservation of spin, momentum, and energy, without implying any other restriction. Finally we note that

$$\sum_{k''\alpha''} \int d\omega''\,\delta(\xi^-, \xi'')\,\delta(\xi'', \xi') = \delta(\xi, \xi') \qquad \textit{(4–146)}$$

[This is why we introduced an index $-\alpha$ in the definition (4–141).] By collecting these results, we can put (4–144) in the very simple form

$$(k^2/2m - \omega - \mu)\,\tilde{G}_{\sigma\sigma'}(p) - \sum_{\sigma''} M_{\sigma\sigma''}(p\,;\,-\alpha)\,\tilde{G}_{\sigma''\sigma'}\left(\frac{\sigma'' p}{\sigma}\right) = \delta_{\sigma\sigma'} \qquad \textit{(4–147)}$$

(4–147) constitutes the final version of our dynamical equation, with which we are going to work in the following section.

Solution of the dynamical equation. The normal approach consists in calculating M from the two-particle Green's function K, then inserting the result into (4–147) to obtain G. A detailed study of the symmetries with respect to reflection, rotation of spins, and time reversal allows us to obtain directly the properties of M. Actually, we shall adopt the inverse approach: knowing the form of G, studied at the beginning of this section, we deduce from it the form of M. Later on, we shall express the coefficients of G as a function of those of M.

First we remark that, according to (4–147), $M_{\sigma\sigma''}(p, -\alpha)$ is independent of α. Thus M has the same structure as G; like the latter, it will be represented by a 2×2 matrix, denoted by $\mathbf{M}(p)$.

In order to solve (4–147) let us put it into matrix form. For this purpose, we separate G into an even and an odd part with respect to ω,

$$\tilde{\mathbf{G}}(p) = \tilde{\mathbf{G}}^{+}(k, \omega) + \tilde{\mathbf{G}}^{-}(k, \omega) \tag{4–148}$$

Equation (4–147) can then be written as

$$(k^2/2m - \omega - \mu)\,\tilde{\mathbf{G}} - \boldsymbol{\beta}_2 \mathbf{M} \boldsymbol{\beta}_2 \tilde{\mathbf{G}}^{-} - \mathbf{M}\tilde{\mathbf{G}}^{+} = 1 \tag{4–149}$$

where $\boldsymbol{\beta}_2$ is defined by (4–124).

Let us replace $\tilde{\mathbf{G}}$ by its explicit form (4–125). We then easily verify that the solution $\mathbf{M}$ of Eq. (4–149) is given by

$$\mathbf{M}(p) = M_1(p)\,\mathbf{1} + M_2(p)\,\boldsymbol{\beta}_1 \tag{4–150}$$

where the functions M_1 and M_2 are given by

$$M_2(p) = \frac{F_2(p)}{F_1(p)\,F_1(-p) + F_2(p)^2}$$
$$k^2/2m - \omega - \mu - M_1(p) = \frac{F_1(-p)}{F_1(p)\,F_1(-p) + F_2(p)^2} \tag{4–151}$$

Note that, since F_2 is even, M_2 is also.

Actually, we are interested in the inverse relation, giving F_1 and F_2 as a function of M_1 and M_2. By using reflection invariance, we finally obtain

$$F_1(k, \omega) = \frac{k^2/2m - \mu + \omega - M_1(k, -\omega)}{D}$$
$$F_2(k, \omega) = \frac{M_2(k, \omega)}{D} \tag{4–152}$$
$$D = [k^2/2m - \mu + \omega - M_1(k, -\omega)]\,[k^2/2m - \mu - \omega - M_1(k, \omega)] + [M_2(k, \omega)]^2$$

If we set $M_2 = 0$, the pair of equations (4–125) and (4–152) reduces to the simple equation (4–8); we are brought back to the usual normal systems. The characteristic of superfluids is thus $M_2 \neq 0$.

We see from (4–139) that $F_1(k,\omega)$ and $F_2(k,\omega)$ have *the same singularities* in ω, whose frequencies satisfy the equation

$$D = 0$$

Since D is even, the poles of F_1 and F_2 are grouped in pairs of *opposite* values; this is a fundamental difference from normal systems. We shall assume that, for each value of k, there exists only a single pair of poles in the neighborhood of the real axis. This hypothesis is reasonable, as M_1 and M_2 are regular functions of ω. The general discussion of Chap. 3 indicates that these poles are located in the second and fourth quadrants. Their frequencies are thus equal to

$$\pm(\omega_k - i\Gamma_k)$$

where ω_k and Γ_k are positive. ω_k and Γ_k determine the excitation energy and the damping of the elementary excitations, which we shall study in the following section.

Let $-z^+$ and $-z^-$ be the residues of $F_1(k,\omega)$ at the poles $\pm(\omega_k - i\Gamma_k)$, $-\hat{z}$ and $+\hat{z}$ the corresponding residues of F_2 (which, we recall, is even). z^+ and z^- are necessarily positive, whereas the sign of $\hat{z}$ depends on the orientation of the axes. Equation (C–11) implies

$$z^+ + z^- \leqslant 1 \tag{4–153}$$

[If the functions $A_\pm(k,\omega)$ did not have a continuous background, we would rigorously have $z^+ + z^- = 1$).] These residues are not independent; in fact, the combination

$$F_1(\omega)\, F_1(-\omega) + (F_2(\omega))^2 = 1/D$$

has only first-order poles. From this we deduce the important relation

$$z^+ z^- = \hat{z}^2 \tag{4–154}$$

As a consequence, the condition $M_2 \neq 0$ ($\hat{z} \neq 0$) implies the existence of two opposite poles of F_1, and vice versa; this gives us two equivalent characteristics of the superfluid state. For a normal gas, one of the residues z^+ and z^- is necessarily zero; there is only one pole per value of k.

Knowing the poles of $F_1(k,\omega)$ and $F_2(k,\omega)$, we find the asymptotic form of $F_1(k,t)$ and $F_2(k,t)$ when t is large (yet less than Γ_k^{-1}),

$$\begin{cases} F_1(k,t) \sim \begin{cases} i z^+ \exp(-i\omega_k t) & \text{if} \quad t > 0 \\ -i z^- \exp(+i\omega_k t) & \text{if} \quad t < 0 \end{cases} \\ F_2(k,t) \sim i\hat{z} \exp(-i\omega_k |t|) \end{cases} \tag{4–155}$$

Note that $F_1(k,t)$ is now long-ranged in both directions in time.

d. The structure of the elementary excitations

Variation of the different parameters as a function of k. The existence of two opposite poles of G leads to elementary excitations which

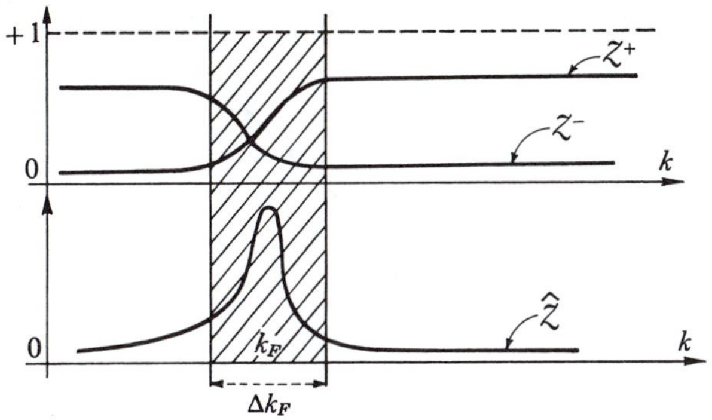

Figure 7

look both like a particle and like a hole. (We shall analyze this rather paradoxical conclusion at the end of this section.) The relative importance of these two characteristics is measured by z^+/z^-. For k very large, the elementary excitations are almost pure particles, and we have $z^+ \gg z^-$. For k very small, on the other hand, the excitation is almost a hole, and we have $z^+ \ll z^-$. Between these two limits, z^+ and z^- vary continuously, M_1 and M_2 being continuous functions of k and ω (Fig. 7.) The variation of $\hat{z}$ is obtained from (4–154). $\hat{z}$ presents a more or less sharp maximum, centered at $k = k_F$, of width Δk_F. This maximum corresponds to the transition region in which z^+ is increasing rapidly, while z^- is decreasing. This behavior should be compared with that characteristic of a normal gas, indicated in Fig. 8. By analogy with the classical case, we can say that k_F represents the Fermi wave vector of the system. But whereas for a normal gas k_F is defined sharply (since it corresponds to a discontinuity), in a superfluid k_F acquires an uncertainty Δk_F; the Fermi level has become *smeared out*. This spreading out arises from the fact that the

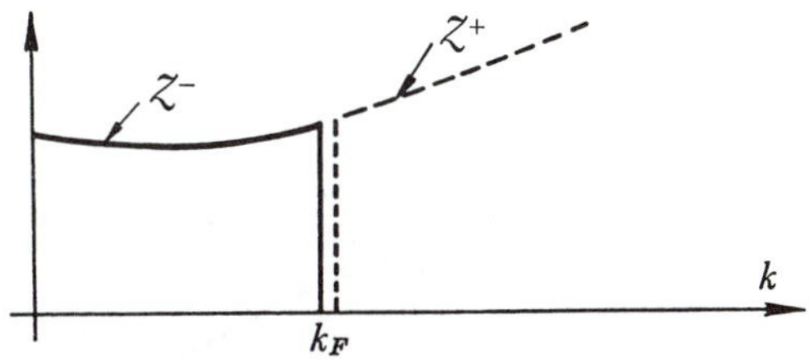

Figure 8

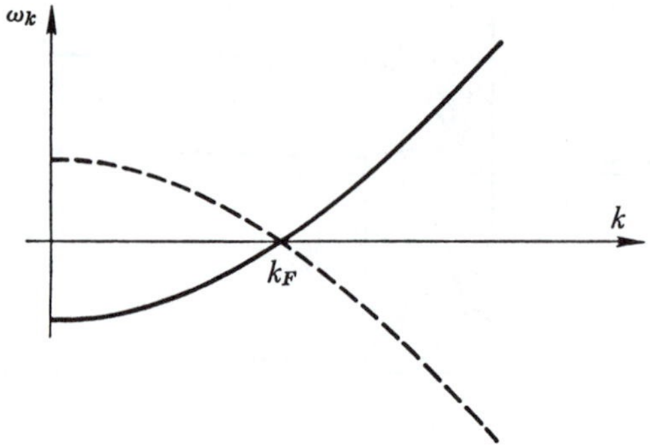

Figure 9

excitations are no longer *either* particles *or* holes, as in a normal gas, but, on the contrary, involve both at once.

In the same way as z^+, z^- and $\hat{z}$, the energy $\pm(\omega_k - i\Gamma_k)$ of the elementary excitations is a continuous function of k (this again arising from the continuity of M_1 and M_2.) Let us first consider the real part ω_k. For a normal system, $\pm\,\omega_k$ has the behavior indicated in Fig. 9. In this case the dotted curve corresponds to artificial singularities of G. For a superfluid system, Fig. 9 is replaced by Fig. 10. The two branches of this curve correspond to true singularities. The energy ω_k has a nonzero minimum ω_0, occurring near the "pseudo" Fermi level k_F. This minimum gives rise to a *forbidden band* in the elementary excitation spectrum. This is a very important new phenomenon. The gap ω_0 has numerous physical consequences, which are experimentally observed in superconductivity.

It still remains to discuss the origin of the damping term Γ_k. The states into which a quasi particle can decay can be classified into two categories:

(a) States containing only quasi particles (quasi holes), whose

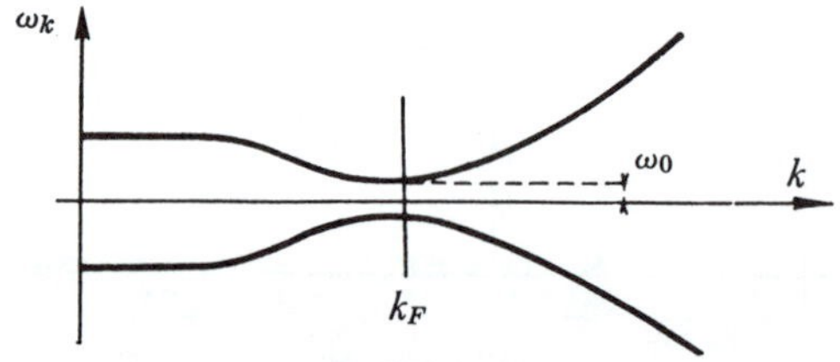

Figure 10

number must be odd to obtain a final state of $(N + 1)$ particles. The energy of these states is greater than $3\omega_0$.

(b) States containing quasi particles and collective excitations, for example, phonons; the energy spectrum thus starts from an energy near ω_0.

If we neglect states of type (b), quasi particles such that

$$\omega_0 < \omega_k < 3\,\omega_0$$

are not damped; Γ_k is then zero. Actually, the collective states give a weak damping. When ω_k exceeds $3\omega_0$, type (a) processes become possible; Γ_k grows rapidly (while, however, remaining of the same order as for a normal gas).

As for a normal system, the energy of the state containing one quasi particle added to $|\varphi_0(N)\rangle$ is equal to

$$E_0(N) + \mu + \omega_k > E_0(N) + \mu + \omega_0$$

This remark leads us to clarify a difficulty which we have deliberately ignored until now. Since a superfluid system contains a large number of condensed pairs, the wave function $|\varphi_0(N)\rangle$ must have a different structure according to whether N is even or odd. This difference has an effect on the spectrum of excited states, thus on the Lehmann representation of the Green's functions. To what extent are the latter affected by this complication?

Until now, we have assumed that N is even. In this case, if we add a particle, there is no other with which it can be paired. $E_0(N + 1)$ is thus greater than $[E_0(N) + \mu]$ (Fig. 11). The difference is just the width ω_0 of the forbidden band. The spectrum of $(N \pm 1)$-particle excited states starts from $E_0(N \pm 1)$ and is indicated by shading in Fig. 11.

Let us now turn to the case where N is odd. The energy spectrum then

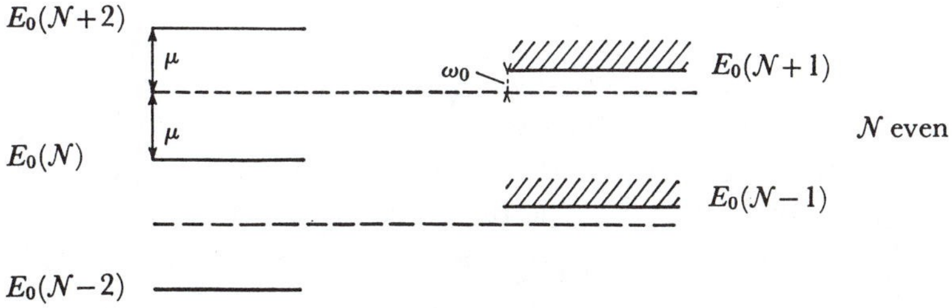

Figure 11

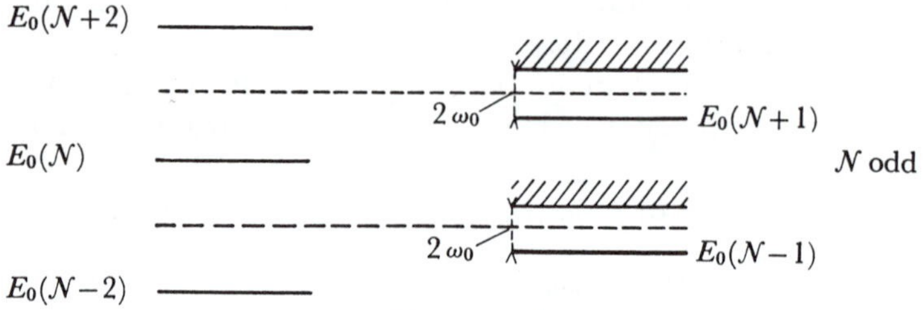

Figure 12

takes the form indicated in Fig. 12. At first glance it seems that the energy ξ_{n0} appearing in the Lehmann representation can take on negative values! Most fortunately, this is only an illusion. In fact, we obtain the ground state $|\boldsymbol{\varphi}_0(N+1)\rangle$ by grouping the $(N+1)$st particle with the single particle of $|\boldsymbol{\varphi}_0(N)\rangle$ in the form of a condensed pair. This state has zero momentum and does not appear in the Lehmann representation of $G(k,t)$. Only the excited states can be involved; these are obtained by adding to $|\boldsymbol{\varphi}_0(N+1)\rangle$ a quasi particle and a quasi hole. The corresponding continuous spectrum (shaded region in Fig. 12) is separated from $E_0(N+1)$ by a forbidden band of width $2\omega_0$ and is identical to that obtained for N even. In conclusion, we do not have to concern ourselves with the parity of N.

Wave function of the elementary excitations. Let us consider the state vector defined by (4–16),

$$|\varphi_{k\sigma}\rangle = C\int_{-\infty}^{0} dt' \exp[-i(\omega_k + \mu)\,t']\exp(\alpha t')\,\mathbf{a}^*_{k\sigma}(t')\,|\varphi_0(N)\rangle \qquad \textit{(4–156)}$$

This state contains $(N+1)$ particles and has total wave vector k and total spin σ. The discussion given for the normal gas still applies: $|\boldsymbol{\varphi}_{k\sigma}\rangle$ represents an *approximate eigenstate*, of energy

$$E_0(N) + \mu + \omega_k$$

The lifetime of this state is equal to Γ_k^{-1}; the only states which make sense are those with a small Γ_k. The spectral distribution of $|\boldsymbol{\varphi}_{k\sigma}\rangle$ is given by (4–18); only the pole $+(\omega_k - i\Gamma_k)$ of the propagator $G_{\sigma\sigma}(k,\omega)$ is involved. The normalization constant C is therefore equal to

$$C = \alpha/\sqrt{z^+}$$

Let us now consider the state

$$| \varphi'_{k\sigma} > =$$

$$(\alpha\sigma/\sqrt{z^-}) \int_{-\infty}^{0} dt' \exp[-i(\omega_k - \mu)\, t'] \exp(\alpha t')\, \mathbf{a}_{-k,-\sigma}(t')\, | \varphi_0(N + 2) > \tag{4–157}$$

(where $\sigma = \pm 1$). Like $|\varphi_{k\sigma}\rangle$, $|\varphi'_{k\sigma}\rangle$ contains $(N + 1)$ particles, of total wave vector k and total spin σ. We easily verify that $|\varphi'_{k\sigma}\rangle$ is normalized. The density of norm at the frequency

$$E_0(N + 2) + \omega - \mu = E_0(N) + \mu + \omega$$

is roughly equal to $\delta(\omega - \omega_k)$; it is identical to that for $|\varphi_{k\sigma}\rangle$ [only the pole $-(\omega_k - i\Gamma_k)$ of the propagator $\bar{G}_{-\sigma,\,-\sigma}(-k,\omega)$ enters the calculation]. Let us calculate the scalar product $\langle\varphi'_{k\sigma}|\varphi_{k\sigma}\rangle$, which reduces to an integral involving the function $\check{G}$. An elementary calculation shows that

$$< \varphi'_{k\sigma} | \varphi_{k\sigma} > = \hat{z}/\sqrt{z^+z^-} = 1$$

(we assume the axis to be oriented in such a way that $\hat{z}$ is positive). The states $|\varphi_{k\sigma}\rangle$ and $|\varphi'_{k\sigma}\rangle$ are *identical.* We can thus create the excitation in two different ways:

(a) By adding a particle $k\sigma$ to the state $|\varphi_0(N)\rangle$.

(b) By removing a particle $(-k, -\sigma)$ from the state $|\varphi_0(N + 2)\rangle$.

These processes are equivalent; if we break up a condensed pair by removing a particle $(-k, -\sigma)$, a particle (k,σ) remains.

Thus all excited states partake simultaneously of the nature of a quasi particle and a quasi hole. The relative importance of these two aspects depends on the ratio z^+/z^-. Let us note that the excited states have a well-defined number of particles, the difference between a hole and a particle being compensated by the addition of a condensed pair (it is only if we ignore the latter that the total number of particles seems not to be conserved).

If in (4–157) we replace $|\varphi_0(N + 2)\rangle$ by $|\varphi_0(N)\rangle$, we obtain a state analogous to $|\varphi_{k\sigma}\rangle$, but containing one less pair; the structure of the elementary excitation is not changed. Note that the two poles of $\bar{G}_{\sigma\sigma}(k)$ correspond to two different elementary excitations, of opposite spin and wave vector. Nevertheless, there is only *one* elementary excitation for each value of $k\sigma$ as shown by the diagram below:

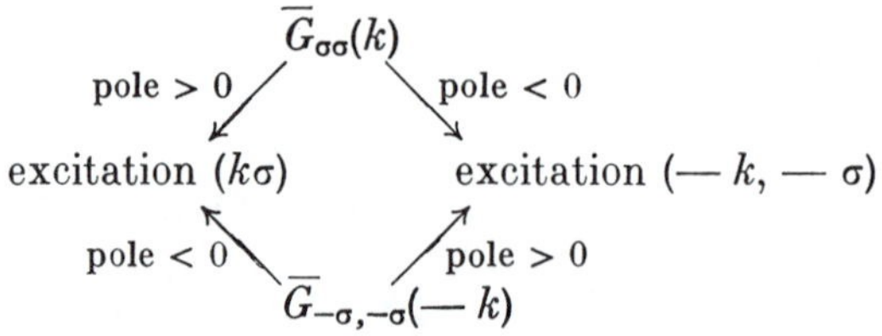

Effects of an elementary excitation on the Green's functions. The excitations corresponding to the states $|\varphi_{k\sigma}\rangle$ will be "elementary" only if they are independent of one another. As for a normal gas, we approach this problem by calculating the Green's functions in the state $|\varphi_{k\sigma}\rangle$. We limit ourselves to the study of $\bar{G}$ (the considerations which follow are easily extended to $\hat{G}$ and $\check{G}$). We set

$$\bar{G}(k'\sigma', t) + \delta_{k\sigma}\bar{G}(k'\sigma', t) = i < \varphi_{k\sigma} \mid T\{ a_{k'\sigma'}(t)\, a^*_{k'\sigma'} \} \mid \varphi_{k\sigma} > e^{i\mu t} \qquad \textit{(4–158)}$$

$\delta_{k\sigma}\bar{G}$ describes the correction arising from the quasi particle $k\sigma$. Let us replace the state $|\varphi_{k\sigma}\rangle$ in (4–158) by its expression (4–146). Integrals related to the two-particle Green's function $K(k_i,t_i)$ appear in the second term. We know that K is composed of a free part (4–130), independent of the volume Ω, and a bound part of order $1/\Omega$. The calculation is analogous to that carried out for a normal gas [Eqs. (4–23) to (4–31)]. We shall content ourselves with giving the results, without going into tedious details. Three cases are possible:

(a) $k', \sigma' \neq (k,\sigma)$ or $(-k, -\sigma)$: $\delta_{k\sigma}\bar{G}(k'\sigma')$ is then of order $1/\Omega$ and therefore negligible. The poles of $\bar{G}$ are the same as before. The excitations of different $|k,\sigma|$ are therefore independent.

(b) $k',\sigma' = k, \sigma$. We then find

$$\delta_{k\sigma}\bar{G}(k, \sigma, t) = -iz^+ \exp(-i\omega_k t) \qquad \textit{(4–159)}$$

If we carry out a Fourier transformation with respect to t, we see that the introduction of the quasi particle (k,σ) has the *sole effect* of causing the positive pole of $\bar{G}(k,\sigma;\omega)$ to move from the lower half-plane to the upper half-plane (Fig. 13a). Except for this, $\bar{G}$ does not change.

(c) $k',\sigma' = -k, -\sigma$. (4–159) is replaced by

$$\delta_{k\sigma}\bar{G}(-k, -\sigma, t) = +i\frac{\hat{z}^2}{z^+}\exp(i\omega_k t) = iz^- \exp(i\omega_k t) \qquad \textit{(4–160)}$$

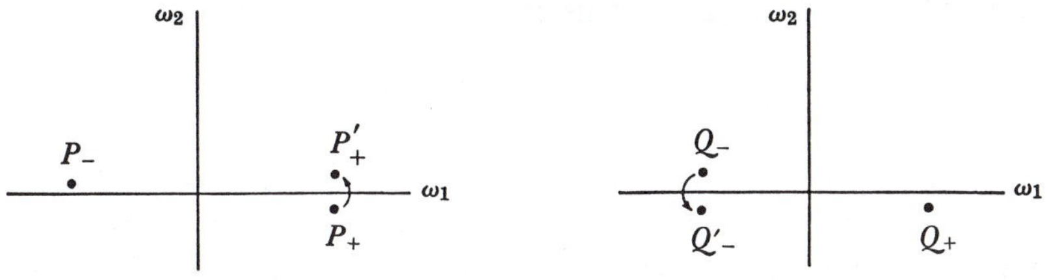

(a) $\bar{G}(k, \sigma; \omega)$ (b) $\bar{G}(-k, -\sigma; \omega)$

Figure 13

In the Fourier transform, the negative pole of $\bar{G}(-k, -\sigma;\omega)$ moves from the upper half-plane to the lower half plane without any other modification (Fig. 13b).

We established above that the poles Q_- and P_+ give rise to the excitation $k\sigma$, whereas Q_+ and P_- correspond to the excitation $(-k' - \sigma)$. The above results then lead to the following conclusions:

(a) The quasi particle $(-k, -\sigma)$ is not affected by the transition from $|\varphi_0\rangle$ to $|\varphi_{k\sigma}\rangle$ (P_- and Q_+ do not change). This assures the *total independence* of the elementary excitations.

(b) The elementary excitation $k\sigma$ becomes reversed in type. As a consequence, one cannot create two excitations of the same type; the quasi particles are *fermions*. The new poles Q'_- and P'_+ correspond to the transition which takes $|\varphi_{k\sigma}\rangle$ back into $|\varphi_0\rangle$.

In this regard, superfluid systems thus have the same properties as normal systems. The existence of two opposite poles does not create any new complications.

Quasi-particle creation and destruction operators. We can put the state $|\varphi_{k\sigma}\rangle$ into the form

$$|\varphi_{k\sigma}\rangle = A^*_{k\sigma} |\varphi_0(N)\rangle \tag{4-161}$$

the operator $A^*_{k\sigma}$ being defined as for a normal gas [cf. (4–16)],

$$A^*_{k\sigma} = (\alpha/\sqrt{z^+}) \int_{-\infty}^{0} dt' \exp[-i(\omega_k + \mu)\, t'] \exp(\alpha t') a^*_{k\sigma}(t') \tag{4-162}$$

$A^*_{k\sigma}$ can be written in a completely different way if we refer to the definition (4–157) of $|\varphi_{k\sigma}\rangle$. Let us set

$$|\varphi_0(N+2)\rangle = P_0^*|\varphi_0(N)\rangle \tag{4–163}$$

The operator P_0^* creates a "condensed pair." P_0^* commutes with all the operators $A_{k\sigma}^*$, to an accuracy of order $1/\Omega$ (the structure of the quasi particles is a continuous function of N). Let us put (4–163) back into (4–157); we see that $A_{k\sigma}^*$ takes the form

$$A_{k\sigma}^* = (\alpha\sigma/\sqrt{z}) \int_{-\infty}^{0} dt' \exp[-i(\omega_k - \mu)t'] \exp(\alpha t')\, a_{-k,-\sigma}(t')\, P_0^* \tag{4–164}$$

The two definitions (4–162) and (4–164) are equivalent.

Let $A_{k\sigma}$ be the operator conjugate to $A_{k\sigma}^*$. We can verify from either of its two expressions that

$$A_{k\sigma}|\varphi_0(N)\rangle = 0 \tag{4–165}$$

(The filtering lets pass only a frequency band for which the density of norm is zero.) Using (4–165) and the conclusions stated in the preceding section, we see that

$$\begin{aligned} [A_{k\sigma}, A_{k'\sigma'}^*]_+ &= \delta_{kk'}\,\delta_{\sigma\sigma'} \\ [A_{k\sigma}, A_{k'\sigma'}]_+ &= 0 \\ [A_{k\sigma}^*, A_{k'\sigma'}^*]_+ &= 0 \end{aligned} \tag{4–166}$$

[For more details, refer to the discussion given for normal systems, Eqs. (4–36) to (4–42).] Therefore $A_{k\sigma}$ and $A_{k\sigma}^*$ have all the properties of quasi-particle creation and destruction operators.

Actually, (4–166) is subject to several limitations:

(a) These operator relations are valid only if they are applied to the ground state or to weakly excited states. Otherwise, the structure of the quasi particle changes, and it is necessary to revise the definition of the operators $A_{k\sigma}$ and $A_{k\sigma}^*$.

(b) These relations are exact only in the limit $\Gamma_k \to 0$. If Γ_k is small, they remain approximately true; they lose their meaning if Γ_k is too large.

Let us now try to compare these "dressed" operators with the "bare" operators $a_{k\sigma}$ and $a_{k\sigma}^*$. For this purpose, let us write the destruction operator $A_{-k,-\sigma}$ using (4–164),

$$A_{-k,-\sigma} = -(\alpha\sigma/\sqrt{z}) \int_{-\infty}^{0} dt' \exp[i(\omega_k - \mu)t'] \exp(\alpha t')\, a_{k\sigma}^*(t')\, P_0 \tag{4–167}$$

(P_0 is the operator conjugate to $\overset{*}{P_0}$, destroying a condensed pair; to an accuracy of order $1/\Omega$, we have $P_0\overset{*}{P_0} = \overset{*}{P_0}P_0 = 1$.) By comparing (4–167) with (4–162), we see that the same operator $\overset{*}{a}_{k\sigma}$ enters the two expressions. This leads us to separate $\overset{*}{a}_{k\sigma}$ into three parts:

(a) A part oscillating at the frequency $(\mu + \omega_k)$, which, from (4–162) gives rise to the creation operator $\overset{*}{A}_{k\sigma}$.

(b) A part oscillating at the frequency $(\mu - \omega_k)$, which gives $A_{-k,-\sigma}$.

(c) Finally, an incoherent part whose frequency spectrum is continuous.

Let us assume that we knew how to eliminate this incoherent part, and let $\overset{c*}{a}_{k\sigma}$ be the coherent part combining terms (a) and (b). (4–162) and (4–167) allow us to write

$$\overset{c*}{a}_{k\sigma} = \sqrt{z_+}\,\overset{*}{A}_{k\sigma} - \sigma\sqrt{z_-}\,A_{-k,-\sigma}\overset{*}{P_0} \tag{4–168}$$

By combining (4–158) with the complex conjugate equation, we easily invert this relation, obtaining

$$\overset{*}{A}_{k\sigma} = \frac{\sqrt{z_+}\,\overset{c*}{a}_{k\sigma} + \sigma\sqrt{z_-}\,\overset{c}{a}_{-k,-\sigma}\,\overset{*}{P_0}}{z_+ + z_-} \tag{4–169}$$

(4–169) gives an instructive idea of the structure of $\overset{*}{A}_{k\sigma}$.

Relations (4–168) and (4–169) resemble a canonical transformation. Actually this is not one, since the anticommutation relations are not preserved; we can verify from (4–168) that

$$[\overset{c}{a}_{k\sigma}, \overset{c*}{a}_{k\sigma}]_+ = z_+ + z_- \tag{4–170}$$

which, in general, is different from 1. The operators $\overset{c*}{a}_{k\sigma}$ thus must be manipulated with great caution. However, it can happen, as a consequence of certain approximations, that the incoherent part of $\overset{*}{a}_{k\sigma}$ is zero. We then have $\overset{c*}{a}_{k\sigma} = \overset{*}{a}_{k\sigma}$, the preceding transformation being canonical. (In this case, $z_+ + z_-$ is necesssarily equal to 1.) The direct definition of the quasi-particle operators by a transformation of the type (4–169) constitutes the essence of Bogoliubov's method in the problem of superconductivity. Our discussion points out the difficulties encountered when one wants to take account of the incoherent part of the excitation spectrum.

In the ideal case, where

$$\begin{cases} \overset{*}{a}_{k\sigma} = \overset{c*}{a}_{k\sigma} \\ z_+ + z_- = 1 \end{cases} \tag{4–171}$$

these problems can be attacked by another method, of a variational nature, due to Bardeen, Cooper, and Schrieffer. The problem is then to determine the wave function $|\varphi_0\rangle$ such that

$$A_{k\sigma} \mid \varphi_0 > = 0 \tag{4–172}$$

Let us start from a state $|\mathbf{0}\rangle$ such that, for a certain value of $k\sigma$, we have

(4–172) is true if we choose

$$\mid \varphi_0 > = \{ \sqrt{z_+} - \sigma \sqrt{z_-}\, a^*_{k\sigma}\, a^*_{-k,-\sigma}\, P_0 \} \mid \mathbf{0} > \tag{4–173}$$

The generalization of (4–173) to the set of all values of k presents some difficulties, which we shall not go into here. Like that of Bogoliubov, this method does not lend itself to a generalization to the real case, for which (4–171) is false. In contrast, the method which we have used is very flexible.

With the exception of certain pathological bound states (excitons, phonons, etc.), the excited states of the system are obtained by making an arbitrary number of operators $A^*_{k\sigma}, P^*_0$, or P_0 act on $|\varphi_0(N)\rangle$. Note that the operators $A^*_{k\sigma}$ do not form a complete set; the state $|\varphi_0(N+2)\rangle$, for example, can be obtained only by direct action of P^*_0.

The distribution function of particles and pairs. In the ground state, the distribution function for bare particles m_k is again given by

$$m_k = \mathrm{i}\, [\overline{G}_{\sigma\sigma}(k, t)]_{t=-0} \tag{4–174}$$

But, in contrast with the normal gas, m_k is now continuous (Fig. 14). We observe, however, a rapid drop near the "Fermi level" k_F, which, we recall, is not sharply defined.

Another interesting quantity is the probability of finding a condensed pair in the state $(k,\sigma,-k,-\sigma)$. This probability is measured by

$$\nu_k = < \varphi_0(N) \mid a_{k\sigma}\, a_{-k,-\sigma} \mid \varphi_0(N+2) > = \mathrm{i}\, [\widehat{G}_{\sigma,-\sigma}(k, t)]_{t=0} \tag{4–175}$$

The quantity ν_k has its maximum in the neighborhood of k_F (Fig. 14).

What happens to the functions $m_{k'\sigma'}$ and $\nu_{k'\sigma'}$ when a quasi particle (k,σ) is introduced? To find out, we go back to Eq. (4–159), etc. We thus find that:

(a) For $k'\sigma' \neq k\sigma$ or $(-k,-\sigma)$, $m_{k'\sigma'}$ and $\nu_{k'\sigma'}$ suffer only a negligible

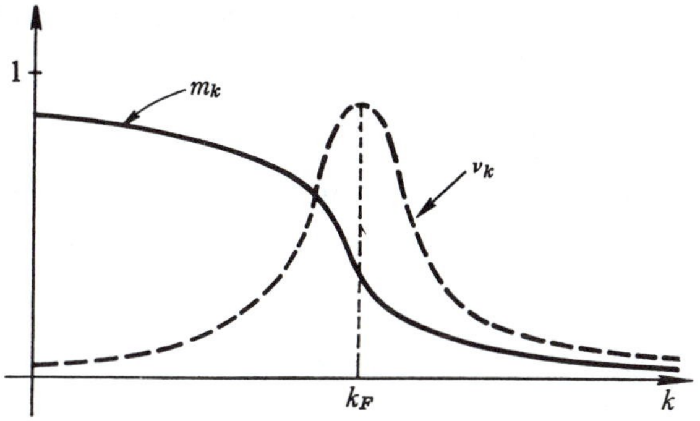

Figure 14

correction of order $1/\Omega$. (The sum of these corrections, is, however, of order unity.)

(b) The population of the state $k\sigma$ is increased by z_+, while that of the state $(-k, -\sigma)$ is diminished by z_-.

(c) $|\nu_{k\sigma}|$ decreases by an amount $\hat{z}$.

The total number of bare particles in states other than $k\sigma$ and $(-k, -\sigma)$ varies by an amount $(1 + z_- - z_+)$, so as to make up the deficit left by these states alone. This deficit has two origins, physically very different. In the first place, it is due to a growth of the number of condensed pairs, according to the following picture:

(a) $m_{k\sigma}$ increases by an amount $(z_+ + z_-)$.

(b) A fraction z_- of the particles $k\sigma$ combine with an equal number of particles $(-k, -\sigma)$, taken from the ground state $|\varphi_0(N)\rangle$, to form bound pairs which drop into the condensed phase. The rest of the deficit, equal to $(1 - z_- - z_+)$, arises from the self-energy cloud carried by the free particles and the pairs. In the approximation of Bardeen et al., this second cause disappears, whereas the first remains.

We thus see that the introduction of a quasi particle is a phenomenon connected with the condensed phase (whose population varies). This connection also appears in the variation of $\nu_{k\sigma}$: when filling the state $k\sigma$, we remove all condensed pairs from it. The corresponding distortion, negligible for a single pair, becomes very important for the whole set.

The deficit of particles, $(1 + z_- - z_+)$, cannot be zero for a superfluid. The terms which make up this deficit all arise from the bound part of K, of order $1/\Omega$. The approximation which consists of neglecting this bound part is therefore not *consistent*; it contains internal contradictions. If we decide nevertheless to accept it, we abandon ipso facto conservation of the total number of particles. Simplification of the calculations is obtained at the cost of increased difficulty of interpretation.

e. Renormalization of the propagators

By analogy with $\bar{G}$, $\hat{G}$, $\check{G}$, we define the quasi-particle propagators by the relations

$$\begin{cases} \bar{\mathcal{G}}_{\sigma\sigma'}(k, t) = \mathrm{i} < \varphi_0(N) \mid T\{ \mathbf{A}_{k\sigma}(t)\, \mathbf{A}^*_{k\sigma'}(0) \} \mid \varphi_0(N) > e^{\mathrm{i}\mu t} \\ \hat{\mathcal{G}}_{\sigma\sigma'}(k, t) = \mathrm{i} < \varphi_0(N) \mid T\{ \mathbf{A}_{k\sigma}(t)\, \mathbf{A}_{-k\sigma'}(0) \} \mid \varphi_0(N+2) > e^{\mathrm{i}\mu t} \\ \check{\mathcal{G}}_{\sigma\sigma'}(k, t) = \mathrm{i} < \varphi_0(N+2) \mid T\{ \mathbf{A}^*_{k\sigma}(t)\, \mathbf{A}^*_{-k\sigma'}(0) \} \mid \varphi_0(N) > e^{-\mathrm{i}\mu t} \end{cases} \tag{4–176}$$

As a quasi particle corresponds to an eigenstate of energy $(E_0(N) + \mu + \omega_k)$, we have, in a reasonable interval of time,

$$\mathbf{A}^*_{k\sigma}(t) = A^*_{k\sigma} \exp[\mathrm{i}(\mu + \omega_k)\, t] \tag{4–177}$$

Using (4–165) and (4–166), we see that

$$\begin{aligned} &\hat{\mathcal{G}} = \check{\mathcal{G}} = 0 \\ &\bar{\mathcal{G}}_{\sigma\sigma'}(k, t) = \begin{cases} \mathrm{i} \exp(-\mathrm{i}\omega_k t)\, \delta_{\sigma,\sigma'} & \text{if } t > 0 \\ 0 & \text{if } t < 0 \end{cases} \end{aligned} \tag{4–178}$$

The *renormalized* propagators $\mathcal{G}$ thus have the same structure as for a normal gas of independent particles. Let us point out the disappearance of the terms $\hat{\mathcal{G}}$ and $\check{\mathcal{G}}$, which, before the transformation $a_{k\sigma} \to A_{k\sigma}$, characterized the superfluid state: we have "diagonalized" the single-particle Green's function.

The Fourier transform of (4–178) with respect to t is simply written as

$$\bar{\mathcal{G}}_{\sigma,\sigma'}(k, \omega) = \frac{1}{\omega_k - \omega}\, \delta_{\sigma,\sigma'} \tag{4–179}$$

it being understood that ω_k has a small *negative* imaginary part.

Let us turn now to the two quasi-particle propagator. Again we define

$$A^\alpha_{k\sigma} = \begin{cases} A_{k\sigma} & \text{if} \quad \alpha = +1 \\ A^*_{k\sigma} & \text{if} \quad \alpha = -1 \end{cases} \tag{4–180}$$

By analogy with (4–128), we set

$$\mathcal{K}(k_i, t_i, \sigma_i, \alpha_i) = \\ < \varphi_0 \mid T\left\{ \mathbf{A}_{k_1\sigma_1}^{\alpha_1}(t_1) \ldots \mathbf{A}_{k_4\sigma_4}^{\alpha_4}(t_4) \right\} \mid \varphi_0 > \exp\left(i\mu \sum_i \alpha_i t_i\right) \tag{4–181}$$

The function $\mathscr{K}$ is different from zero only if we have

$$\sum_i \alpha_i \sigma_i = 0 \qquad \text{and} \qquad \sum_i \alpha_i k_i = 0$$

The physical interpretation of $\mathscr{K}$ varies with the values of α_i. If, for example, $\alpha_1 = \alpha_2 = +1$, $\alpha_3 = \alpha_4 = -1$, $\mathscr{K}$ describes the scattering of two quasi particles. If $\alpha_1 = \pm 1$, $\alpha_2 = \alpha_3 = \alpha_4 = \pm 1$, we are dealing with the reaction

$$1 \text{ quasi particle} \rightleftharpoons 3 \text{ quasi particles}$$

Finally, the case $\alpha_1 = \alpha_2 = \alpha_3 = \alpha_4 = \pm 1$ corresponds to the destruction or creation of four excitations (note that the number of quasi particles is not conserved).

In the same way as K, $\mathscr{K}$ separates into a free part and a bound part. The free part is given by (4–130), except that the propagators G are replaced by $\mathscr{G}$; this introduces nothing new. The bound part is analyzed as for a normal gas [Eqs. (4–59) to (4–61).]

We start from the Fourier transform of (4–132) with respect to the space variables,

$$\Delta K(k_i t_i \sigma_i \alpha_i) = \int dt'_i \sum_{k'_i, \alpha'_i, \sigma'_i} \gamma(k'_i, t'_i, \sigma'_i, \alpha'_i) \prod_{i=1}^{4} G(k_i t_i \sigma_i \alpha_i, k'_i t'_i \sigma'_i \alpha'_i) \tag{4–182}$$

To go from ΔK to $\Delta\mathscr{K}$, we must replace the four operators

$$\mathbf{a}_{k_i \sigma_i}^{\alpha_i}(t_i) \qquad \text{by} \qquad \mathbf{A}_{k_i \sigma_i}^{\alpha_i}(t_i)$$

As a consequence,

$$\Delta\mathcal{K}(k_i \ldots \alpha_i) = \int dt'_i \sum_{k'_i, \alpha'_i, \sigma'_i} \gamma(k'_i \ldots \alpha'_i) \prod_{i=1}^{4} g(k_i \ldots \alpha_i, k'_i \ldots \alpha'_i) \tag{4–183}$$

where the function g is defined by

$$g(kt\sigma\alpha, k't'\sigma'\alpha') = \mathrm{i} < \varphi_0 \mid T \{ \mathbf{A}^{\alpha}_{k\sigma}(t) \, \mathbf{a}^{\alpha'}_{k'\sigma'}(t') \} \mid \varphi_0 > \qquad (4\text{–}184)$$

To calculate these functions g, we use the decomposition of $\overset{*}{a}_{k\sigma}$ given above. The incoherent part of $a^{\alpha}_{k\sigma}$ does not contribute to (4–184). Its continuous spectral distribution has only a negligible overlap with the characteristic peaks of $A^{\alpha}_{k\sigma}$. Only the coherent part given by (4–168) contributes. It is then easily verified that

$$g(kt\sigma\alpha, k'0\sigma'\alpha') = \begin{cases} \sqrt{z_+}\,\overline{\mathcal{G}}_{\sigma\sigma}(k, t) & \text{if } \alpha = -\alpha' = +1 \\ -\sqrt{z_+}\,\overline{\mathcal{G}}_{\sigma\sigma}(k, -t) & \text{if } \alpha = -\alpha' = -1 \\ \sigma\sqrt{z_-}\,\overline{\mathcal{G}}_{\sigma\sigma}(k, t) & \text{if } \alpha = \alpha' = +1 \\ -\sigma\sqrt{z_-}\,\overline{\mathcal{G}}_{\sigma\sigma}(k, -t) & \text{if } \alpha = \alpha' = -1 \end{cases} \qquad (4\text{–}185)$$

These expressions simplify after Fourier transformation with respect to time. By using (4–120) and (4–179), we obtain

$$g(\xi, \xi') = \mathcal{G}(p)\; U(\xi, \xi') \qquad (4\text{–}186)$$

where the quantities $\mathcal{G}$ and U are defined by

$$\mathcal{G}(P) = \frac{1}{\omega_k - \omega} \qquad (4\text{–}187)$$

$$U(\xi, \xi') = \delta(\xi, \xi') \{ \alpha\sqrt{z_+}\,\delta_{\sigma\sigma'} + \alpha\sigma\sqrt{z_-}\,\delta_{\sigma,-\sigma'} \} \qquad (4\text{–}188)$$

The Fourier transform of (4–183) with respect to time is

$$\Delta\mathcal{K}(\xi_i) = \sum_{\xi'_i} \gamma(\xi'_i) \prod_{i=1}^{4} g(\xi_i, \xi'_i) \qquad (4\text{–}189)$$

By using (4–186), we finally put this expression into the form

$$\Delta\mathcal{K}(\xi_i) = \gamma_R(\xi_i) \prod_{i=1}^{4} \mathcal{G}(p_i) \qquad (4\text{–}190)$$

where the interaction γ_R is given by

$$\gamma_R(\xi_i) = \sum_{\xi'_i} \gamma(\xi'_i)\; U(\xi_i, \xi'_i) \qquad (4\text{–}191)$$

We can give an explicit expression for γ_R by referring to (4–188),

$$\gamma_R(p_i\sigma_i\alpha_i) =$$

$$\Big|\sum_{\sigma_i'} \gamma\left(\frac{\sigma_i p_i}{\sigma_i'}, \sigma_i', -\frac{\sigma_i\alpha_i}{\sigma_i'}\right) \prod_{i=1}^{4} \left(\alpha_i\sqrt{z_+}\,\delta_{\sigma_i,\sigma_i'} + \alpha_i\sigma_i\sqrt{z_-}\,\delta_{\sigma_i,-\sigma_i'}\right) \tag{4–192}$$

γ_R is the *renormalized* interaction operator, which describes the interaction between quasi particles, while γ describes that between bare particles. Let us point out that the relation between $\Delta\mathscr{K}$ and γ_R is "scalar," while that between ΔK and γ was of a matrix nature. This important simplification results from the diagonalization of the propagator.

Knowledge of γ_R opens the door to numerous applications: calculation of real scattering amplitudes, study of the energy of interaction between quasi particles in the sense of Landau, etc. Since (4–64) and (4–188) have identical structures, one may transpose, with little modification, the calculations carried out for normal systems. We leave to the reader the task of generalizing the results obtained in the first part of this chapter.

f. Conclusion

We have limited ourselves in this chapter to a purely *descriptive* study of normal and superfluid systems. Actually a complete theory must include two steps:

(a) An analytic step, whose object is to find out under what conditions a gas is normal, superfluid, etc. In other words, why does the ground state $|\varphi_0\rangle$ exhibit certain characteristics rather than others? This is a basic problem, unfortunately a very delicate one, which we have not explored. It is necessary first to determine what type of states are compatible with a given law of interaction. We find, for example, that the normal state can always exist, whereas the isotropic superfluid state which we have considered requires an attraction. Once this selection is made, we must find out to which type the ground state belongs. This final choice can be based on *energy* considerations, ($|\varphi_0\rangle$ must have a minimal energy) or on considerations of *stability* with respect to small deformations of $|\varphi_0\rangle$. (An unstable system contains collective modes of pure imaginary frequency.) These questions have not yet received definite answers, although Thouless has very recently clarified many points.

(b) Once the nature of the system is known, its properties must still be described. It is to this second step that we have devoted this chapter. Our aim has been to develop a formalism, not to resolve difficulties of principle.

The formalism presented is general and rigorous. Of course, we can carry out quantitative calculations of the functions M, γ, etc., only by approximate methods. The simplest of these approximations consists in neglecting the bound part of the two-particle Green's function K. The dynamical equation can then be solved without any further hypothesis. In the case of normal gases, this simplification leads us to the well-known Hartree-Fock approximation. For a superfluid, the results thus obtained are equivalent to those of Bardeen, Cooper, and Schrieffer and of Bogoliubov. Although this approximation is *consistent* in the case of a normal gas, it is no longer so for a superfluid (since it causes conservation of the total number of particles to be violated). In any case, we cannot treat in this way correlations between quasi particles, which are automatically excluded by the hypothesis $\gamma = 0$.

In the present state of our knowledge, we can do little better than the above approximation. However, our formalism lends itself to a generalization, in contrast to variational or other treatments. This is an example of the flexibility of methods using Green's functions. Another nonnegligible advantage in the study of superfluids is the explicit conservation of the total number of particles at all steps of the calculation.

We have considered only *normal* and *isotropic superfluid* states. More complex states can exist. We know, for example, of anisotropic superfluid states. This discussion makes no pretense of being complete. Moreover, even within the restricted framework to which we have confined ourselves, many problems remain unsolved: structure of the interaction operator γ, demonstration of the existence of bound states, etc. In order to investigate these, we need a convenient formalism for calculating the functions M, γ, etc. It is to this formalism, based on a diagrammatic representation, that the next chapter is devoted.

Chapter 5

Perturbation Methods

Until now we have limited ourselves to a formal study of Green's functions: we have no method for calculating them. We are going to fill this gap by developing a "perturbation" technique, based on an expansion in powers of the interaction between particles. This is a very difficult procedure to justify, as the convergence of the series obtained is in doubt. As a matter of fact, we are incapable of calculating *all* the terms in the series. Thus, we cannot draw any conclusions concerning its convergence, and we are therefore reduced to *assuming* the mathematical validity of the procedure used.

For this expansion to make sense, it is necessary for the properties of the system to vary in a continuous way when the coupling constant increases from zero to its actual value. The structure of the ground state cannot change; this automatically limits us to normal systems, having the same Fermi surface as the noninteracting gas.

Actually, one can often avoid this restriction by choosing a suitable unperturbed Hamiltonian. We shall see at the end of this chapter how to take account of a change in the Fermi surface for a normal gas. The extension to superfluid systems is treated in Chap. 7.

For the moment, we limit ourselves to the simplest case: that of a normal gas with a spherical Fermi surface. We first discuss the mathematical aspect of the problem, by studying in detail the "algebra" of perturbations. Then we introduce the fundamental physical hypothesis of *adiabatic* turning on of the interaction, which will give us the energy and the wave function of the *real* ground state. We apply these results to the study of the one- and two-particle Green's functions, thus establishing a solid foundation for Chaps. 3 and 4 . Next we extend these results to the excited states, and we show that the adiabatic turning on of the interaction leads to the same elementary excitations as the poles of G. We emphasize the elegance of diagrammatic methods in establishing diverse relations between the Green's functions and the energy. Finally, we show that the ground-state energy is a stationary functional of the Green's function G, and we discuss the consequences of this important theorem.

1. Mathematical Formulation

a. Setting up of the problem

Let $\mathbf{H}$ be the Hamiltonian of the system, which we write in the form

$$\mathbf{H} = \mathbf{H}_0 + \mathbf{H}_1$$

In our case, $\mathbf{H}_0$ reduces to the kinetic energy of the particles. We assume its eigenvectors and eigenvalues to be known, designating them, respectively, by $|\mathbf{n}\rangle$ and E_n° (as opposed to the eigenvectors and eigenvalues of $\mathbf{H}$, which we write $|\boldsymbol{\varphi}_n\rangle$ and E_n). $\mathbf{H}_0$ is a time-independent operator, which constitutes the "unperturbed" Hamiltonian.

The interaction between particles $\mathbf{H}_1$ acts as a "perturbation" which transforms the states $|\mathbf{n}\rangle$ into states $|\boldsymbol{\varphi}_n\rangle$. Our object is to expand all the properties of the system in powers of $\mathbf{H}_1$. Although physically $\mathbf{H}_1$ is independent of time, we shall be led to study a fictitious problem in which $\mathbf{H}_1$ is turned on "adiabatically" between the times $-\infty$ and 0, by means, for example, of an exponential factor $e^{\eta t}$. The system thus obtained is no longer conservative, which from the start poses some difficulties of representation.

Interaction representation. Until now we have used the Heisenberg representation, characterized by a state vector which is constant in time. Dynamical development is then incorporated entirely into the operators, according to the scheme

$$\begin{cases} \dfrac{d}{dt} | \varphi \rangle = 0 \\ \dfrac{d}{dt} \mathbf{A} = i[\mathbf{H}, \mathbf{A}] + \dfrac{\partial \mathbf{A}}{\partial t} \end{cases} \tag{5-1}$$

(We denote the Heisenberg representation by using bold face letters; the derivative $\partial\mathbf{A}/\partial t$ expresses the *explicit* dependence of $\mathbf{A}$ on the time t.) For a conservative system, $\partial\mathbf{H}/\partial t = 0$, which immediately implies $d\mathbf{H}/dt = 0$.

This representation proves to be inconvenient for a nonconservative system. It is better to return to the Schrödinger representation, in which dynamical development appears only in the wave function. Equations (5–1) are then replaced by

$$\begin{cases} \mathrm{i}\dfrac{\mathrm{d}}{\mathrm{d}t} \mid \varphi_s > = H_s \mid \varphi_s > \\ \dfrac{\mathrm{d}A_s}{\mathrm{d}t} = \dfrac{\partial A_s}{\partial t} \end{cases} \tag{5–2}$$

(the index s refers to Schrödinger). We pass from one representation to the other by a canonical transformation. Its form is simple only in conservative systems, for which

$$\begin{cases} \mathbf{H} = H_s \\ \mid \varphi > = \mathrm{e}^{\mathrm{i}\mathbf{H}t} \mid \varphi_s(t) > = \mid \varphi_s(0) > \\ \mathbf{A} = \mathrm{e}^{\mathrm{i}\mathbf{H}t} A_s \, \mathrm{e}^{-\mathrm{i}\mathbf{H}t} \end{cases} \tag{5–3}$$

Equations (5–3) are no longer valid when $\partial\mathbf{H}/\partial t = 0$.

In practice, neither of these two representations is adapted to the perturbation problem. To demonstrate the special role played by $\mathbf{H}_0$, we need a mixed representation, which is "Heisenberg" as far as $\mathbf{H}_0$ is concerned and "Schrödinger" as far as $\mathbf{H}_1$ is concerned. To do this, we make the canonical transformation

$$\begin{cases} \mid \varphi > = \mathrm{e}^{\mathrm{i}H_0 t} \mid \varphi_s > \\ A = \mathrm{e}^{\mathrm{i}H_0 t} A_s \, \mathrm{e}^{-\mathrm{i}H_0 t} \end{cases} \tag{5–4}$$

The new representation thus defined is known as the *interaction representation.* The dynamics of the operators are governed by H_0 alone, H_1 acting only on the state vector. More precisely, (5–2) becomes

$$\begin{cases} \mathrm{i}\dfrac{\mathrm{d}}{\mathrm{d}t} \mid \varphi > = H_1(t) \mid \varphi > \\ \dfrac{\mathrm{d}}{\mathrm{d}t} A = \dfrac{\partial A}{\partial t} + \mathrm{i}[H_0, A] \end{cases} \tag{5–5}$$

The first of Eqs. (5–5) readily lends itself to solution by iteration, which forms the basis of the perturbation calculation. Equations (5–5) are valid for any H_1 (even if it depends on time). Let us point out that H_1 is itself expressed in the interaction representation, according to the formula

$$H_1(t) = \mathrm{e}^{\mathrm{i}H_0 t} H_{1s} \, \mathrm{e}^{-\mathrm{i}H_0 t}$$

In the particular case where H_1 is independent of time, we can, by going back to (5–3) and (5–4), relate the interaction and Heisenberg representations,

$$\begin{cases} |\varphi> = e^{iH_0t}\, e^{-i\mathbf{H}t}\, |\boldsymbol{\varphi}> \\ A \quad = e^{iH_0t}\, e^{-i\mathbf{H}t}\, \mathbf{A}\, e^{i\mathbf{H}t}\, e^{-iH_0t} \end{cases} \tag{5–6}$$

(5–6) constitutes the "formal" solution of (5–5).

Unless specifically stated to the contrary, we shall use the interaction representation exclusively in this chapter.

The time-development operator. We can characterize the time development of $|\varphi\rangle$ by means of an operator $U(t,t')$, defined by

$$|\varphi(t')> = U(t', t)\, |\varphi(t)> \tag{5–7}$$

U is the *time-development operator* of the system. It is a unitary operator; this important property is a result of the hermiticity of H_1 (the scalar product of any two state vectors is a constant of the motion). A result of the definition (5–7) is that

$$U(t, t')\, U(t', t'') = U(t, t'') \tag{5–8}$$

In particular

$$U(t, t')\, U(t', t) = U(t, t) = 1 \tag{5–9}$$

The operators $U(t,t')$ and $U(t',t)$ are thus conjugate to each other.

The dynamical equation satisfied by U follows from (5–5) and can be written as

$$i \frac{\partial}{\partial t'} U(t't) = H_1(t')\, U(t', t) \tag{5–10}$$

We must supplement (5–10) by the boundary condition $U(t,t) = 1$. It is convenient to combine these two pieces of information in the form of an integral equation,

$$U(t', t) = 1 - i \int_t^{t'} H_1(t'')\, U(t'', t)\, dt'' \tag{5–11}$$

(5–11) constitutes the point of departure for all the perturbation formalism.

Equations (5–7) and (5–11) are general. In the particular case of a conservative system, we can also give a formal expression for U by going back to (5–6); we find at once

$$U(t', t) = e^{iH_0t'} e^{iH(t-t')} e^{-iH_0t} \tag{5–12}$$

[Let us emphasize that (5–12) is no longer valid if $\partial H_1/\partial t \neq 0$.] Note that U does not depend only on $(t - t')$; this is not surprising, as the definition of $|\varphi\rangle$ causes the time $t = 0$ to play a special role.

Solution of Eq. (5–11) by iteration. If H_1 is relatively small, we are led to solve (5–11) by iteration. The result can be put in the form

$$\begin{cases} U(t', t) = \sum_{p=0}^{\infty} U_p(t',t) \\ U_0(t', t) = 1 \\ U_p(t', t) = (-i)^p \int_t^{t'} dt_1 \int_t^{t_1} dt_2 \ldots \int_t^{t_{p-1}} dt_p \; H_1(t_1) \ldots H_1(t_p) \end{cases} \tag{5–13}$$

[By putting the solution (5–13) back into (5–11), one can easily verify that the integral equation is satisfied term by term.] (5–13) is just the expansion in powers of H_1 for which we are looking. It only remains to develop a simple method of calculation of the different terms of the series.

The operators $H_1(t_i)$ do not commute with each other. It is therefore important to preserve their order carefully. Let us assume that $t' > t$; we see from (5–13) that the times t_i increase monotonically from right to left. We obtain the same result by writing U_p in the form

$$U_p(t', t) = \frac{(-i)^p}{p\,!} \int_t^{t'} dt_1 \ldots \int_t^{t'} dt_p \; P\{ H_1(t_1) \ldots H_1(t_p) \} \tag{5–14}$$

where $P[\ldots]$ is Dyson's chronological product, or P product, whose factors are arranged according to increasing time from right to left (the permutation of two operators with respect to the order indicated is not accompanied by any sign change, in contrast to the operator T defined in Chap. 3). There are $p!$ ways of ordering the t_i, each giving the same contribution to $U_p(t',t)$, hence the factor $p!$ in the denominator of (5–14). This equation is of a much more symmetric nature than (5–13).

Explicit form of the solution. The interaction Hamiltonian H_1 was calculated in Appendix B [Eq. (B–16)]. We rewrite it, using the interaction representation,

$$H_1(t) = \frac{1}{2} \sum_{\sigma,\sigma'} \sum_{k,k',q} \frac{V_q}{\Omega} \overset{*}{a}_{k+q,\sigma}(t)\, \overset{*}{a}_{k'-q,\sigma'}(t)\, a_{k',\sigma'}(t)\, a_{k,\sigma}(t) \tag{5-15}$$

Thanks to the interaction representation, the operators $\overset{*}{a}_{k\sigma}(t)$, $a_{k\sigma}(t)$ have a very simple form,

$$\begin{cases} \overset{*}{a}_{k\sigma}(t) = e^{iH_0 t}\, \overset{*}{a}_{k\sigma}\, e^{-iH_0 t} = \exp(i\varepsilon_k^0 t)\, \overset{*}{a}_{k\sigma} \\ a_{k\sigma}(t) = e^{iH_0 t}\, a_{k\sigma}\, e^{-iH_0 t} = \exp(-\, i\varepsilon_k^0 t)\, a_{k\sigma} \end{cases} \tag{5-16}$$

where $\varepsilon_k^\circ = k^2/2m$ is the kinetic energy of the particle k. It is advantageous to put (5–15) into a more symmetric form by setting

$$\begin{cases} H_1 = \frac{1}{4} \sum_{\mathbf{k}_i} a^*_{\mathbf{k}_1} a^*_{\mathbf{k}_2} a_{\mathbf{k}_3} a_{\mathbf{k}_4} V(\mathbf{k}_1, \mathbf{k}_2, \mathbf{k}_3, \mathbf{k}_4) \\ V(\mathbf{k}_1, \mathbf{k}_2, \mathbf{k}_3, \mathbf{k}_4) = \frac{1}{\Omega} \left[V_{k_1-k_4}\, \delta_{\sigma_1,\sigma_4} - V_{k_1-k_3}\, \delta_{\sigma_1,\sigma_3} \right] \delta_{\mathbf{k}_1+\mathbf{k}_2,\mathbf{k}_3+\mathbf{k}_4} \end{cases} \tag{5-17}$$

[$\mathbf{k}_i$ is an abbreviated notation for (k_i,σ_i)]. The equivalence of (5–15) and (5–17) is easily verified. Note that, by definition, $V(\mathbf{k}_i)$ changes sign when $\mathbf{k}_1$, $\mathbf{k}_2$ or $\mathbf{k}_3$, $\mathbf{k}_4$ are permuted. If we want to make the law of interaction vary explicitly in time, we shall modify $V(\mathbf{k}_i)$.

A term of U of the pth order is thus represented by the product of $2p$ creation and $2p$ destruction operators, chronologically ordered from right to left. There is an ambiguity for simultaneous operators corresponding to the same factor H_1; in this case, the creation operators are found at the left of the destruction operators, arranged in the same order as the variables $\mathbf{k}_i$ in $V(\mathbf{k}_1,\mathbf{k}_2,\mathbf{k}_3,\mathbf{k}_4)$ [see (5–17)].

In order to make the expression for U more compact, we introduce the chronological product of creation or destruction operators, or T product, already used in the definition of the Green's functions. The T product is related to Dyson's P product by the relation

$$T(UV \ldots Z) = \lambda P(UV \ldots Z) \tag{5-18}$$

where $\lambda = \pm 1$ has the sign of the permutation which takes the indicated order $UV \ldots Z$ into the chronological order. For example, we have, for a product of two factors,

$$T\{\, U(t)\, V(t')\,\} = \begin{cases} U(t)\;\; V(t') & \text{if} \quad t > t' \\ -\, V(t')\;\; U(t) & \text{if} \quad t < t' \end{cases} \tag{5-19}$$

If the operators U and V anticommute, we always have

$$T(U\,V) = U\,V$$

(This is precisely why we prefer the T product to the P product.) When $t = t'$, the value of the T product [(5–19)] is defined only if U and V anticommute; otherwise there is an ambiguity. By *definition*, we obtain the T product of several simultaneous factors by placing the creation operators to the left of the destruction operators. (The order in which we arrange the $a_{\mathbf{k}}^*$ and the $a_{\mathbf{k}}$ separately is of no importance, since in each of these groups the operators anticommute.) For example, by making use of (5–16) and the anticommutation relations, we can write

$$\left\{\begin{aligned} &T\left\{ a_{\mathbf{k}}^*(t)\, a_{\mathbf{k}'}^*(t') \right\} = a_{\mathbf{k}}^*(t)\, a_{\mathbf{k}'}^*(t') \\ &T\left\{ a_{\mathbf{k}}(t)\, a_{\mathbf{k}'}(t') \right\} = a_{\mathbf{k}}(t)\, a_{\mathbf{k}'}(t') \\ &T\left\{ a_{\mathbf{k}}^*(t)\, a_{\mathbf{k}'}(t') \right\} = a_{\mathbf{k}}^*(t)\, a_{\mathbf{k}'}(t') - \begin{cases} 0 \text{ if } t \geqslant t' \\ \exp\left[i\varepsilon_k^0(t - t')\right] \delta_{\mathbf{k},\mathbf{k}'}, \text{ if } t < t' \end{cases} \end{aligned}\right. \tag{5–20}$$

which illustrates the preceding definition.

We can now give a more specific form to the expansion (5–14) (valid for $t' > t$),

$$U_p(t', t) = \left(\frac{i}{4}\right)^p \frac{1}{p\,!} \int_t^{t'} dt_1 \ldots \int_t^{t'} dt_p \sum_{\mathbf{k}_{1i}, \ldots, \mathbf{k}_{pi}} V(\mathbf{k}_{1i}, t_1) \ldots V(\mathbf{k}_{pi}, t_p)$$
$$T\left\{ a_{\mathbf{k}_{11}}^*(t_1) \ldots a_{\mathbf{k}_{14}}(t_1) \ldots a_{\mathbf{k}_{p1}}^*(t_p) \ldots a_{\mathbf{k}_{p4}}(t_p) \right\} \tag{5–21}$$

The T product of (5–21) reduces to a P product of factors H_1: by permuting two H_1's, we interchange two sets of four operators, which is an even permutation. We have allowed for an eventual variation of the law of interaction in time. [Note that (5–21) is invariant with respect, for example, to permutation of $\mathbf{k}_{11}$ and $\mathbf{k}_{12}$: the T product and the factor $V(\mathbf{k}_{1i}, t_1)$ both change sign.]

b. The matrix elements of U; reduction of chronological products

Equation (5–21) in principle gives the complete solution of the problem. Actually it is quite cumbersome. We can help to remedy this by separate calculation of the different matrix elements of U.

We choose as a basis the eigenstates of H_0, which we shall refer to the ground state $|0\rangle$. The latter has a very simple distribution function in

k space, equal to $+1$ if $|k| < k_F$ and to 0 if $|k| > k_F$ [the Fermi wave vector is given by (4–47)]. Let us introduce the notation

$$b_{\mathbf{k}}^* = \begin{cases} a_{\mathbf{k}}^* & \text{if} \quad |k| > k_F \\ a_{\mathbf{k}} & \text{if} \quad |k| < k_F \end{cases} \tag{5–22}$$

$b_{\mathbf{k}}^*$ is the creation operator of an elementary excitation. We can then write

$$\begin{cases} b_{\mathbf{k}} \,|\, 0 > = 0 \\ |\, n > = \displaystyle\prod_{i=1}^{i=2m} b_{\mathbf{k}_i}^* \,|\, 0 > \end{cases} \tag{5–23}$$

In order to calculate the matrix element $\langle n'|U|n\rangle$, we should permute the factors of U so as to place all the $b_{\mathbf{k}}^*$ to the left of the $b_{\mathbf{k}}$. This leads us to the notion of the normal product.

Definition of the normal product. Let us consider a product $UV \ldots Z$, whose factors can be either of the type $b_{\mathbf{k}}$ or of the type $b_{\mathbf{k}}^*$. The *normal product* $N(UV \ldots Z)$ is obtained by placing all the $b_{\mathbf{k}}^*$ to the left of the $b_{\mathbf{k}}$ and giving the resulting product the sign of the permutation made to obtain this result. Again, the relative order of the $b_{\mathbf{k}}^*$ (or the $b_{\mathbf{k}}$) is of no importance, since, from (5–16),

$$\begin{cases} [b_{\mathbf{k}}(t),\, b_{\mathbf{k}'}(t')]_+ = 0 \\ [b_{\mathbf{k}}^*(t),\, b_{\mathbf{k}'}^*(t')]_+ = 0 \end{cases}$$

If U and V anticommute, we always have

$$N(UV) = UV \tag{5–24}$$

(5–24) becomes false if U and V do not anticommute. For example,

$$N\{\, a_{\mathbf{k}}^*(t)\, a_{\mathbf{k}}(t')\,\} = a_{\mathbf{k}}^*(t)\, a_{\mathbf{k}}(t') - \begin{cases} 0 & \text{if} \quad |k| > k_F \\ \exp[i\varepsilon_k^0(t - t')] & \text{if} \quad |k| < k_F \end{cases} \tag{5–25}$$

Let us assume that we know how to decompose the operator $U_p(t,t')$ into a sum of normal products, containing 0, 2, 4, . . ., $4p$ operators. A typical term of this expansion can be written as

$$\prod_{j'} b_{\mathbf{k}_{j'}}^* \prod_{j} b_{\mathbf{k}_j}$$

By going back to (5–23), we see that the corresponding contribution to the matrix element $\langle n'|U|n\rangle$ is proportional to

$$<0\,|\prod_{i'} b_{\mathbf{k}_{i'}} \left\{\prod_{j'} b^*_{\mathbf{k}_{j'}} \prod_{j} b_{\mathbf{k}_j}\right\} \prod_{i} b^*_{\mathbf{k}_i}\,|\,0> \tag{5-26}$$

Using (5–23) and the anticommutation relations for the $b_{\mathbf{k}}$, we see that (5–26) is different from zero only if the set of the $\mathbf{k}_j$ is contained in that of the $\mathbf{k}_i$ and the set of the $\mathbf{k}'_j$ is contained in the $\mathbf{k}'_i$. As a consequence, a few types of normal product contribute to the matrix element. If, for example, we look for the diagonal element $\langle 0|U|0\rangle$, the only contribution comes from the scalar term of the decomposition of U_p, containing no operators. We thus see the considerable interest there is in transforming the chronological product (5–21) into a sum of normal products.

The notion of contraction of two operators. We call the *contraction* $\overparen{UV}$ of two operators U and V the difference

$$\overparen{UV} = T(UV) - N(UV) \tag{5-27}$$

If U and V anticommute, $\overparen{UV} = 0$. The contraction of two creation operators, or of two destruction operators, is therefore zero. More generally, the relations

$$\begin{cases} T(UV) = -\,T(VU) \\ N(UV) = -\,N(VU) \end{cases}$$

imply $\overparen{UV} = -\overparen{VU}$.

Let us consider the contraction $\overparen{a^*_{\mathbf{k}}(t)a_{\mathbf{k}'}(t')}$; it is a scalar. Using the definition of the normal product, we thus find

$$\overparen{UV} = <0\,|\,\overparen{UV}\,|\,0> = <0\,|\,T(UV)\,|\,0> \tag{5-28}$$

By going back to the definition (3–8), we see that

$$\overparen{a^*_{\mathbf{k}}(t)\; a_{\mathbf{k}'}(t')} = \mathrm{i}\; G_0(k, t' - t)\; \delta_{\mathbf{k},\mathbf{k}'} \tag{5-29}$$

where G_0 is the single-particle Green's function of the *noninteracting system*, given by relations (3–17). The notion of contraction thus leads us to the techniques developed in Chap. 3.

It remains to discuss the case where $t' = t$. Our definition of the T product then places $a_{\mathbf{k}}^*$ to the left of $a_{\mathbf{k}}$ and corresponds to the limit $t' \to t - 0$. The contraction (5–29) is in this case equal to $iG_0(k, -0)$, which removes the ambiguity.

Wick's theorem. The notion of contraction allows us to reduce a chronological product of two factors into a sum of normal products. We shall now generalize this result to a product containing an arbitrary number of factors; this is the object of *Wick's theorem.* This theorem is purely mathematical and depends on combinatorial analysis. We indicate its proof in Appendix E; for now, we merely state it.

The chronological product of N factors $U_1 \ldots, U_N$ can be decomposed in the form

$$\begin{aligned} T(U_1, \ldots, U_N) &= N(U_1, \ldots, U_N) + \sum_{i_1<j_1} \lambda_{i_1 j_1} \overbrace{U_{i_1} U_{j_1}} N_{i_1 j_1}(U_1, \ldots, U_N) \\ &+ \sum_{\substack{i_1<j_1 \\ i_2<j_2 \\ i_1<i_2}} \lambda_{i_1 i_2 j_1 j_2} \overbrace{U_{i_1} U_{j_1}}\, \overbrace{U_{i_2} U_{j_2}} N_{i_1 i_2 j_1 j_2}(U_1, \ldots, U_N) \\ &+ \cdots + \sum_{\substack{i_1<j_1 \\ i_p<j_p \\ i_1<i_2<\cdots<i_p}} \lambda_{i_1 \cdots i_p j_1 \cdots j_p} \overbrace{U_{i_1} U_{j_1}} \cdots \overbrace{U_{i_p} U_{j_p}} N_{i_1 \cdots i_p j_1 \cdots j_p}(U_1, \ldots, U_N) + \cdots \end{aligned} \tag{5–30}$$

In this expression

$$\lambda_{i_1 \cdots i_p j_1 \cdots j_p} = \pm 1$$

has the sign of the permutation required to bring together the different pairs of operators which are contracted, while

$$N_{i_1 \cdots i_p j_1 \cdots j_p}(U_1, \ldots, U_N)$$

is the normal product which remains when the contracted operators have been removed.

In other words, we contract any number of pairs of operators, after having brought them together by an appropriate permutation, which is reflected in the overall sign. The remainder is put in the form of a normal product. We sum over all sets of possible contractions, containing $1, \ldots,$

$N/2$ contractions, taking care that each set of contractions *is accounted for only once* (which we have expressed by restrictions on the summations). This theorem, apparently complex, lends itself to a very simple graphical interpretation, which we shall demonstrate a little later.

Let us first apply this theorem to the calculation of the average value $\langle 0|U(t,t')|0\rangle$. Since the average value of a normal product is zero, only the terms of the Wick expansion in which *all* the factors have been contracted contribute; this makes for considerable simplification. Note that U_p contains $2p$ destruction operators and $2p$ creation operators; the expansion of $\langle 0|U_p|0\rangle$ contains $(2p)!$ independent sets of contractions.

Let us turn now to the general matrix element $\langle n'|U|n\rangle$. We see from (5–26) that the terms of the Wick expansion giving a nonzero contribution must contain a well-defined normal product. The sole arbitrariness remains in the order of the factors within the normal product. In practice, we can always reduce the calculation to that of an average value in the state $|0\rangle$. Let us write

$$\begin{cases} |\, n > = b_n^* \,|\, 0 > \\ |\, n' > = b_{n'}^* \,|\, 0 > \end{cases} \tag{5–31}$$

where b_n^*, $b_{n'}^*$ are products of operators $b_{\mathbf{k}}^*$, given by Eq. (5–26). We can write

$$\begin{aligned} < n' \,|\, U(t', t) \,|\, n > &= < 0 \,|\, b_{n'} U(t', t)\, b_n^* \,|\, 0 > \\ &= \exp\left[i(\omega_{n'}^0\, t' - \omega_n^0 t)\right] < 0 \,|\, b_{n'}(t')\, U(t', t)\, b_n^*(t) \,|\, 0 > \end{aligned} \tag{5–32}$$

where ω_n° and $\omega_{n'}^\circ$ are the excitation energies of the states $|n\rangle$ and $|n'\rangle$. When $t' > t$, the operator $b_{n'}(t')U(t',t)b_n^*(t)$ is a chronological product, amenable to a Wick expansion. Its average value in the state $|0\rangle$ is obtained by contracting *all* the factors, including the b_n^*, $b_{n'}$.

c. Graphical representation of the expansion; Feynman diagrams

For the moment we limit ourselves to the calculation of $\langle 0|U(t',t)|0\rangle$. We propose to represent any term of the corresponding expansion by a diagram. We first draw a vertical axis, on which we mark off a time scale (see Fig. 1). We next represent each of the factors H_1 by a set of four points: two "positive" points, representing the creation operators $a_{\mathbf{k}_1}^*$ and $a_{\mathbf{k}_2}^*$, and two "negative" points, representing $a_{\mathbf{k}_3}$ and $a_{\mathbf{k}_4}$ [see Eq. (5–17)]. We call this group an *interaction vertex*. A typical term of $\langle 0|U_p(t',t)|0\rangle$ will contain p vertices at times in the interval between t and t' (Fig. 1). Each of the operators $a_{\mathbf{k}}$ or $a_{\mathbf{k}}^*$ appearing in U_p thus corresponds to a well-determined point.

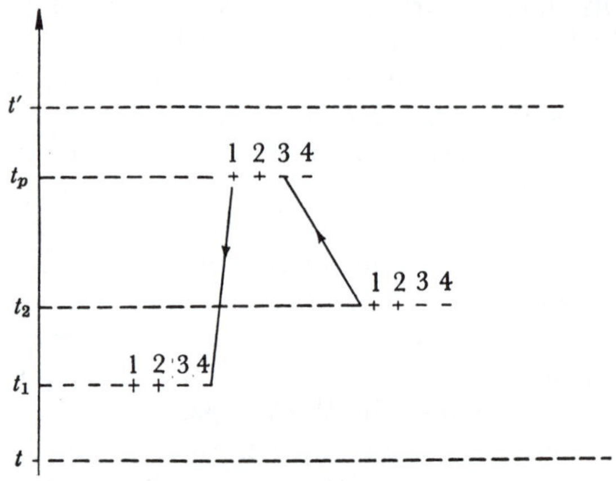

Figure 1

A contraction between two operators can now be represented by a line connecting the two corresponding points. Only nonzero contractions are of interest; they connect a positive point to a negative point. We shall orient the line from the point + toward the point − . We indicate several examples of this in Fig. 1. Such a line is called a *propagation line.*

A complete set of contractions is obtained by connecting the $4p$ points in pairs by propagation lines. There is a one-to-one correspondence between the "diagram" thus obtained and the set of contractions. In order to exhaust all sets of contractions contributing to $\langle 0|U_p(t',t)|0\rangle$, we must draw all possible diagrams containing p vertices, each diagram being counted only once.

It now remains to establish the formula allowing us to evaluate the contribution of a given diagram to $\langle 0|U(t',t)|0\rangle$:

(a) With each vertex we associate the factor

$$\tfrac{1}{4}\, V(\mathbf{k}_1, \mathbf{k}_2, \mathbf{k}_3, \mathbf{k}_4) \tag{5–33}$$

resulting from (5–17).

(b) With each line we associate the factor

$$\overbrace{a_{\mathbf{k}'}(t')\, a^*_{\mathbf{k}}(t)} = -\,\mathrm{i}G_0(\mathbf{k}, t'-t)\, \delta_{\mathbf{k}\mathbf{k}'} \tag{5–34}$$

in accordance with (5–29). Note that the two ends of a line correspond to the same **k** and that the line is oriented from t to t'. By referring to (3–17), we see that, if $|k| > k_F$, the line points upward, whereas it points

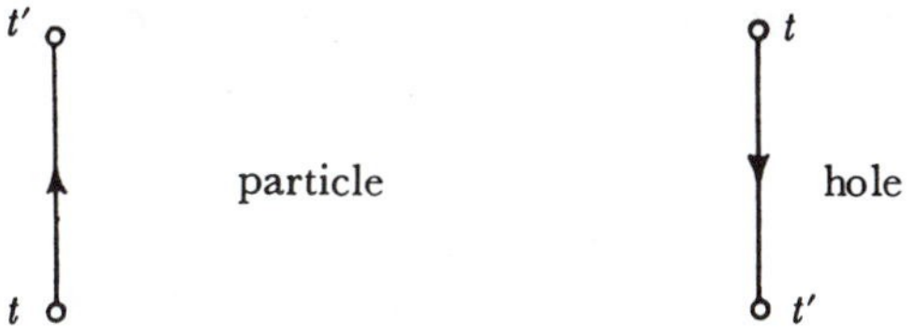

Figure 2

downward if $|k| < k_F$ (Fig. 2). We can think of a hole as a particle which propagates backward in time. This interpretation, due to Feynman, allows us to treat holes and particles equivalently.

(c) There remains the question of the sign. For this purpose we, slightly modify the structure of the interaction vertices, placing $a_{\mathbf{k}_4}$ next to $a^*_{\mathbf{k}_1}$ (Fig. 3). This permutation is even and does not affect the sign. The operators (1,4) and (2,3) now form *blocks* which can be displaced at will, with no change in sign. One propagation line enters each block, and another leaves it (Fig. 3). We thus form *closed loops*, as indicated in Fig. 3. Let us consider one of these loops, and let us arrange the blocks of operators in the order corresponding to the loop; we thus arrive at a system of contractions

$$a^*a \;.\; a^*a \;.\; a^*a \;\ldots\; a^*a$$

The interior contractions carry no change of sign. On the other hand, the contraction between the first and the last operators gives a factor $+\, iG_0$ instead of $-\, iG_0$. As a consequence, if a diagram contains l closed loops, we must add to the contributions (a) and (b) a factor $(-\,1)^l$.

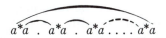

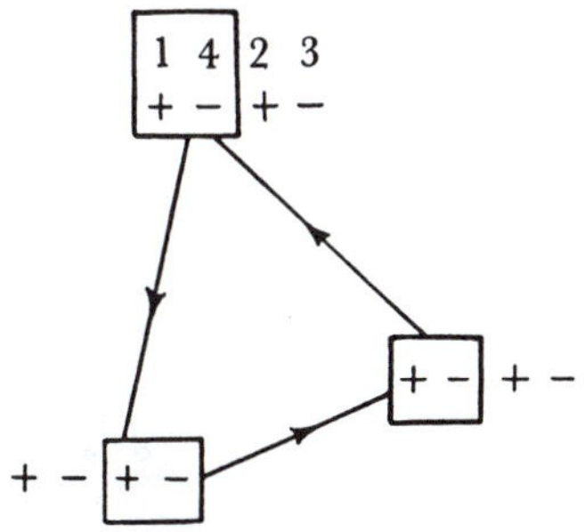

Figure 3

(d) After all these factors have been collected, we add a factor $(-i)^p/p!$, sum over all wave vectors, and integrate over the t_i, in accordance with Eq. (5–14).

Note: the rule for evaluating the signs applies to loops containing a single vertex:  : we have seen that a contraction of this type is equal to $+iG_0(k, -0)$

We can thus calculate *explicitly* the different terms of the expansion of $\langle 0|U|0\rangle$. Note that we have not had to concern ourselves with the exclusion principle. Several lines of a diagram can correspond to the same wave vector **k**; it is easily verified that the corresponding contribution is always zero (by interchanging the end points of the two lines we obtain opposite terms). Therefore there is no difficulty in this aspect, *on condition that we neglect the exclusion principle everywhere*; in particular, we should not reinstate it after having made partial summations.

The graphical representation which we have just described is convenient, but still quite complicated. We are now going to simplify it, step by step.

Simplification of the interaction vertices. The vertices defined above have a structure, which allows localization of the four operators a or a^*. Actually it is desirable to define the vertex only by the two lines which enter and the two lines which leave it, of wave vectors, respectively, $\mathbf{k}_3\mathbf{k}_4$ and $\mathbf{k}_1\mathbf{k}_2$. There are four ways of attaching these lines to the vertex, diagramed in Fig. 4. We indicate with regard to each combination the contribution which it gives to the diagram, taking into account the change of sign required by the permutation of two lines. Because of the antisymmetry of $V(\mathbf{k}_i)$, these four contributions are identical.

We can thus simplify the definition of the interaction vertex; the latter will hereafter be designated by a point, the origin of two lines, 1

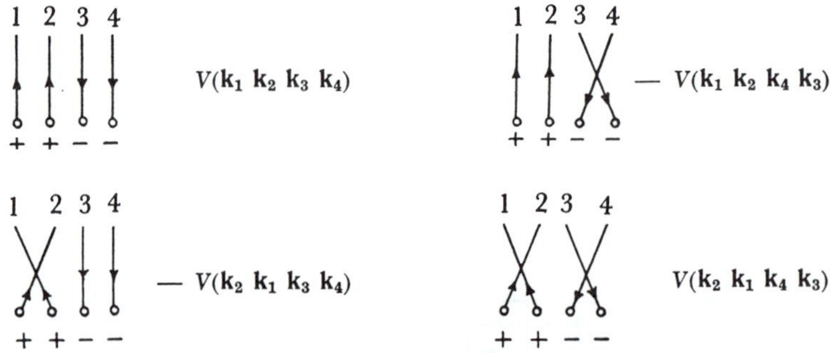

Figure 4

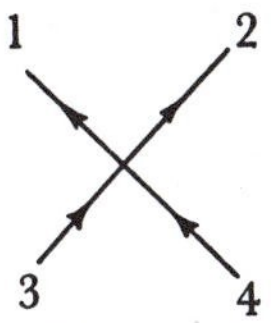

Figure 5

and 2, and the end of two others, 3 and 4, arranged as indicated in Fig. 5. With this vertex we associate the factor $V(\mathbf{k}_1\mathbf{k}_2\mathbf{k}_3\mathbf{k}_4)$, instead of (5–33). To avoid sign troubles it is essential to represent the vertex as a *crossing*, the line 4 becoming 1 and the line 3 becoming 2. The continuity of the lines on passing through the vertex determines the position of the $\mathbf{k}_i$ in $V(\mathbf{k}_i)$, that is to say, the sign of the term. By following a line continuously across different vertices, one can reconstruct the closed loops. With the above conventions, the sign to be assigned to the total product is $(-1)^l$, where l is the number of closed loops.

There is a case in which the preceding regrouping is invalid: when the two lines coming from a given vertex meet again at a neighboring vertex (Fig. 6). The lines are then said to be *equivalent*. We see in Fig. 6 that there are only two independent sets of contractions and not four. It is therefore necessary to correct the rules given above by dividing by a factor 2^m, where m is the number of pairs of equivalent lines.

Reduction of labeled diagrams into unlabeled diagrams. The diagrams with which we are working for the moment are "labeled." By this we understand that with each vertex there is associated a well-defined variable of integration t_i. Actually, this is of no interest, as we integrate over the t_i. We can therefore regroup all diagrams having the same form and differing only by the distribution of the t_i among the various vertices. We thus obtain an "unlabeled" diagram, whose vertices are unmarked. A priori, the contribution of an unlabeled diagram seems to be $p!$ times greater than that of a labeled diagram, since there are $p!$ possible permutations of the t_i among the different vertices. This

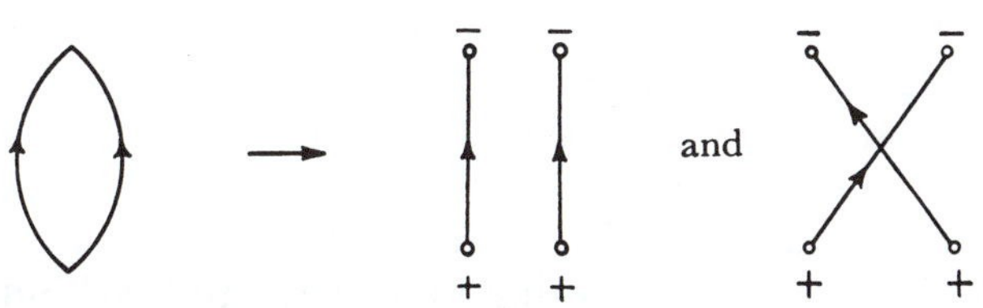

Figure 6

conclusion is premature; actually a certain number r of permutations of the t_i can give the same labeled diagram. For example, the labeled diagram 7*a* is invariant with respect to any permutation of the t_i, Fig. 7*b* is invariant only with respect to cyclic permutations, and Fig. 7*c* is not invariant with respect to any permutation. Thus, we must multiply the contribution of a labeled diagram by $p!/r$ to obtain that of an unlabeled diagram.

All these regroupings considerably reduce the extent of the calculations and make the method more useful. Before going further, we shall summarize the above results.

Rules for the use of unlabeled Feynman diagrams. We want to calculate the matrix element $\langle 0|U_p(t',t)|0\rangle$, where U_p is the term of order p in the expansion of U (we assume that $t' > t$). For this purpose, we draw all possible unlabeled diagrams containing p vertices. To each line is attributed a definite wave vector. With each diagram we associate a term calculated according to the following rules:

(5–35)

(a) To each vertex

corresponds a factor $V(\mathbf{k}_1,\mathbf{k}_2,\mathbf{k}_3,\mathbf{k}_4)$

(b) A line of wave vector $\mathbf{k}$ going from t to t' gives a factor

$$G_0(k, t'-t)$$

(c) The whole term is multiplied by

$$(-1)^l \, i^p \, \frac{1}{2^m r}$$

where l is the number of closed loops, m the number of pairs of equivalent lines, and r the number of permutations leaving the labeled diagram invariant.

(d) Finally, we sum over all wave vectors, and we integrate over the times of the p vertices (in the interval t, t').

In the rules we have regrouped all constant factors, relying on the fact that a diagram of p vertices contains $2p$ lines. By adding the contributions from the different diagrams, we obtain $\langle 0|U_p(t,t')|0\rangle$.

In practice, the two principal difficulties are, on the one hand, not to forget diagrams and, on the other hand, to calculate r correctly. We can make a simple check by noting that there are $(2p)!$ independent sets of contractions, that is to say, $(2p)!$ diagrams of the initial type (labeled

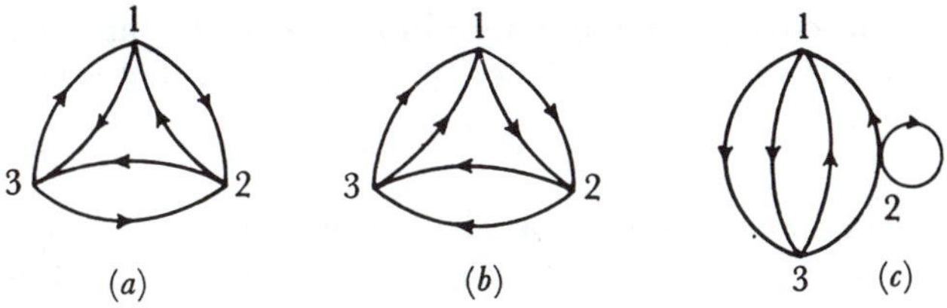

Figure 7

diagrams with "decomposed" vertices). By following the reduction process, we can ascertain that an unlabeled diagram D_p corresponds to $4^p p!/2^m r$ decomposed labeled diagrams. Hence the following rule, which constitutes a valuable check:

$$\sum_{D_p} \frac{1}{r\,2^m} = \frac{(2p)\,!}{4^p\,p\,!} \tag{5-36}$$

By way of example, we enumerate in Fig. 8 all the unlabeled

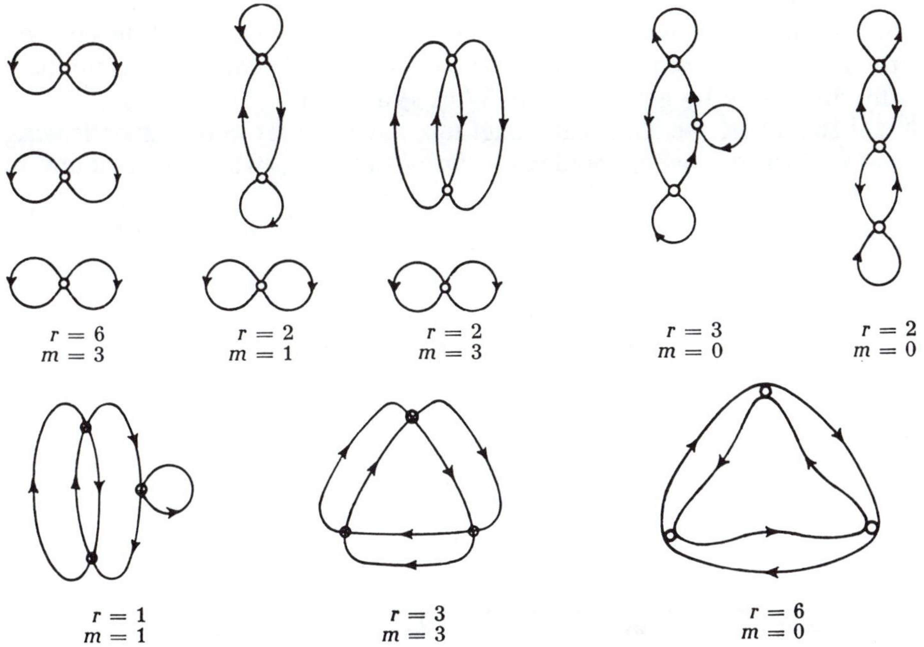

Figure 8

diagrams D_3, indicating the corresponding values of r and m. It is easily verified that (5–36) is satisfied. (Note that the factor $1/2^{m_r}$ is unfavorable to diagrams which are too symmetric.)

Calculation of other matrix elements of $U_p(t',t)$. From Eq. (5–32), we see that this calculation reduces to the evaluation of the chronological product

$$< 0 \mid b_{n'}(t')\; U(t', t)\; b_n^*(t) \mid 0 >$$

where the operators b_n^* and $b_{n'}^*$ are products of factors $a^*_{\mathbf{k}\,>\,k_F}$ or $a_{\mathbf{k}'\,<\,k_F}$. In order to extend to this problem the techniques developed for $\langle 0|U|0\rangle$, we represent the different factors of b_n^* and $b_{n'}$ by points located, respectively, at times t and t', "positive" if we are concerned with a creation operator, "negative" for a destruction operator. A complete set of contractions of the product $b_{n'}Ub_n^*$ is represented by a diagram in which the vertices *and* the external points are all related to each other, that is to say, by a diagram having lines entering and leaving from external points (see Fig. 9). The rules yielding the contribution of these diagrams are analogous to those established for $\langle 0|U|0\rangle$. In particular, we are led in the same way to unlabeled diagrams. The only change concerns the numerical factor. To begin with, if a diagram of order p contains $(2p + q)$ lines, we must add a factor $(-i)^q$, arising from the q additional lines. Furthermore, the overall sign is difficult to determine, since it depends on the exact structure of the operators b_n and $b_{n'}$. We can always eliminate this ambiguity by going back to Wick's theorem.

Hereafter, we shall call diagrams having lines entering or leaving *linked diagrams*, as opposed to *unlinked diagrams*, which contribute to $\langle 0|U|0\rangle$.

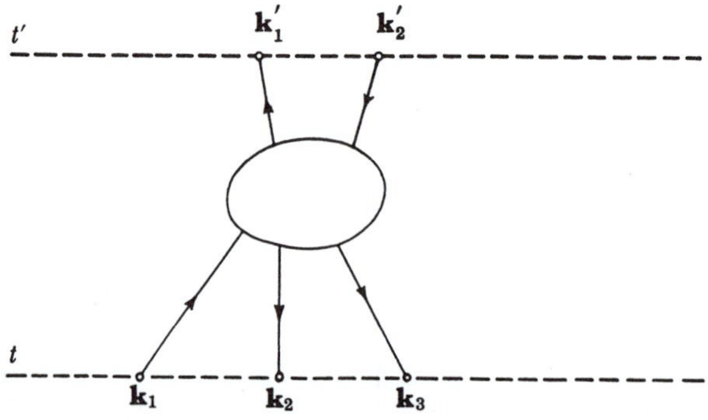

Figure 9

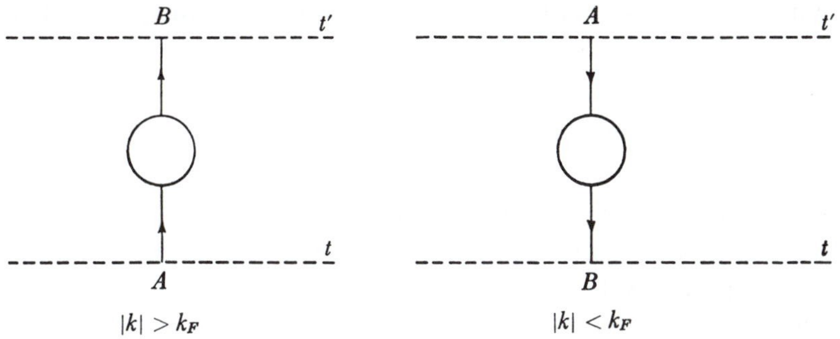

Figure 10

We shall illustrate this discussion by two particular cases:

(1) *Calculation of* $\langle 0|b_{\mathbf{k}}(t')U(t',t)b_{\mathbf{k}}^*(t)|0\rangle$. The corresponding diagrams are indicated in Fig. 10. They contain $(2p + 1)$ lines. The only point in doubt concerns the sign. These diagrams contain, in addition to a certain number l of closed loops, an open line which goes from A to B. This line yields the set of contractions

$$\begin{cases} a_{\mathbf{k}}(a^*a \ldots\ldots a^*a)\, a_{\mathbf{k}}^* & \text{when } k > k_F \\ a_{\mathbf{k}}^*(a^*a \ldots\ldots a^*\, a)a_{\mathbf{k}} & \text{when } k < k_F \end{cases}$$

which implies an additional minus sign when $k < k_F$. In summary, the numerical factor which supplements the contribution of the lines and the vertices is in this case given by

$$\mp(-1)^l\, i^{p+1}\, \frac{1}{r2^m} \qquad \text{according to whether } k \gtrless k_F \tag{5-37}$$

[compare with (5–35).] Here again, we can verify the calculation by noting that there are $(2p + 1)!$ independent sets of contractions of order p. (5–36) must then be replaced by

$$\sum_{D_p} \frac{1}{r2^m} = \frac{(2p+1)\,!}{4^p\, p\,!} \tag{5-38}$$

(5–38) will prove to be useful in the calculation of Green's functions.

(2) *Calculation of the state vector* $U(t',t)|0\rangle$. We can write

$$U(t', t)\,|\,0\rangle = \sum_n |\,n\rangle\langle n\,|\,U(t', t)\,|\,0\rangle \tag{5-39}$$

The only states $\langle n|$ which contribute are those containing the same number $q \geqslant 2$ of excited particles and holes, of the type

$$< n \mid = < 0 \mid a_{\mathbf{k}_1} \dots a_{\mathbf{k}_q} \dots a^{*\prime}_{\mathbf{k}_1} \dots a^{*\prime}_{\mathbf{k}_q}$$

with $|k_i| > k_F$ and $|k_i'| < k_F$. The matrix element $\langle n|U|0\rangle$ is obtained, to within a factor $e^{i\omega_n^\circ t'}$, with the help of the diagrams indicated in Fig. 11. In addition to the closed loops, these diagrams contain q open lines, relating the points $\mathbf{k}_i'$ to the points $\mathbf{k}_i$. We shall define the state $\langle n|$ so that the operators corresponding to the two end points $\mathbf{k}_i$ and $\mathbf{k}_i'$ of a single open line are found next to each other, according to the scheme

$$< n \mid = < 0 \mid a^{*\prime}_{\mathbf{k}_1} a_{\mathbf{k}_1} \dots a^{*\prime}_{\mathbf{k}_q} a_{\mathbf{k}_q} \qquad (5\text{–}40)$$

With this convention, each open line contributes a factor (-1). By combining this with the factor $(\pm i)^q$ due to the presence of $(2n + q)$ lines, we arrive at the multiplicative factor replacing (5–37),

$$(-1)^l \, i^{p+q} \frac{1}{r2^m} \qquad (5\text{–}41)$$

These formulas will be used to calculate the wave function of the ground state of the real system.

We have dwelt upon these questions of expansion and graphical representation at the risk of becoming tiresome. However, it is useful to perfect a tool which we shall use constantly in the rest of our study.

d. The elimination of disconnected diagrams

The average value $\langle 0|U|0\rangle$ is given by the set of *unlinked* diagrams. Let us refer to Fig. 8; we see that two quite distinct types of diagrams appear. The first three diagrams are composed of several independent parts, "disconnected" from each other. On the other hand, the other

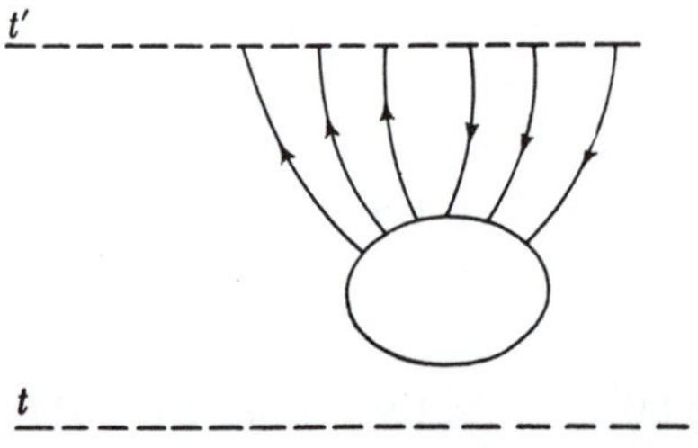

Figure 11

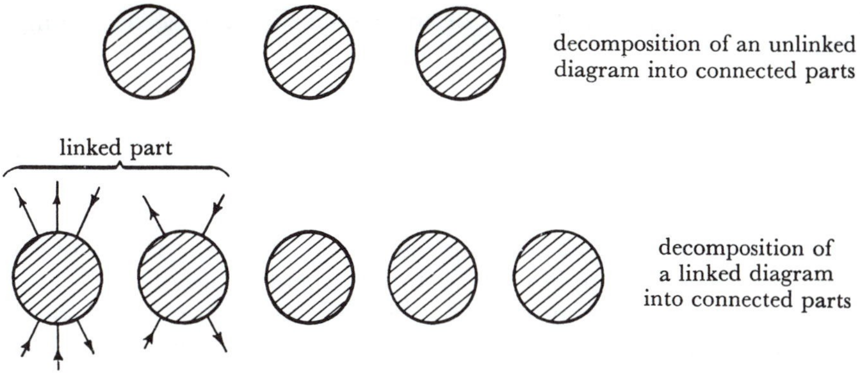

Figure 12

five are formed from a single block. We shall call the latter *connected.* Every unlinked diagram reduces to a combination of some number of *unlinked connected parts.* Similarly, a linked diagram can be decomposed into a *linked part* comprised of all the external lines, supplemented by a certain number of connected unlinked parts (Fig. 12). Let us emphasize that the linked part of a diagram must include *all* the external lines. This linked part can eventually be decomposed itself into several independent parts (see Fig. 12); such a reduction at the moment is beyond the scope of this study; at present, we shall consider the linked part as a whole, disregarding its structure.

In the interior of a single connected part, the times t_i and the wave vectors $\mathbf{k}_i$ of the different interaction vertices are coupled to each other by the propagators G_0. On the other hand, the variables t_i or $\mathbf{k}_i$ corresponding to different connected parts are totally independent. By referring to (5–21), we see that the contribution of a disconnected labeled diagram is equal to the *product* of factors arising from each of the connected parts. This conclusion must not be extended to unlabeled diagrams without some precautions. In fact, let us take a diagram made up of a linked part D_L and of n_1 unlinked connected parts $D_{c1}, \ldots, n_p$ unlinked connected parts D_{cp}. The permutations of vertices which leave the "labeled" version of this diagram invariant are of two types:

(a) Permutations within a single connected part.

(b) Permutations interchanging connected parts among themselves, whose number is

$$n_1 ! \ldots n_p ! \qquad (5\text{–}42)$$

Let $d_L, d_{c1} \ldots, d_{cp}$ be the contributions to U from the diagrams D_L,

$D_{c1}, \ldots, D_{cp}$. The contribution of the whole diagram is, according to (5–42),

$$\frac{d_L(d_{c1})^{n_1} \ldots (d_{cp})^{n_p}}{n_1 ! \ldots n_p !} \tag{5–43}$$

The combination of unlabeled connected diagrams is therefore not done by simple multiplication, in contrast to the labeled diagrams.

Let us group together all diagrams having the same linked part D_L; we must sum over the number n_i of each of the unlinked connected parts D_{ci}. We thus obtain, from (5–43), a contribution

$$d_L \prod_i \sum_{n_i} \frac{(d_{ci})^{n_i}}{n_i !} = d_L \exp\left\{\sum_i d_{ci}\right\} \tag{5–44}$$

(5–44) leads us at once to the fundamental relation

$$\boxed{U = U_L \exp(U_{0c})} \tag{5–45}$$

where U_L is an *operator* which sums the contributions to U from all the *linked* parts and U_{0c} a *scalar*, the sum of the contributions to $\langle 0|U|0\rangle$ from all *unlinked connected* diagrams.

Relation (5–45) is of major importance. As a matter of fact, we shall see in a moment that terms including n unlinked connected parts are proportional to Ω^n (Ω is the volume of the system). If Ω is large, the expansion diverges and we cannot be satisfied with the first terms. Fortunately, (5–45) resolves the difficulty by summing the whole series. In practice, the divergent exponential factor will disappear from all observable physical properties. Without (5–45), perturbation theory cannot be used.

Note: In writing (5–45), we assume that the diagram of order 0 (corresponding to the first term of the expansion of U) is a linked diagram and is not an unlinked connected diagram.

Relation between the different terms of U and the volume Ω of the system. From rules (5–35), a term of order p of U contains a factor $1/\Omega^p$, arising from the p factors V. Actually, the volume Ω also appears in an implicit way in the summations over the wave vectors $\mathbf{k}_i$,

$$\sum_{\mathbf{k}} = \sum_{\sigma} \frac{\Omega}{(2\pi)^3} \int d^3k \tag{5-46}$$

To know how the contribution from a diagram depends on Ω, we must therefore count the number of *independent* variables $\mathbf{k}_i$.

Let us first consider an unlinked connected diagram; it contains $2p$ lines, and thus $2p$ variables $\mathbf{k}_i$. Each vertex gives a relation between these $\mathbf{k}_i$, expressing conservation of momentum. Actually, the last of these relations results from the $(p - 1)$ preceding ones (the total wave vector before the first interaction and after the last is zero, its conservation being automatic). There are therefore $(p + 1)$ independent variables; the contribution from the diagram is proportional to Ω. Note that

$$< 0 \mid U \mid 0 > = \exp(U_{oc}) \tag{5-47}$$

depends exponentially on the volume. This result is not at all surprising. In fact, if we place next to each other macroscopic fractions of the system, say, 1 cm^3, we can neglect "contact effects"; the wave function is simply the product of the wave functions of each cm^3. $\langle 0|U|0\rangle$ will therefore be the product of equal factors corresponding to each volume element; it must vary exponentially with Ω.

Let us now turn to the study of U_L. Here the situation is more complex. In general, a linked diagram can be reduced to several independent parts ("disconnected"), according to the scheme of Fig. 13. Let us consider one of these parts, having q entering lines and q emerging lines whose wave vectors, given in advance, must sum to zero. There are $(2p - q)$ interior lines, and $(p - 1)$ independent conservation relations; hence a result proportional to $\Omega^{(1-q)}$. For example, diagrams 14a and 14b are independent of Ω, whereas Fig. 14c is of order $1/\Omega$. We shall see later that these conclusions apply directly to the Green's functions. They allow us to distinguish between intensive and extensive properties.

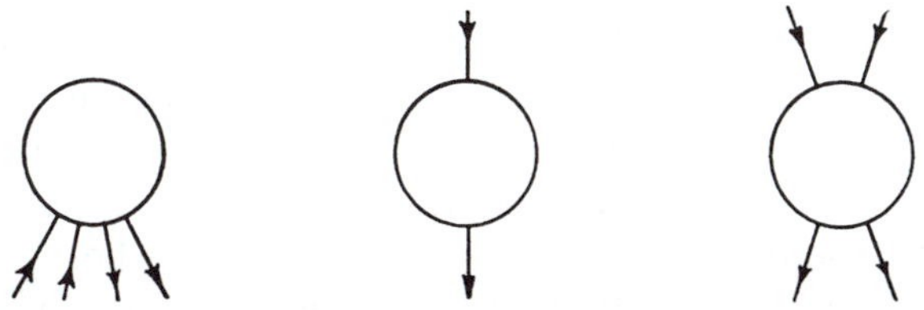

Figure 13

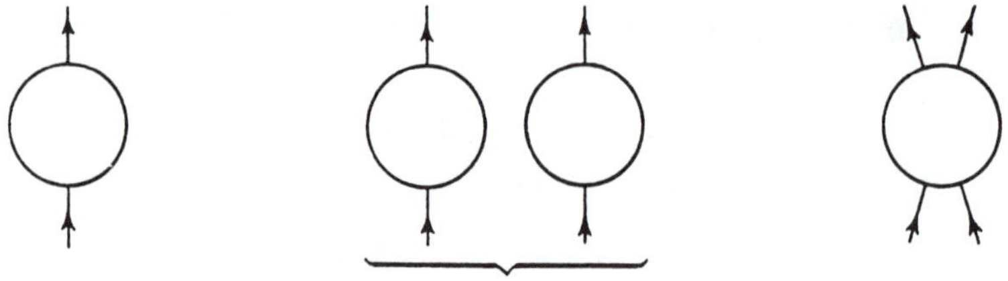

(a) (b) (c)

Figure 14

2. Adiabatic Hypothesis—Perturbed Ground State

The preceding considerations are of a formal nature. In order to draw practical conclusions from them, it is necessary to add a physical hypothesis. We shall assume that the interaction, zero when $t \to -\infty$, is turned on slowly up to its actual value at $t = 0$. We add to the actual interaction Hamiltonian H_1 a factor $e^{\eta t}$, where η is a very small positive constant, which we shall eventually let tend to 0; the interaction is thus established "adiabatically."

When $t \to -\infty$ the "instantaneous" eigenstates of the system are those of H_0. Thanks to the choice of the interaction representation, the corresponding state vectors are constant. Let us consider a system for which $t = -\infty$ is in the state $|n\rangle$. When t increases, $H_1 e^{\eta t}$ becomes appreciable and the state vector changes and takes on more and more complex values $|\varphi_{\eta n}(t)\rangle$. At time $t = 0$, we obtain a state $|\varphi_{\eta n}(0)\rangle$ of the *real system*. We are going to show that, when $\eta \to 0$, this state tends toward an eigenstate of the real system, whose energy we shall calculate. We thus obtain the one-to-one correspondence between perturbed and unperturbed states discussed in Chap. 1; the ground state $|0\rangle$ is transformed into a state $|\varphi_0\rangle$ which we *assume* to be the perturbed ground state; the excited states give rise to a spectrum of "quasi particles" and "quasi holes," characteristic of a *normal* system, having the same Fermi surface as the noninteracting gas.

For these conclusions to correspond to physical reality and not to a mathematical fiction, the state $|\varphi_0\rangle$ thus obtained must be the ground state. It must then remain stable with respect to any deformation whatever. This considerably restricts the field of application of perturbation methods. In practice, several types of instability are possible. It can happen that the energy of certain quasi particles is less than that of quasi holes; we then have spontaneous formation of "particle-hole" pairs, which amounts to an instability of the Fermi surface (see the end of this Chapter). In other cases, the instability arises from collective modes (phonons, for example); this is what occurs in superfluid gases (see Chap. 7). We limit ourselves in this chapter to the study of simple systems for which $|0\rangle$ adiabatically generates the ground state $|\varphi_0\rangle$. We shall see that this hypothesis is essential to the mathematical development of our theory.

We propose to study the state vector $|\varphi_{\eta 0}(0)\rangle$ as a function of η. We write

$$|\varphi_{\eta 0}(0)\rangle = U_\eta(0, -\infty)\,|0\rangle \tag{5-48}$$

where U_η is the transformation function corresponding to the interaction $H_1 e^{\eta t}$ [we reserve the symbol $U(t',t)$ for the physical case $\eta = 0$]. In order to characterize $|\varphi_{\eta 0}(0)\rangle$, we project it on the different unperturbed states $|n\rangle$. Each component $\langle n|\varphi_{\eta 0}(0)\rangle$ is given by the whole set of diagrams, connected or not, having external lines corresponding to the state $|n\rangle$, directed upward [see Eq. (5–39) and (5–41)]. These diagrams can extend over all negative values of time. Using (5–45), we find

$$|\varphi_{\eta 0}(0)\rangle = U_{\eta L}(0, -\infty)\,|0\rangle \exp\left[U_{\eta 0c}(0, -\infty)\right] \tag{5-49}$$

(let us recall that $U_{\eta L}$ and $U_{\eta 0c}$ arise, respectively, from linked diagrams and from unlinked connected diagrams).

When η is different from zero, expression (5–49) is well defined: with each interaction vertex, at time t_i, is associated a factor $e^{\eta t_i}$ which assures the convergence of the integration

$$\int_{-\infty}^{0} dt_i$$

This is no longer so when $\eta \to 0$. Those integrations which do not converge give a result of order $1/\eta$. This leads us to study in more detail the extension in time of the various diagrams.

a. The extension in time of the diagrams

Let us consider any connected diagram (linked or unlinked), and let us separate it into two parts, related by a certain number of ascending lines $\mathbf{k}_i$ and of descending lines $\mathbf{k}'_i$ (Fig. 15). Let us displace the lower half downward by an amount δt without changing the structure of parts A and B. We assume η to be small, so that $\eta\,\delta t \ll 1$. The contribution from the diagram changes by a factor

$$\exp\left[i\,\delta t\left(\sum_i \varepsilon^0_{k'_i} - \sum_i \varepsilon^0_{k_i}\right)\right] \tag{5-50}$$

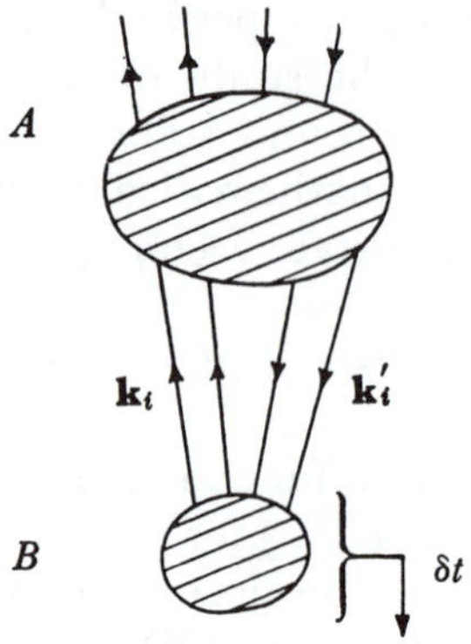

Figure 15

A priori, the separation of parts A and B does not seem to diminish the contribution from the diagram. Actually, in most cases, at least one of the variables $\mathbf{k}_i$ or $\mathbf{k}_i'$ is subject to a summation over all reciprocal space. In this case, the frequency of the phase factor (5–50) varies continuously over a certain finite interval. When the sum over the $\mathbf{k}_i$ and the $\mathbf{k}_i'$ is taken, these phase factors interfere and give a result close to zero as soon as δt is large enough for the phases to be incoherent. By repeating this reasoning for all other groups of interior lines, we conclude that only diagrams *localized in time* contribute.

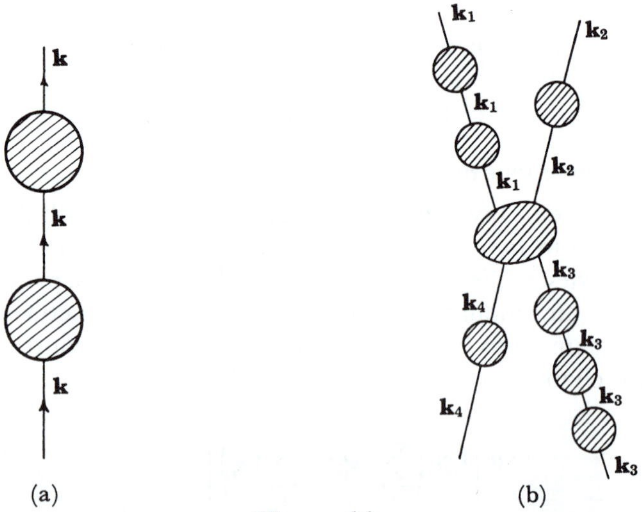

Figure 16

The only exception to this rule arises when two parts of a diagram are connected by lines all of whose wave vectors are fixed by momentum conservation; the frequency of the phase factor (5–50) is then unique, and there is no destructive interference. These "pathological" cases all come from the diagram shown in Fig. 16*a*; we indicate in Fig. 16*b* a more complex combination all of whose elements can be separated from one another by an arbitrarily large amount.

Let us take an unlinked connected diagram; it is necessarily localized in time, of average extent Δ, where Δ plays the role of a characteristic correlation time for the physical system. Only its average time

$$t_D = \frac{1}{p}[t_1 + \cdots + t_p] \tag{5-51}$$

can vary from $-\infty$ to 0; all the t_i must be located in the region $t_D \pm \Delta$. Let $q_\eta(t_D)$ be the contribution from all diagrams having a fixed t_D (obtained by inserting a factor

$$\delta\left(t_D - \frac{t_1 + \cdots + t_p}{p}\right)$$

in the rules (5–35).) If $\eta\Delta \ll 1$, the coupling constant $g = e^{\eta t}$ remains practically constant over the whole extent of the diagram, equal to $e^{\eta t_D}$. We can therefore write

$$q_\eta(t_D) = q_0(g) + O(\eta) \tag{5-52}$$

where $q_0(g)$ is obtained by associating with each vertex the same interaction gH_1, independent of time. It is easily verified that $q_0(g)$ no longer depends on the time t_D; thus, instead of introducing the δ function and integrating over all t_i, it is sufficient to integrate over all t_i except one.

Note: We have been careful, in the definition of q_0, to fix the *average* time of the diagram t_D; we thus preserve the symmetry between different vertices, and the notion of an unlabeled diagram still makes sense. We could, on the other hand, have fixed the time of one of the vertices, for example, t. The preceding discussion remains valid, but it is necessary to pay careful attention to the counting of diagrams. By fixing t_1, we "mark" the corresponding vertex. Two approaches are possible:

(a) We preserve the rules (5–35). We must then assume that two diagrams which differ only by the choice of the fixed vertex are actually identical.

(b) If we prefer to specify the fixed vertex, it is necessary to add to the prescription (5–35) a factor $1/p$ (where p is the order of the diagram).

We shall return to this question in section 5 of the present chapter.

Let us return to the expansion of $U_{\eta 0c}$, and let us separate out explicitly the integration over t_D, by setting

$$U_{\eta oc}(0, -\infty) = \int_{-\infty}^{0} dt_D\, f(t_D)$$

If $t_D \ll -\Delta$, the different vertices of the diagram never reach the limit $t = 0$ and as a consequence $f(t_D) = q_\eta(t_D)$. This is no longer so when $|t_D| \lesssim \Delta$; there appear corrections due to the limits of integration, which we shall call in general "boundary effects." This correction, arising only in a finite interval of t_D, gives a result b_1 independent of η; hence

$$U_{\eta oc}(0, -\infty) = \int_{-\infty}^{0} q_\eta(t_D)\, dt_D + b_1 \tag{5-53}$$

Let us go back to (5–52); since the integration over t_D runs over an interval of order $1/\eta$, the correction $0(\eta)$ leads to a new contribution b_2 independent of η. In contrast, $q_0(g)$ gives a "divergent" term of order $1/\eta$. We finally obtain

$$\begin{cases} U_{\eta oc}(0, -\infty) = a/\eta + b \\ b = b_1 + b_2 \\ a = \displaystyle\int_0^1 \frac{dg'}{g'}\, q_0(g') \end{cases} \tag{5-54}$$

The coefficient a is pure imaginary. In fact, $U_\eta(0, -\infty)$ is a unitary operator. According to (5–45), we must have

$$U^*_{\eta L}(0, -\infty)\, U_{\eta L}(0, -\infty) \exp\left(\frac{a + a^*}{\eta} + b + b^*\right) = 1$$

We shall see later that $U_{\eta L}$ is regular when $\eta \to 0$. The above relation therefore implies that $a + a^* = 0$. This property expresses the stability of the ground state and is no longer true for an excited state. The coefficient b, on the other hand, can have a real part.

In summary, the set of all unlinked diagrams gives a contribution

$$U_{\eta o}(0, -\infty) = \exp(U_{\eta oc}) = \exp(a/\eta + b) \tag{5-55}$$

which contains a divergent phase factor when $\eta \to 0$. We shall later see the physical consequences of this phenomenon.

It is interesting to compare (5–54) with the expression for $U_{0c}(t',t)$, calculated by taking $\eta = 0$. The diagrams have the same structure as for $U_{\eta 0c}(0, -\infty)$, but the integrations over t_D [Eq. (5–53)] are now limited by construction to the interval (t,t') and no longer need an adiabatic factor for convergence. If $(t' - t) \gg \Delta$, boundary effects give a relatively unimportant contribution b' and we can write, by analogy with (5–53),

$$U_{oc}(t', t) = b' + q_0(1)\,(t' - t) = b' + \left[g\,\frac{da}{dg}\right]_{g=1}(t' - t) \qquad (5\text{–}56)$$

Again, we see that the phase of $U_0(t',t)$ does not have a limit when $(t' - t) \to \infty$. Note that (5–56) involves the derivative da/dg; this is normal, since the adiabatic turning-on amounts to integrating over all values of the coupling constant g.

Let us now turn to the study of linked diagrams of the type indicated in Fig. 17. As in Fig. 16*b*, we isolate the "insertions" carried by each of the $2q$ external lines. We thus obtain a core A, whose position we fix by the coordinate t_A of some vertex. This core is necessarily localized. From the core extend q ascending "dressed" propagators, of wave vectors $\mathbf{k}_i > k_F$, attached at points $t_A + \delta_i$, and q descending propagators, of wave vectors $\mathbf{k}'_j < k_F$, attached at points $t_A + \delta'_j$. For a given structure of A, we must calculate the integral

$$\int_{-\infty}^{0} A(g)\, Q(\mathbf{k}_1, t_A + \delta_1) \cdots Q(\mathbf{k}'_q, t_A + \delta'_q)\, dt_A \qquad (5\text{–}57)$$

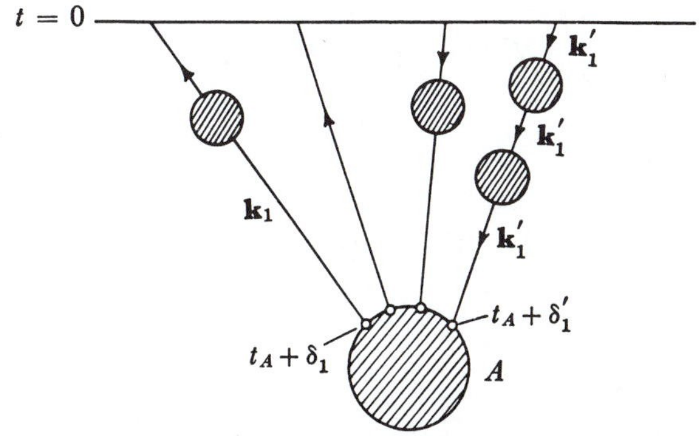

Figure 17

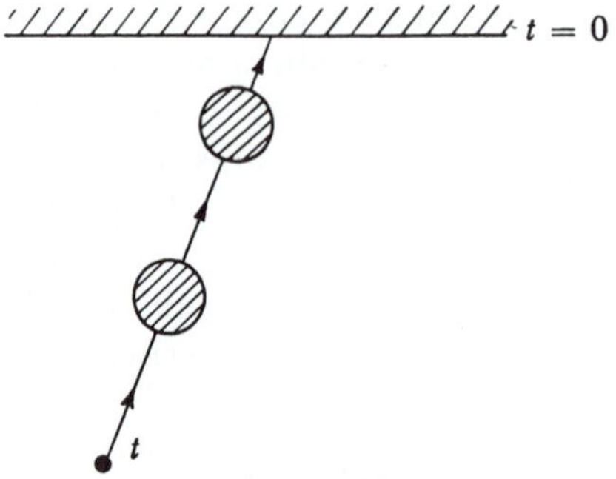

Figure 18

where $g = e^{\eta t_A}$, $A(g)$ is the contribution of the core A, and $Q(k,t)$ is given by the set of diagrams of Fig. 18, having an arbitrary number of insertions. In writing (5–57), we neglect "boundary effects," occurring when t_A is small. The corresponding corrections are independent of η when $\eta \to 0$; they will cause us no trouble.

We shall see later that $Q(k,t)$ has the same asymptotic form as the Green's function $G(k,t)$. More precisely, $Q(k,t)$ is of the form

$$Q(\mathbf{k}, t) = f(\mathbf{k}, g) \exp[\pm \mathrm{i}\varepsilon_{\mathbf{k}}(g)\, t] \tag{5–58}$$

where $\varepsilon_{\mathbf{k}}(g)$ is the energy of the elementary excitation $\mathbf{k}$ for the value g of the coupling constant. The signs $\pm$ correspond, respectively, to $k \gtrless k_F$. Here again, we neglect the boundary effects which arise for small t. We see in (5–58) a rapid oscillation superposed on a slow variation due to the time dependence of g.

We can now write (5–57) in the form

$$\int_{-\infty}^{0} F(g) \exp[\mathrm{i}w(g)\, t_A]\, \mathrm{d}t_A \tag{5–59}$$

where we have defined

$$\begin{cases} F(g) = A(g)\, f(\mathbf{k}_1, g) \cdots f(\mathbf{k}'_q, g) \exp[\mathrm{i}(\varepsilon_{\mathbf{k}_1}\, \delta_1 \cdots - \varepsilon_{\mathbf{k}'_q}\, \delta'_q)] \\ w(g) = \sum\limits_{i=1}^{q} \varepsilon_{\mathbf{k}_i}(g) - \sum\limits_{j=1}^{q} \varepsilon_{\mathbf{k}'_j}(g) \end{cases} \tag{5–60}$$

$w(g)$ is just the *excitation energy* of the state containing the q quasi particles $\mathbf{k}_i$ and the q quasi holes $\mathbf{k}'_j$, for the value g of the coupling constant.

According to our hypothesis, $|\varphi_0(g)\rangle$ is nondegenerate. $w(g)$ is therefore positive and can *never be zero*. This simple fact is the key to the perturbation calculation; it is because of this that the linked diagrams give a regular contribution when $\eta \to 0$.

Let us return to Eq. (5–59). The integral is always defined, since F decreases exponentially as $t_A \to -\infty$. We can expand it in powers of η, using the fact that g varies slowly with t_A. We thus obtain

$$\int F(g) \exp[iw(g)\, t_A]\, dt_A = \frac{F(g)}{iw(g)} \exp[iw(g)\, t_A] + O(\eta) \tag{5–61}$$

The expansion (5–61) *is valid only if* $w(g) \neq 0$. Under this condition (5–59) tends to a finite limit as $\eta \to 0$, equal to

$$\frac{F(1)}{iw(1)} \tag{5–62}$$

(5–62) is *false* if $w(g)$ goes through zero at any point. (We shall study this other case in detail in the discussion of excited states.)

Once the integration over t_A has been performed, we can integrate over the internal variables of the core A and sum over all possible cores; the result is regular as $\eta \to 0$. As a consequence, the vector $U_{\eta L}(0, -\infty)|0\rangle$ has a well-defined limit for $\eta \to 0$. This basic property assumes that $|\varphi_0(g)\rangle$ is nondegenerate and thus is rigorously stable. It is no longer true for an excited state.

b. The perturbed ground state

Let us return to (5–49). According to our conjectures, $|\varphi_{\eta 0}(0)\rangle$ must tend to the perturbed ground state $|\varphi_0\rangle$ as $\eta \to 0$. Actually, the limit is not defined, because of the divergent term a/η present in $U_{\eta 0c}$. This divergence is limited to a constant phase factor, which has no physical meaning. The "structure" of the state, given by $U_{\eta L}$, is perfectly regular. We shall therefore write

$$|\varphi_0\rangle = C \lim_{\eta \to 0} \frac{U_\eta(0, -\infty)\,|0\rangle}{\langle 0|\, U_\eta(0, -\infty)\,|0\rangle} = C U_{\eta L}(0, -\infty)\,|0\rangle \tag{5–63}$$

where C is a normalization constant. The passage to the limit $\eta \to 0$ is now made without difficulty, the divergent phase factor having disappeared. According to (5–55), C is given by

$$|C|^2 = |U_{\eta 0}(0, -\infty)|^2 = \exp(b + b^*) \tag{5–64}$$

We know that $C < 1$. As a matter of fact, C depends exponentially on the volume Ω, which makes it extremely small for a macroscopic system. Let us note that this artifice will not work if the singular term a/η contains a real part; in such a case, C will not have a limit as $\eta \to 0$, and we no longer know how to define $|\varphi_0\rangle$ explicitly.

We still have to prove that (5–63) represents an eigenstate of the real system, and to calculate the corresponding energy. For this purpose let us go back to (5–48), and let us replace U_η by its series expansion (5–14); we find

$$U_\eta(0, -\infty) \,|\, 0> = \sum_{p=0}^{\infty} \frac{(-\mathrm{i})^p}{p\,!} \int_{-\infty}^{0} \mathrm{d}t_1 \ldots \mathrm{d}t_p \; P[H_\mathrm{I}(t_1) \ldots H_\mathrm{I}(t_p)] \exp[\eta(t_1 + \cdots + t_p)] \,|\, 0> \tag{5-65}$$

Let us now apply to the state vector (5–65) the operator $(H_0 - E_0^0)$, where E_0^0 is the ground-state energy of the unperturbed system. We clearly have

$$(H_0 - E_0^0)\, U_\eta(0, -\infty) \,|\, 0> = [H_0, U_\eta(0, -\infty)] \,|\, 0> \tag{5-66}$$

In order to evaluate the commutator in (5–66), we rely on the identity

$$[H_0, A_1 \ldots A_p] = \sum_{i=1}^{p} A_1 \ldots A_{i-1}[H_0, A_i]\, A_{i+1} \ldots A_p \tag{5-67}$$

Furthermore if we note that by definition

$$[H_0, H_\mathrm{I}(t)] = -\mathrm{i} \frac{\mathrm{d}}{\mathrm{d}t} H_\mathrm{I}(t) \tag{5-68}$$

we can put (5–66) in the form

$$(H_0 - E_0^0)\, U_\eta(0, -\infty) \,|\, 0> = \sum_{p=1}^{\infty} \frac{(-\mathrm{i})^{p+1}}{p\,!} \int \mathrm{d}t_1 \ldots \mathrm{d}t_p \exp[\eta(t_1 + \cdots + t_p)] \times \sum_{j=1}^{p} P\left[H_\mathrm{I}(t_1) \ldots \frac{\mathrm{d}}{\mathrm{d}t_j} H_\mathrm{I}(t_j) \ldots H_\mathrm{I}(t_p)\right] |\, 0> \tag{5-69}$$

Let us first note that all the t_j enter symmetrically; we can reduce the summation to the term $j = 1$ and multiply the result by p. On the other hand,

$$P\left[\frac{\mathrm{d}H_\mathrm{I}(t_1)}{\mathrm{d}t_1} \ldots H_\mathrm{I}(t_p)\right] = \frac{\mathrm{d}}{\mathrm{d}t_1} P[H_\mathrm{I}(t_1) \ldots H_\mathrm{I}(t_p)] \tag{5-70}$$

(the chronological product $P[\ldots]$ being continuous when $t_1 = t_i$).

We can then integrate (5–69) by parts with respect to t_1,

$$(H_0 - E_0^0)\, U_\eta(0, -\infty)\,|\,0> = \\ -\sum_{p=1}^{\infty} \frac{(-\mathrm{i})^{p-1}}{(p-1)!} \int_{-\infty}^{0} \mathrm{d}t_2 \ldots \mathrm{d}t_p \exp[\eta(t_2 + \cdots + t_p)] \\ \times \left\{ \exp(\eta t_1)\, P[H_\mathrm{I}(t_1) \ldots H_\mathrm{I}(t_p)] \right\}_{t_1=-\infty}^{t_1=0} |\,0> \\ + \sum_{p=0}^{\infty} \mathrm{i}p\eta \frac{(-\mathrm{i})^p}{p\,!} \int_{-\infty}^{0} \mathrm{d}t_1 \ldots \mathrm{d}t_p \exp[\eta(t_1 + \cdots + t_p)]\, P[H_\mathrm{I}(t_1) \ldots H_\mathrm{I}(t_p)]\,|\,0> \tag{5–71}$$

The integrated term in (5–71) is just $-H_\mathrm{I} U_\eta(0, -\infty)|0\rangle$. As a consequence, we find

$$(H - E_0^0)\, U_\eta(0, -\infty)\,|\,0> = \\ \sum_{p=0}^{\infty} \mathrm{i}p\eta \frac{(-\mathrm{i})^p}{p\,!} \int_{-\infty}^{0} \mathrm{d}t_1 \ldots \mathrm{d}t_p \exp[\eta(t_1 + \cdots + t_p)]\, P\,[H_\mathrm{I}(t_1) \ldots H_\mathrm{I}(t_p)]\,|\,0> \tag{5–72}$$

where $H = H_0 + H_\mathrm{I}$ is the complete Hamiltonian.

The second term of (5–72) differs from $U_\eta(0, -\infty)|0\rangle$ only by the presence of the factor $i\eta p$ in the expansion. In order to sum this series, we write the interaction H_I in the form

$$\begin{cases} H_\mathrm{I} = [H_\mathrm{I}(g)]_{g=1} \\ H_\mathrm{I}(g) = gH_\mathrm{I} \end{cases} \tag{5–73}$$

where g is the coupling constant. The factor p in (5–72) can be interpreted as resulting from the operation $g(d/dg)$. This allows us to write (5–72) in the much more compact form

$$(H - E_0^0)\, U_\eta(0, -\infty)\,|\,0> = \mathrm{i}\eta g \frac{\mathrm{d}}{\mathrm{d}g} U_\eta(0, -\infty)\,|\,0> \Big|_{g=1} \tag{5–74}$$

This basic relation was established by Gell-Mann and Low.

When $\eta \to 0$, we are tempted to neglect the second member of (5–74). Actually, we do not have the right to do this, as $U_\eta(0, -\infty)|0\rangle$ contains a divergent phase factor, whose derivative with respect to g is of order $1/\eta$. In order to avoid this difficulty, we go back to (5–63); the ground state $|\varphi_0\rangle$ is actually the limit for $\eta \to 0$ of the quantity

$$|\,\varphi_{\eta 0}> = \exp(-a/\eta)\, U_\eta(0, -\infty)\,|\,0>$$

which does not exhibit any singularity. Let us write (5–74) in the form

$$(H - E_0^0) \mid \varphi_{\eta 0} > = \mathrm{i}\eta g \frac{\mathrm{d}}{\mathrm{d}g} \mid \varphi_{\eta 0} > \Big|_{g=1} + \mathrm{i}g \frac{\mathrm{d}a}{\mathrm{d}g} \mid \varphi_{\eta 0} > \Big|_{g=1} \qquad (5\text{–}75)$$

We can now neglect the term of order η when $\eta \to 0$, which gives us

$$\begin{cases} H \mid \varphi_0 > = E_0 \mid \varphi_0 > \\ E_0 = E_0^0 + \mathrm{i}g \dfrac{\mathrm{d}a}{\mathrm{d}g} \Big|_{g=1} \end{cases} \qquad (5\text{–}76)$$

(5–76) demonstrates that $|\varphi_0\rangle$ is an eigenstate of the real system, of energy E_0. Again, it is necessary to assume that $|\varphi_0\rangle$ is actually the ground state in order for the perturbation method to make sense.

The interaction H_I thus displaces the ground-state energy by a quantity

$$\Delta E_0 = E_0 - E_0^0 = \mathrm{i}g \frac{\mathrm{d}a}{\mathrm{d}g} \Big|_{g=1} \qquad (5\text{–}77)$$

By referring to (5–54), we see that

$$\Delta E_0 = \mathrm{i}q_0(1) \qquad (5\text{–}78)$$

We thus obtain a practical method for calculating ΔE_0,

(5–79)
- Take all unlinked connected unlabeled diagrams, and evaluate them according to the prescription (5–35).
- Integrate all the t_i except *one* from $-\infty$ to $+\infty$.
- Add a factor i to the result.

We can give a more explicit form to (5–79) by referring to the general formula (5–14); we find

$$\Delta E_0 = \sum_{p=1}^{\infty} \frac{(-\mathrm{i})^{p-1}}{p\,!} \int_{-\infty}^{+\infty} \mathrm{d}t_2 \dots \mathrm{d}t_p < 0 \mid P[H_\mathrm{I}(t_1) \dots H_\mathrm{I}(t_p)] \mid 0 >_c \qquad (5\text{–}80)$$

where the index c indicates that we are limited in the expansion to connected diagrams. We obtain a still more compact relation, but a more formal one, by writing

$$\Delta E_0 = \mathrm{i} < 0 \mid U(+\infty, -\infty)\, \delta(t_1) \mid 0 >_c = \frac{\mathrm{i} < 0 \mid U(+\infty, -\infty)\, \delta(t_1) \mid 0 >}{< 0 \mid U(+\infty, -\infty) \mid 0 >} \qquad \textit{(5–81)}$$

In practice (5–81) contributes only an illusory simplification, since, in order to make sense of it, it is necessary to return to (5–80).

We shall see in section 5 various other expressions for ΔE_0, which we shall relate to one another by manipulating diagrams. For the moment, let us note that ΔE_0, given by unlinked connected diagrams, is proportional to the volume Ω. As anticipated, this is an extensive quantity. On the other hand, the wave function $|\varphi_0\rangle$ depends exponentially on the volume.

Before leaving this discussion of the ground state, let us return to (5–56). Expressing $q_0(1)$ with the help of (5–78), we find

$$U_{0c}(t', t) = b' - \mathrm{i}\Delta E_0(t' - t) \qquad \textit{(5–82)}$$

From (5–82) we can obtain the average value of $U(t',t)$ in the state $|0\rangle$,

$$< 0 \mid U(t', t) \mid 0 > = \exp[U_{0c}(t', t)] = \exp[b'] \cdot \exp[-\mathrm{i}\Delta E_0(t' - t)] \qquad \textit{(5–83)}$$

We see that ΔE_0 is just the frequency at which $\langle 0|U|0\rangle$ oscillates for large time intervals. This result can be established directly, without appeal to perturbation methods.

c. The S matrix—reversibility of the adiabatic switching on

Until now we have studied the adiabatic switching on of the interaction H_1 between time $-\infty$ and 0. Let us now assume that we decrease H_1 progressively from $t = 0$ to $t = +\infty$, taking $H_1 e^{-\eta|t|}$ as the instantaneous interaction. What is the state for $t = +\infty$ when the system started from $|0\rangle$ at time $t = -\infty$? This state is simply $U_\eta(+\infty, -\infty)|0\rangle$. The operator $U(+\infty, -\infty)$ is commonly called the *S matrix* of the system. Likewise, we shall put $U_\eta(+\infty, -\infty) = S_\eta$. According to (5–45), we have

$$S_\eta \mid 0 > = U_{\eta L}(+\infty, -\infty) \mid 0 > \exp[U_{\eta 0c}(+\infty, -\infty)] \qquad \textit{(5–84)}$$

We plan to discuss the structure of $U_{\eta L}$ and $U_{\eta 0c}$.

Let us first consider the linked part $U_{\eta L}$. We immediately set aside the term of order zero, equal to 1, and keep only terms containing at least one vertex. The corresponding diagrams have the same structure as that in Fig. 17, except that the exterior lines extend indefinitely upward. We can reproduce point by point the line of argument which led us from (5–57) to (5–59). The contributions from the linked diagrams can therefore be put in the form

$$\int_{-\infty}^{+\infty} F(g) \exp\left[iw(g)\, t_A\right] dt_A \qquad \textit{(5–85)}$$

[where we have used the same notation as for (5–59).] We know that, within the framework of a perturbation method, $w(g)$ is always >0, which allows us to expand the integral (5–85) in powers of η. The first term of this expansion is zero, the intervals $(-\infty,0)$ and $(0,+\infty)$ exactly compensating each other. The contribution from linked diagrams of order $\geqslant 1$ thus tends to zero when $\eta \to 0$, and we can write

$$\lim_{\eta\to 0} U_{\eta L}(+\infty, -\infty)\,|\,0> \,=\, |\,0> \qquad \textit{(5–86)}$$

Let us turn now to the unlinked connected diagrams contributing to $U_{\eta 0c}(+\infty, -\infty)$. The integration over time extends from $-\infty$ to $+\infty$; there are no longer any boundary effects. Relation (5–53) thus becomes

$$U_{\eta 0c}(+\infty, -\infty) = \int_{-\infty}^{+\infty} q_\eta(t_D)\, dt_D \qquad \textit{(5–87)}$$

Let us replace q_η by its expanded form (5–52). The main term $q_0(g)$ gives a contribution equal to $2a/\eta$, which is pure imaginary. It remains to calculate the finite contribution from the term of order η of q_η. When replacing q_η by q_0, we neglect a factor

$$\exp\left[\eta\,\{\, p\,|\,t_D\,| - |\,t_1\,| \cdots - |\,t_p\,|\,\}\right]$$

The correction of order η is thus proportional to

$$\eta\,\{\, p\,|\,t_D\,| - |\,t_1\,| \cdots - |\,t_p\,|\,\}$$

Let us consider two diagrams of opposite t_D, but otherwise identical, and let us assume that $t_D \gg \Delta$; the corrections of order η cancel exactly. We therefore rigorously have

$$\lim_{\eta\to 0} U_{\eta 0c}(+\infty, -\infty) = 2a/\eta \qquad \textit{(5–88)}$$

The correction b_2, present in (5–54), has disappeared as a result of the symmetry between positive and negative times.

Returning to (5–84), we see that

$$S_\eta \mid 0 > \; = \exp(2a/\eta) \mid 0 > \tag{5-89}$$

Up to a constant factor, we recover the ground state $|0\rangle$. (5–89) shows that the adiabatic switching on of the interaction is perfectly *reversible.* $|0\rangle$ is a *stable* state. At no time is there any transition to excited states; the deformation of the wave function is perfectly elastic. This reversibility results from our hypothesis: (5–86) is exact only if $w(g)$ is always $\neq 0$ [see Eq. (5–60)]. If $w(g)$ passed through zero, real transitions would take place, and the reversibility would be destroyed. (We shall see an example of this more complex case in the study of the excited states.)

Note the importance of Eq. (5–89); it is this which gives meaning to the perturbation method. If it were not true, the state at time $t = 0$ would depend on the way in which the interaction was turned on, that is to say, on the function $g(t)$. For example, the two functions indicated in Fig. 19 would lead to different results in the limit $\eta \to 0$. The choice of an exponential g being purely arbitrary, we could only obtain erroneous results. Because of (5–86), the "history" of the interaction is reflected only in the singular phase factor, which has no physical importance.

Equation (5–89) gives us a new expression for the real ground state $|\varphi_0\rangle$. According to (5–8),

$$\begin{aligned} U_\eta(0, -\infty) \mid 0 > \; &= U_\eta(0, +\infty)\, U_\eta(+\infty, -\infty) \mid 0 > \\ &= U_\eta(0, +\infty)\, S_\eta \mid 0 > \end{aligned} \tag{5-90}$$

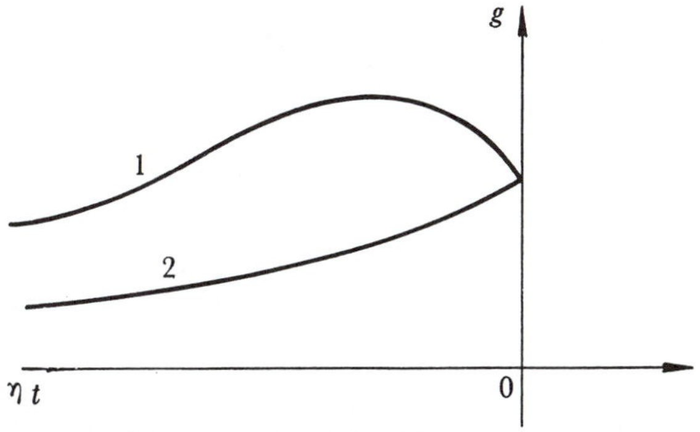

Figure 19

As a consequence, we can write

$$
\begin{aligned}
| \varphi_0 > &= \lim_{\eta\to 0} \exp(-a/\eta)\, U_\eta(0, -\infty)\, | 0 > \\
&= \lim_{\eta\to 0} \exp(a/\eta)\, U_\eta(0, +\infty)\, | 0 >
\end{aligned} \tag{5-91}
$$

We can construct $|\varphi_0\rangle$ starting either from $t = -\infty$ or from $t = +\infty$. This latitude will prove very useful for calculating the Green's functions.

We have available a convenient formalism for studying the characteristics of the ground state. What is its domain of validity? Essentially, $|0\rangle$ must give rise at each instant to the ground state corresponding to the value $g = e^{\eta t}$ of the coupling constant. We shall see later that the adiabatic switching on of the interaction does not change the Fermi surface; as a consequence, the formalism applies only to normal systems whose Fermi surface does not depend on g. This is the case for normal isotropic systems, whose Fermi surface is a sphere of radius

$$
k_F = \left(\frac{3\pi^2 N}{\Omega}\right)^{1/3} \tag{5-92}
$$

We shall see later how to generalize these methods to more complicated systems.

3. The Green's Functions

Knowing the ground state $|\varphi_0\rangle$, we are in a position to calculate the Green's functions of our system; we can thus put into practice the general discussion given in Chaps. 3 and 4 and obtain valuable information concerning the elementary excitations.

First let us consider the single-particle Green's function $G(\mathbf{k}, t - t')$, given by (3–8),

$$
G(\mathbf{k}, t - t') = i < \varphi_0 \,|\, T\{\, \mathbf{a}_\mathbf{k}(t)\, \mathbf{a}^*_\mathbf{k}(t') \,\}\, |\, \varphi_0 > \tag{5-93}
$$

In (5–93), state vectors and operators are expressed in the Heisenberg representation. In order to go back to the interaction representation, we note that

$$
| \varphi_0 > = | \varphi_0(0) > = U(0, t)\, | \varphi_0(t) > \tag{5-94}
$$

which automatically gives

$$
\mathbf{A}(t) = U(0, t)\, A(t)\, U(t, 0) \tag{5-95}
$$

When t is finite, we can replace $U(0,t)$ by

$$\lim_{\eta\to 0} U_\eta(0, t)$$

It remains to specify the state vectors $|\varphi_0\rangle$. We want to leave (5–93) in the form of a chronological product; for this reason we define the $|\varphi_0\rangle$ on the right starting from $t = -\infty$, the $\langle\varphi_0|$ on the left starting from $t = +\infty$ [Eq. (5–91)]. In this way we obtain, using relation (5–8),

$$G(\mathbf{k}, t-t') = \tag{5-96}$$
$$i \lim_{\eta\to 0} e^{-2a/\eta} \begin{cases} <0 \mid U_\eta(+\infty, t)\, a_\mathbf{k}(t)\, U_\eta(t, t')\, a^*_\mathbf{k}(t')\, U_\eta(t', -\infty) \mid 0> & \text{if } t > t' \\ -<0 \mid U_\eta(+\infty, t')\, a^*_\mathbf{k}(t')\, U_\eta(t', t)\, a_\mathbf{k}(t)\, U_\eta(t, -\infty) \mid 0> & \text{if } t < t' \end{cases}$$

(5–96) allows us to calculate G starting from the time-development operator U_η.

Let us replace the various factors U_η of (5–96) by their series expansions (5–14); the expression constitutes a chronological product, with time increasing from right to left. For the present, let us remove the factors $a_\mathbf{k}(t)$ and $a^*_\mathbf{k}(t')$; the product then reduces to

$$U_\eta(+\infty, t)\, U_\eta(t, t')\, U_\eta(t', -\infty) = S_\eta$$

Let us expand $\langle 0|S_\eta|0\rangle$ in a series, according to Eq. (5–14). The interactions located in each of the three intervals $(-\infty, t')$, (t', t), and $(t, +\infty)$ give rise, respectively, to the three factors $U_\eta(t', -\infty)$, $U_\eta(t, t')$, and $U_\eta(+\infty, t)$. To obtain (5–96), we just have to insert into this expansion the operators $a_\mathbf{k}(t)$ and $a^*_\mathbf{k}(t')$ in their proper chronological position (the permutation of a and a^* is accompanied by a change in sign). We can thus write (5–96) in a compact form,

$$G(\mathbf{k}, t-t') = i \lim_{\eta\to 0} \frac{<0 \mid T\{ S_\eta a_\mathbf{k}(t)\, a^*_\mathbf{k}(t') \} \mid 0>}{<0 \mid S_\eta \mid 0>} \tag{5-97}$$

[see Eq. (5–89)]. The T product in the numerator is a formal notation, and acquires a precise meaning only when we replace S_η by its series expansion. The propagation of the additional particle is simply superimposed on the normal development of the ground state.

The same argument yields the two-particle Green's function $K(\mathbf{k}_i, t_i)$, which is given by

$$K(\mathbf{k}_i, t_i) = \lim_{\eta\to 0} \frac{<0 \mid T\{ S_\eta a_{\mathbf{k}_1}(t_1)\, a_{\mathbf{k}_2}(t_2)\, a^*_{\mathbf{k}_3}(t_3)\, a^*_{\mathbf{k}_4}(t_4) \} \mid 0>}{<0 \mid S_\eta \mid 0>} \tag{5-98}$$

where the T product is defined as in (5–97).

a. Expression of the Green's functions in terms of diagrams

The chronological product (5–96) is subject to decomposition by Wick's theorem. Let us represent the operators $a_{\mathbf{k}}(t)$ and $a^*_{\mathbf{k}}(t')$ by points P and P'. The numerator of (5–97) is represented by the set of all diagrams having one open line coming from P' and ending at P (Fig. 20). The vertices of these diagrams can be moved from $t = -\infty$ to $t = +\infty$, without restriction. It is important not to confuse these diagrams with those which appear in the calculation of

$$< 0 \mid b_{\mathbf{k}} U(t', t)\, b^*_{\mathbf{k}} \mid 0 >$$

in this latter case, the vertices are all restricted to the interval (t,t') (see Fig. 11).

The diagrams contributing to the numerator of (5.97) can contain an arbitrary number of unlinked connected parts. These contribute a factor which is just the average value $\langle 0|S|0\rangle$. We thus eliminate the denominator of (5–97), which becomes

$$G(\mathbf{k}, t - t') = \mathrm{i} < 0 \mid T\left(S a_{\mathbf{k}}(t)\, a^*_{\mathbf{k}}(t')\right) \mid 0 >_L \qquad \textit{(5–99)}$$

(the index L means that we are limited to linked diagrams in the expansion). We see that the divergent phase factor has disappeared; the calculation of G thus becomes very simple.

We can similarly decompose Eq. (5–98) into diagrams; $K(\mathbf{k}_i,t_i)$ is given by the set of all *linked* diagrams having two incoming lines proceeding from P_3 and P_4 and two outgoing lines terminating at P_1 and P_2

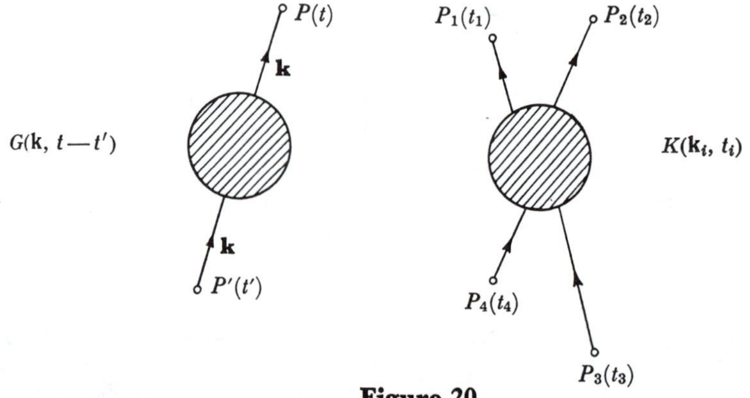

Figure 20

[the points $P_1 \ldots P_4$ represent, respectively, the operators $a_{\mathbf{k}_1}(t_1) \ldots a^*_{\mathbf{k}_4}(t_4)$]. Here again, the vertices of the diagram can be placed anywhere. This graphical representation will allow a simple analysis of the structure of K.

We have not yet specified the rules giving the contribution of each diagram to G or K. We can easily deduce them from (5–35). $G(k, t - t')$ is obtained by the following set of operations:

(5–100)

(a) Take all unlabeled linked diagrams having a single external line, of wave vector $\mathbf{k}$, starting from $P'(t')$, ending on $P(t)$. With each vertex and with each line associate the factors $V(\mathbf{k}_i)$ and G_0 indicated in (5–35).

(b) Multiply the term by $(-1)^l i^p / r 2^m$ [see the definitions of r, l, m, p in (5–35)].

(c) Integrate over the times of all vertices from $-\infty$ to $+\infty$, and sum over the wave vectors of all internal lines.

The sign $(-1)^l$ can be justified in the same way as in (5–37). To zeroth order, the diagram reduces to a line connecting P' to P (Fig. 21*a*); The corresponding contribution is equal to $G_0(k, t-t')$, which was obvious a priori. To zeroth order, propagation can take place only in a single sense (positive if $k > k_F$, negative if $k < k_F$); in contrast, the true propagator G exists for either sign of $(t - t')$ (Fig. 21*b*).

The rules for the calculation of K are identical to (5–100). The only "tricky" point concerns the sign we must give to the result. We can verify without difficulty that the sign is $(-1)^l$ if the open line starting at P_3 ends on P_1 but that it becomes $(-1)^{l+1}$ if this line ends on P_2. There is a small complication here which we must not forget.

We have seen that a linked diagram can be decomposed into several disconnected parts. This leads us to separate out of $K(\mathbf{k}_i,t_i)$ the terms corresponding to diagrams 22*a* and *b*. The contribution of these diagrams

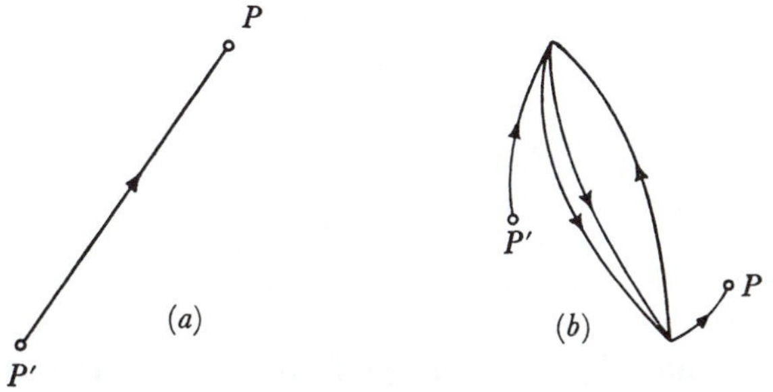

Figure 21

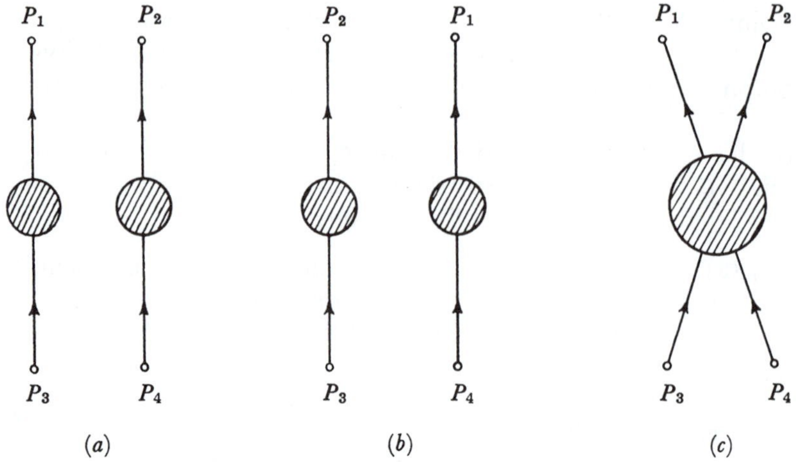

Figure 22

to K is equal to

$$G(1, 3)\, G(2, 4)\, \delta_{\mathbf{k}_1\mathbf{k}_3}\, \delta_{\mathbf{k}_2\mathbf{k}_4} - G(1, 4)\, G(2, 3)\, \delta_{\mathbf{k}_1\mathbf{k}_4}\, \delta_{\mathbf{k}_2\mathbf{k}_3}$$

This is just the *free part* of the two-particle Green's function, defined in Chap. 3. The rest of K, which we have called the *bound part*, is given by the *connected* diagrams, of Fig. 22*c*. We thus give simple physical support to the somewhat arbitrary decomposition carried out in Chap. 3.

Finally, the general rules established in the preceding section (Fig. 14) show that G is independent of the volume Ω, whereas the bound part of K is of order $1/\Omega$; this confirms the conclusions reached in Chap. 3.

b. Fourier transformation of the diagrammatic expansion

Until now, we have used a time-dependent formalism, which enters naturally within the framework of the interaction representation. This formulation is valuable, since it allows us to eliminate the disconnected diagrams. Once this basic step is taken, it is better to make a Fourier transformation of the diagrammatic expansion in order to calculate the various physical quantities, such as ΔE_0, G, K. We thus obtain very compact and elegant calculations, which permit a detailed analysis of the different diagrams.

With each line of a diagram is associated a factor $G_0(\mathbf{k}, t - t')$, where t' is the origin and t the terminus of the line. Let us express this factor in the form of a Fourier transform [see Eq. (3–16)],

$$G_0(\mathbf{k}, t - t') = \frac{1}{2\pi} \int_{-\infty}^{+\infty} d\omega \; G_0(\mathbf{k}, \omega) \exp[-i\omega(t - t')] \qquad (5\text{–}101)$$

The function $G_0(\mathbf{k},\omega)$ is given by (3–18). Before integration over ω, (5–101) appears as a product $e^{i\omega t'}\, e^{-i\omega t}$; the two end points of the line are therefore decoupled, which will allow us to integrate over the variables t_i. This is the great advantage of the Fourier transform; it makes the different vertices independent of one another.

Let us carry out the operation (5–101) on all the lines of the diagram; to each line, of wave vector $\mathbf{k}_i$, corresponds an energy ω_i. Let us consider a vertex, at time t, at which the lines $\mathbf{k}_1\omega_1$ and $\mathbf{k}_2\omega_2$ enter and from which the lines $\mathbf{k}_3\omega_3$ and $\mathbf{k}_4\omega_4$ leave. The time t is involved only in the quantity

$$\exp[i(\omega_3 + \omega_4 - \omega_1 - \omega_2)\, t] \exp(-\eta \mid t \mid) \qquad (5\text{–}102)$$

The integration of (5–102) from $t = -\infty$ to $+\infty$ is trivial and gives the result

$$\begin{cases} \dfrac{2\eta}{\Delta\omega^2 + \eta^2} \\ \Delta\omega = \omega_3 + \omega_4 - \omega_1 - \omega_2 \end{cases} \qquad (5\text{–}103)$$

It is easily verified that when $\eta \to 0$

$$\frac{2\eta}{\Delta\omega^2 + \eta^2} \to 2\pi\, \delta(\Delta\omega) \qquad (5\text{–}104)$$

There is thus conservation of energy at each vertex.

Let us first apply this transformation to the calculation of $G(\mathbf{k}, t - t')$. Each line has its associated energy; in particular, the entering and leaving lines have, respectively, the energies ω' and ω. With the various parts of the diagram there are associated the following contributions:

(5–105)
- (a) For each vertex, a factor $V(\mathbf{k}_i)\,\delta(\omega_1 + \omega_2 - \omega_3 - \omega_4)$
- (b) For each line a factor $G_0(\mathbf{k},\omega)$

All the ω_i must be integrated from $-\infty$ to $+\infty$. Usually, these integrals converge, except for graphs such as  in which a line closes on itself, we know that such a diagram corresponds to the limit $(t - t') \to -0$. Therefore, if there is any doubt, the integration over ω

must be made over the contour formed by the real axis and the infinite semicircle in the upper half-plane [see (3–24)].

In addition to the above contributions, we must include a factor $e^{i(\omega' t' - \omega t)}$, arising from the two end points of the diagram, and a factor $1/(2\pi)^{p+1}$, coming from the p vertices [(5–104)] and the $(2p + 1)$ lines [(5–101)]. In practice, it is much simpler to calculate directly the Fourier transform $G(\mathbf{k},\omega,\omega')$. According to (3–14), we have

$$G(\mathbf{k}, t - t') = \frac{1}{2\pi} \iint d\omega \, d\omega' \, G(\mathbf{k}, \omega, \omega') \exp[i(\omega' t' - \omega t)] \qquad (5\text{–}106)$$

In order to pass from $G(\mathbf{k}, t - t')$ to $G(\mathbf{k},\omega,\omega')$, we suppress the factor

$$\frac{\exp[i(\omega' t' - \omega t)]}{2\pi}$$

and we do not integrate over ω and ω'. To get $G(\mathbf{k},\omega,\omega')$, we must therefore add the following rules to (5–105):

(5–107)
- (c) Add the numerical factor $(-1)^l \left(\dfrac{i}{2\pi}\right)^p \dfrac{1}{r2^m}$
- (d) Fix the energies of the entering and leaving lines, respectively, at ω' and ω. Sum over the internal $\mathbf{k}$, integrate over the internal ω.

In practice, we know that $G(\mathbf{k},\omega,\omega')$ is of the form

$$G(\mathbf{k}, \omega, \omega') = G(\mathbf{k}, \omega) \, \delta(\omega - \omega')$$

[see Eq. (3–15)]. This result also follows from the structure of the diagrams. As energy is conserved at each vertex, it must be also between the end points of the diagram—hence the factor $\delta(\omega - \omega')$. In order to obtain $G(\mathbf{k},\omega)$ directly, we take the frequencies of the two external lines to be equal, and we omit the function $\delta(\omega_1 \ldots - \omega_4)$ at any one of the vertices of the diagram.

The rules (5–107) are easily generalized to the calculation of $K(\mathbf{k}_i,\omega_i)$, defined by (3–48). The lines coming from $P_1 \ldots P_4$ carry energies $\omega_1 \ldots \omega_4$. The numerical factor is the same as in (5–107), to within an overall sign. [We have seen that this sign is $\pm (-1)^l$ according to whether the line coming from P_1 ends on P_3 or on P_4.] Conservation of energy requires a factor $\delta(\omega_1 + \omega_2 - \omega_3 - \omega_4)$ for the linked diagrams (Fig. 22*c*) and a product of two δ functions for the "free" diagrams (Fig. 22*a* and *b*).

To conclude, let us apply this transformation to the variation in the ground-state energy ΔE_0, given by (5–80). Integration over the variables $t_2 \ldots t_p$ gives $(p-1)$ energy-conservation relations at the corresponding vertices. The conservation of ω at the last vertex t_1 is then automatic (the total energy at times before or after the set of t_i is zero; the whole diagram thus conserves energy by construction). The factor (5–102) at the vertex t_1 reduces to 1. Hence the following rules for calculating ΔE_0:

(5–108)
- (a) Take the set of unlabeled unlinked connected diagrams, associate with the vertices and with the lines the factors indicated in (5–105) but omit the factor $\delta(\omega_1 \ldots - \omega_4)$ at any one of the vertices of the diagram.
- (b) Add a factor $\left(\frac{i}{2\pi}\right)^{p+1} \cdot \frac{(-1)^l}{r2^m}$
- (c) Sum over all $\mathbf{k}$; integrate over all ω.

By comparing (5–105), (5–107) and (5–108), we become aware that ΔE_0 is intimately related to the Green's functions. We shall demonstrate this relation in section 5, with the help of a detailed analysis of the structure of the diagrams.

c. *Reduction of the Green's functions*

We propose to decompose the diagrams contributing to G and K into elements "localized" in time, that is to say, spreading over a finite time interval Δ. Δ plays the role of a "correlation time." We studied this question in section 2, and we saw that a diagram can extend over a time $\gg \Delta$ only if it contains lines whose wave vector is fixed, equal to that of one of the external lines (see Fig. 16).

a. Single-particle Green's function. In general, all the lines of a diagram contributing to $G(\mathbf{k},\omega)$ have variable wave vector, except in the case indicated in Fig. 23*a*; the central line must then have the same wave vector $\mathbf{k}$ and the same energy ω as the external lines. Let us cut the central line (Fig. 23*b*), and let $G_I(\mathbf{k},\omega)$ and $G_{II}(\mathbf{k},\omega)$ be the contributions to G arising from the two subdiagrams I and II. The contribution from the whole diagram is simply

$$\frac{G_{\mathrm{I}}(\mathbf{k}\omega)\, G_{\mathrm{II}}(\mathbf{k}\omega)}{G_0(\mathbf{k}\omega)} \qquad (5\text{–}109)$$

(By cutting the diagram, we count the propagator G_0 corresponding to the central line twice.)

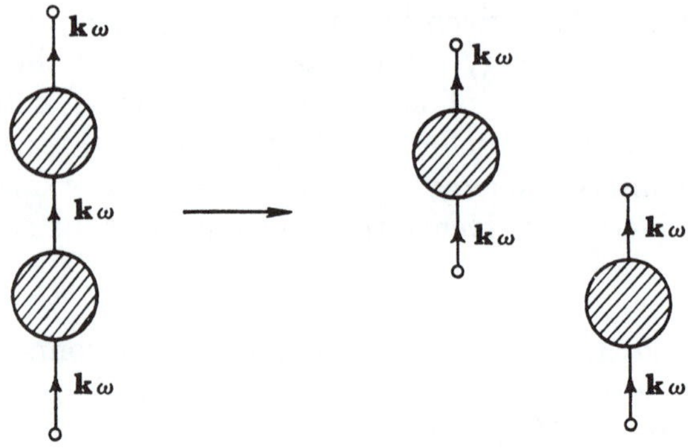

(*a*) (*b*)

Figure 23

Let us now consider all the diagrams of order $\geqslant 1$ which cannot be cut up like Fig. 23*a*. We shall call these diagrams with the two external lines removed *irreducible self-energy parts*. Let $M(\mathbf{k},\omega)$ be the corresponding contribution. The diagrams containing a single self-energy part lead to the term

$$[G_0(\mathbf{k},\omega)]^2\, M(\mathbf{k},\omega)$$

We obtain all the diagrams of G by inserting into the principal line any number of self-energy parts (0, 1, 2, etc.). G is therefore given by

$$G = G_0 + G_0^2 M + G_0^3 M^2 + \cdots = \frac{G_0}{1 - G_0 M} \tag{5-110}$$

Let us represent the complete propagator G by a double line and G_0 by a single line. Relation (5–110) just expresses the graphical equation indicated in Fig. 24. If we replace $G_0(\mathbf{k},\omega)$ by its value (3–18), (5–110) takes the very simple form

$$\begin{cases} G(\mathbf{k},\omega) = \dfrac{1}{\varepsilon_k^0 - \omega - M(\mathbf{k},\omega) - i\eta} \\ \eta = \begin{cases} +0 & \text{if} \quad k > k_F \\ -0 & \text{if} \quad k < k_F \end{cases} \end{cases} \tag{5-111}$$

The effect of the interaction H_1 is thus expressed by the introduction of $M(\mathbf{k},\omega)$ into the denominator of (5–111).

Let us compare (5–111) with (4–8). $M(\mathbf{k},\omega)$ is just the *self-energy*

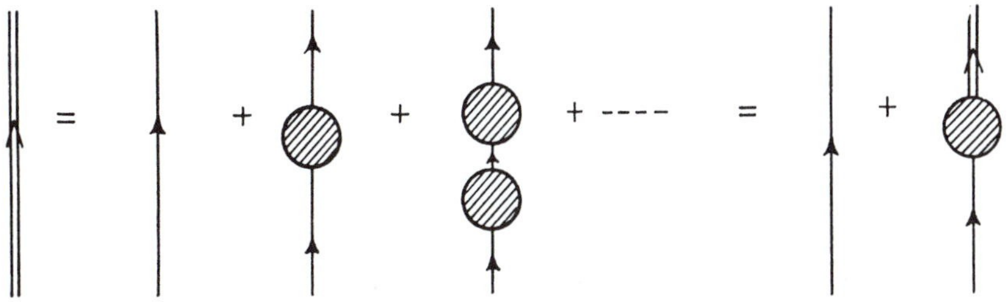

Figure 24

operator which we introduced in analyzing the dynamical equation satisfied by the Green's functions; we know how to calculate it starting from G and K [Eqs. (4–8), and (4–7)]. But whereas, in Chap. 4, M appears only as an intermediate step in the calculation, it acquires here a very suggestive physical interpretation.

Let $M(\mathbf{k},t)$ be the Fourier transform of $M(\mathbf{k},\omega)$. Equation (5–110) can be written in the form of a convolution product,

$$G(\mathbf{k}, t - t') = G_0(\mathbf{k}, t - t') + \iint_{-\infty}^{+\infty} dt'' \, dt''' \, G_0(\mathbf{k}, t - t'') \, M(\mathbf{k}, t'' - t''') \, G(\mathbf{k}, t''' - t') \tag{5-112}$$

As a matter of fact, (5–112) is obvious when we look at Fig. 24. The great interest in this reduction lies in the fact that $M(\mathbf{k},t)$ is bounded, negligible whenever $t \gg \Delta$ [in other words, $M(\mathbf{k},\omega)$ is a regular function]. The complete Green's function $G(\mathbf{k},t)$ can be extended over large intervals of time t only by elongating the intermediate lines $\mathbf{k}$ or by increasing the number of self-energy parts.

In establishing (5–111) we have "partially" summed the perturbation series. We thus considerably simplify the calculation of G. We shall meet later in this study other examples of partial summations; without claiming to solve the problem rigorously, we shall thus obtain a much more exact idea of various physical phenomena.

By looking at (5–111), we see that M acts like a correction to the energy $\overset{\circ}{\varepsilon}_k = k^2/2m$: hence the name self-energy. The simplest approximation consists in keeping only the first-order diagram, indicated in Fig. 25. The corresponding contribution to $M(\mathbf{k},\omega)$ is equal to

$$M_1(\mathbf{k}, \omega) = \frac{1}{\Omega} \sum_{\sigma', k' < k_F} [V_{\mathbf{k}-\mathbf{k}'} \, \delta_{\sigma,\sigma'} - V_0] \tag{5-113}$$

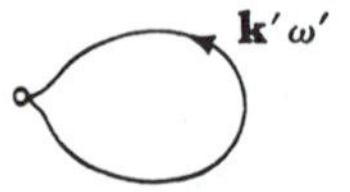

Figure 25

By referring to (4–9), we easily see that (5–113) is equivalent to the Hartree-Fock approximation [which consists of neglecting the bound part of the two-particle Green's function K in (4–1)]. Note that to this order M is independent of ω; we can interpret it rigorously as a correction to $\overset{\circ}{\varepsilon}_k$. Unfortunately, this is no longer true when we go to higher orders; G then acquires the complicated analytic form which we studied in Chap. 3.

b. Two-particle Green's function. We are interested only in the bound part δK, since the free part reduces to products of G's. The reduction of the corresponding diagrams is dictated by Fig. 16*b*, which shows the decomposition of these diagrams into localized elements. Let us fix the central core of this figure, and let us sum over all combinations of self-energy parts in the external lines; the result is represented in Fig. 26, where the double lines correspond to complete propagators $G(\mathbf{k}_i,\omega_i)$. We shall give A the name "interaction core." This core is necessarily localized in time, of extent $\lesssim\Delta$.

Let $\gamma(\mathbf{k}_i,\omega_i)$ be the contribution of all the interaction cores (with the external propagators G removed), computed with the same rules as for K. We can write δK in the form

$$\delta K(\mathbf{k}_i, \omega_i) = \gamma(\mathbf{k}_i, \omega_i)\, G(\mathbf{k}_1, \omega_1) \dots G(\mathbf{k}_4, \omega_4) \tag{5–114}$$

Comparing (5–114) with (3–49), we discover that $\gamma(\mathbf{k}_i,\omega_i)$ is just the

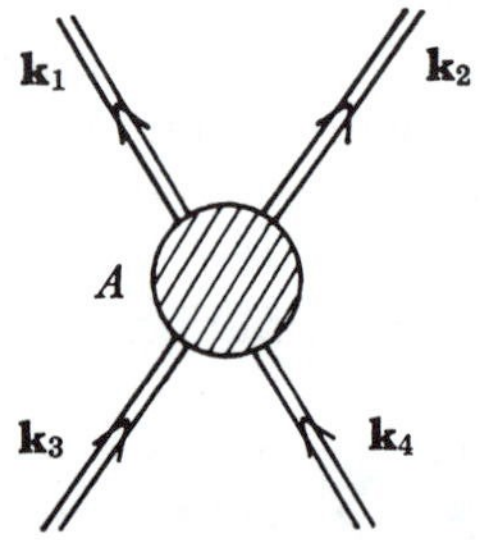

Figure 26

interaction operator defined in Chap. 3. The qualitative arguments used there thus are given a very simple graphical interpretation.

Note: Actually, $\gamma(\mathbf{k}_i,t_i)$ may have long-range terms, arising from the poles of $\gamma(\mathbf{k}_i,\omega_i)$, corresponding to an interaction through the exchange of collective modes. The interaction is then "delocalized", owing to a "resonance" of its elements.

By replacing G_0 by G, we correctly describe the propagation of a single particle. A double line is therefore the "renormalized" version of a single line, the effects of the interaction H_1 on the propagation being taken into account. The additional particle carries along with it a cloud of neighboring particles, leading to an apparent mass correction; in practice, this cloud possesses a certain amount of inertia, which explains why the mass (or self-energy) depends on ω. Similarly, the interaction operator γ is a renormalized version of the two-particle interaction. To lowest order, the interaction core reduces to a single vertex, whose contribution is

$$\gamma_1(\mathbf{k}_i,\, \omega_i) = -\frac{i}{2\pi}\, V(\mathbf{k}_i)\, \delta(\omega_1 + \omega_2 - \omega_3 - \omega_4) \tag{5–115}$$

(the minus sign comes in because line 1 ends up at 4 instead of 3). γ_1 is the first-order interaction between two "isolated" particles, the influence of the embedding medium being disregarded. Actually, the *effective* interaction, measured by γ, is modified by various physical effects; multiple scattering, screening, interaction of the self-energy clouds, etc. In going from V to $2\pi i\gamma$, we renormalize the interaction.

We shall treat these problems in more detail in Chap. 6, which is devoted to a study of γ. For the moment, let us note that the first-order γ is independent of the frequencies ω_i. After Fourier transformation, $\gamma(\mathbf{k}_i,t_i)$ contains a factor

$$\delta(t_1 - t_2)\, \delta(t_1 - t_3)\, \delta(t_1 - t_4)$$

The interaction is *instantaneous*. This is no longer true in higher orders; the interaction becomes *retarded*.

Note: In order to calculate the interaction operator $\Gamma(\mathbf{k}_i,\omega_i)$ defined by (3–50), it is sufficient, in the calculation of γ, to omit the energy δ function at any one of the vertices in the interaction core.

d. Properties of the single-particle Green's function

Using the results of the preceding section, we shall establish a number of properties of $G(\mathbf{k},\omega)$ which will complement those we discussed in Chap. 3. Whereas the latter are rigorously exact, the considerations

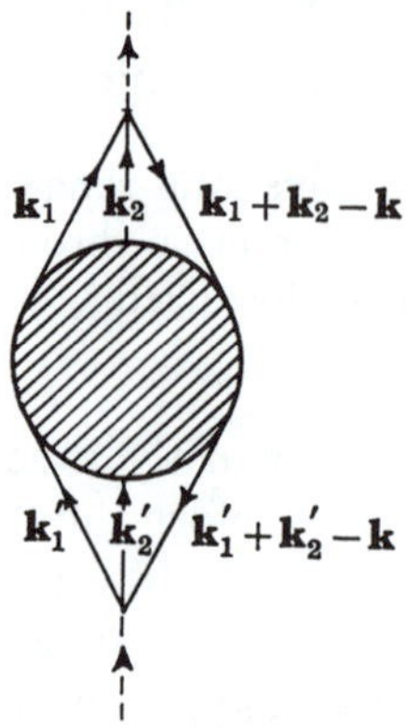

Figure 27

which follow are valid only within the limits imposed by a perturbation expansion. We shall thus justify the hypotheses made at the beginning of Chap. 4.

Let us first turn to the continuity properties of $M(\mathbf{k},\omega)$: is M continuous when k crosses k_F or when ω crosses the chemical potential μ? As a matter of fact, the values k and ω affect the corresponding diagrams only through the energy and momentum conservation relations at the first and last vertices. The structure of a diagram contributing to M is indicated in Fig. 27. The distribution of the vectors $\mathbf{k}_1$, $\mathbf{k}_1'$ . . . does not suffer any discontinuity when $k = k_F$. As a consequence, $M(k,\omega)$, as well as its derivatives with respect to k, is continuous when k crosses k_F; this is what we stated in Chap. 4. [This result can be extended to $\gamma(\mathbf{k}_i,\omega_i)$ when one of the k_i passes through k_F.] We can repeat this reasoning point by point in relation to the energy; $M(k,\omega)$ is therefore continuous at $\omega = \mu$.

Note: There can occur discontinuities in *high*-order derivatives of M and γ when k is equal to a multiple of k_F. These singularities are very weak and are of purely geometric origin.

The analytic structure of the function $M(\omega)$ can easily be deduced from that of $G(\omega)$, thanks to (5–111). Like G, M gives rise to a cut, located above the real axis for $\omega < \mu$, below for $\omega > \mu$ (see Fig. 1 of Appendix C.) From (C–3) we deduce

$$M(\mathbf{k}, \omega^*) = [M(\mathbf{k}, \omega)]^* \tag{5–116}$$

(5–116) is valid off the real axis and enables us to relate the two forms of M.

Let us compare (5–111) with (C–2). Since the functions $A_\pm$ are by nature positive, we see that, when ω is real,

$$\operatorname{Im} M(\mathbf{k}, \omega) \begin{cases} > 0 & \text{if} \quad \omega > \mu \\ < 0 & \text{if} \quad \omega < \mu \end{cases} \tag{5–117}$$

The continuity of M when $\omega = \mu$ then implies

$$\operatorname{Im} M(\mathbf{k}, \mu) = 0 \tag{5–118}$$

(5–118) confirms the conclusions obtained qualitatively in Chap. 4.

In order to determine the asymptotic form of $M(\mathbf{k},\omega)$ when $\omega \to \infty$, let us consider a diagram of M whose end points t' and t are separated. Its contribution can be put into the form

$$\int_{-\infty}^{+\infty} \mathrm{d}(t - t') \exp[\mathrm{i}\omega(t - t')]\, f(t - t') \tag{5–119}$$

The result is of order $1/\omega$ when $\omega \to \infty$ [since $f(t - t')$ is a finite function]. Let us consider, on the other hand, the diagram of Fig. 28*a*. The end points are identical by construction. The corresponding contribution to $M(\mathbf{k},\omega)$ is *independent* of ω. The set of diagrams of this type can be represented by Fig. 28*b* (where the double line represents a propagator G). The corresponding term of M is equal to

$$-\frac{\mathrm{i}}{\Omega}\sum_{k'\sigma'} [V_0 - V_{k-k'}\, \delta_{\sigma,\sigma'}]\, [G(\mathbf{k}', t)]_{t=-0} = -\frac{1}{\Omega}\sum_{k'\sigma'} [V_0 - V_{k-k'}\, \delta_{\sigma\sigma'}]\, m_{k'} \tag{5–120}$$

where $m_{k'}$ is the *real* probability of occupation of the plane wave of wave

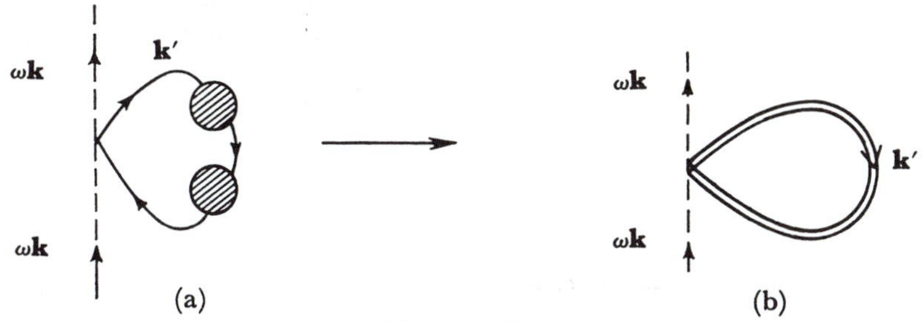

Figure 28

vector $\mathbf{k}'$. (5–120) is just $(k^2/2m - E_k)$, where E_k is the Hartree-Fock energy of a bare particle, already calculated in (C–15). In summary,

$$M(\mathbf{k}, \omega) \underset{\omega\to\infty}{\longrightarrow} k^2/2m - E_k + O(1/\omega)$$

By putting this result back into (5–111), we recover the asymptotic form of $G(\mathbf{k},\omega)$, in accordance with (C–15).

Let us now turn to a more specific property of normal systems by studying the behavior of $G(\mathbf{k},t)$ when t is large. We shall assume that $k > k_F$; our results are easily extended to the case $k < k_F$. We propose to show qualitatively that, in a certain time interval, $G(\mathbf{k},t)$ takes the form

$$G(\mathbf{k}, t) \sim \begin{cases} iz_k \exp(-i\varepsilon_k t) & \text{if} \quad t > 0 \\ 0 & \text{if} \quad t < 0 \end{cases} \tag{5–121}$$

where z_k and ε_k are constants. The imaginary part of ε_k must be negative, since $|G| < 1$ (G is the scalar product of two vectors, each of norm < 1). (5–121) will be true if $|t|$ is much greater than the extent Δ of a self-energy part, but such that $\text{Im}\ \varepsilon_k \cdot t < 1$. This limits us to the immediate neighborhood of the Fermi surface.

A propagator G is obtained by inserting in the "guide" line G_0 an arbitrary number of self-energy parts M. It can easily be verified that the contribution does not change if we displace a self-energy part without deformation along the guide line. By construction, the guide line is *ascending*. G will therefore have very different properties for $t \lessgtr 0$.

Let us first take the case $t \ll -\Delta$. The diagrams then have the behavior indicated in Fig. 29. Two solutions come to mind for going back up the time interval t:

(a) Keep a small number of self-energy parts, of much greater extent than the average Δ. The corresponding contribution is small (it decreases as $1/t^3$ when t increases).

(b) Accumulate a number $\sim t/\Delta$ of self-energy parts of extent $\sim \Delta$. The connecting lines must then be very short, of length less than Δ. If the contribution of a single part is of order M, the whole diagram gives a term

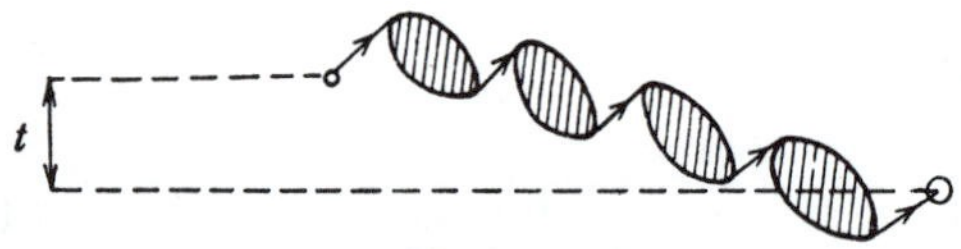

Figure 29

of order $(M\Delta)^{-t/\Delta}$ which is exponentially small. This discussion makes no pretense as to rigor, but it clearly shows why $G(\mathbf{k},t) \to 0$ when $t \ll -\Delta$; it is not possible, using localized self-energy parts, to go very far in a direction opposite to that of the lines G_0.

The situation is completely changed when $t \gg \Delta$. The self-energy parts can then move freely from one end of the guide line to the other (Fig. 30). Let us consider the diagram containing a single part M_1, whose contribution is

$$i^2 \int_{t'}^{\infty} dt''' \int_{-\infty}^{t} dt'' \exp[-i\varepsilon_k^0(t - t'' + t''' - t')]\, M(\mathbf{k}, t'' - t''') \qquad (5\text{–}122)$$

If M occurs at a distance $\gg \Delta$ from each of the end points t and t', we can integrate over $(t'' - t''')$ from $-\infty$ to $+\infty$. (5–122) then reduces to

$$i^2 M(\mathbf{k}, \varepsilon_k^0)\,.\,(t - t')\,.\,\exp[-i\varepsilon_k^0(t - t')] + \delta \qquad (5\text{–}123)$$

where the correction δ expresses the boundary effects when M_1 is brought closer to t or to t'. The calculation involves two steps: integrate over $(t'' - t''')$, then displace M_1, reduced to a point, from one end of the guide line to the other.

In practice we must consider diagrams containing an arbitrary number of self-energy parts. When the contributions of a part M_1 are added up by integrating over $(t'' - t''')$, it is necessary to take account of the fact that neighboring parts exert a certain "pressure" on M_1. If, for example, we integrate over t'', this effect amounts to replacing the phase factor $G_0(\mathbf{k}, t - t'')$ by $G(\mathbf{k}, t - t'')$. Let us anticipate our results and use (5–121); we see that we take account of this effect by replacing

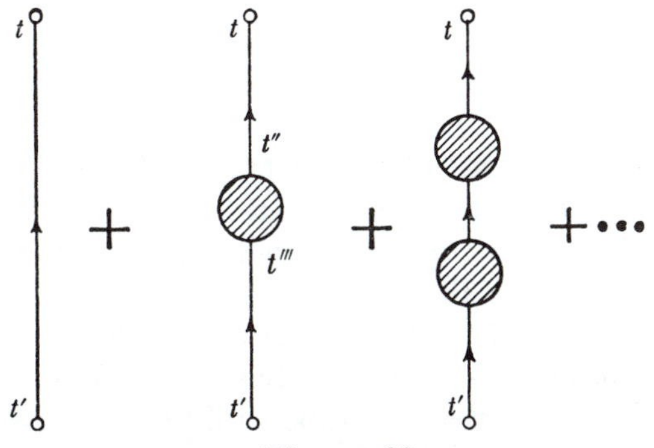

Figure 30

$M(\mathbf{k},\varepsilon_k^\circ)$ by $M(\mathbf{k},\varepsilon_k)$ in (5–123). Once this precaution is taken, we no longer have to concern ourselves with the interaction between self-energy parts.

Let us first neglect boundary effects near t and t'. The contribution of the diagrams containing p parts M_i is equal to

$$\mathrm{i}\,\frac{[\mathrm{i}M(\mathbf{k},\varepsilon_k)\,(t-t')]^p}{p\,!}\exp[-\mathrm{i}\varepsilon_k^0(t-t')] \tag{5–124}$$

(account being taken of the fact that a permutation of the M_i does not change the diagram). The corresponding series is easily summed and leads to the following value of G, which is just a first approximation:

$$\mathrm{i}\exp\{-\mathrm{i}[\varepsilon_k^0 - M(\mathbf{k},\varepsilon_k)]\,(t-t')\} \tag{5–125}$$

To finally obtain the proper result, it is necessary to take account of boundary effects. The latter limit the integration over the length $(t'' - t''')$ of the self-energy parts too near to t and t'; they introduce the factor z_k of (5–121). We shall not attempt to establish this result rigorously, contenting ourselves with learning two lessons from this discussion:

(a) The perturbation expansion amounts to expanding the exponential of (5–121) in a series.

(b) For this expansion to make sense, it is necessary for Im $M(\mathbf{k},\varepsilon_k)$ to be > 0.

The energy ε_k is determined by the equation

$$\varepsilon_k = \varepsilon_k^0 - M(\mathbf{k},\varepsilon_k) \tag{5–126}$$

We see from (5–111) that ε_k is a pole of the Fourier transform $G(\mathbf{k},\omega)$ (or rather of its analytic continuation across the cut). This agrees with the conjectures of Chap. 3. The residue at this pole is just $-\ z_k$. Within the limits of a perturbation method, we have

$$\mathrm{Im}\,\varepsilon_k < 0 \tag{5–127}$$

(5–127) shows that the elementary excitation of wave vector $k > k_F$ is a *quasi particle* which can propagate for a long distance only in the positive time direction. When $k < k_F$, this conclusion must be reversed. The nature of the excitation is therefore the same for G as for G_0. We thus prove that *the Fermi surface of the real system is identical to that of the unperturbed system.* This basic result is true only within the limits of a perturbation treatment and expresses the impossibility of reversing the sense of propagation by using localized self-energy parts.

We have seen that M is a continuous function of k and ω which varies slowly with ω. ε_k is thus a continuous function of k. In particular, $\text{Im}\,\varepsilon_{k_F} = 0$; excitations near the Fermi surface are weakly damped. If ε_k is located near the real axis, we see from (5–117) that

$$\text{Re}\,\varepsilon_k \begin{cases} > \mu & \text{if} \quad k > k_F \\ < \mu & \text{if} \quad k < k_F \end{cases} \tag{5–128}$$

(5–128) expresses the stability of the ground state. The continuity of M then implies $\varepsilon_{k_F} = \mu$; we find again the Van Hove theorem stated in Chap. 4.

This discussion remains very qualitative. We have presented it because it shows the physical origin of the hypotheses proposed in Chap. 4 more clearly than a rigorous mathematical study. We see better the limitations imposed by the perturbation treatment. Let us emphasize the importance of these results, which constitute the basis of the following section.

4. Adiabatic Generation of Excited States

We have limited ourselves until now to the study of the ground state. This allows us to calculate the Green's functions. According to Chap. 3 and 4, we thus obtain all necessary information about the elementary excitations. This approach is consistent and complete.

In this chapter, we shall adopt a completely different attitude; we shall extend to the excited states the technique of "adiabatic generation" used for the ground state in section 2. We shall show that the quasi particles thus obtained are identical to those which result from the poles of the Green's function, in terms of their wave functions as well as their energies. This new approach emphasizes the irreversibility introduced by the finite lifetime of the particles; the real transitions which the system undergoes appear clearly. This formalism applies only if the lifetime of the elementary excitation is much greater than the width Δ of the self-energy parts; again, this limits us to the immediate neighborhood of the Fermi surface.

We characterize the time development of the system by the operator $U_\eta(t',t)$. Let us start from the ground state $|0\rangle$ at time $t = -\infty$, and at time t let us add an extra bare particle of wave vector $\mathbf{k}$. At time $t = 0$ the state vector is given by

$$|\varphi_{\eta\mathbf{k}t}\rangle = U_\eta(0, t)\, a^*_{\mathbf{k}}(t)\, U_\eta(t, -\infty)\,|0\rangle \tag{5–129}$$

If $t = 0$, we add a *bare* particle to the ground state $|\varphi_0\rangle$. On the other hand, if $t \to -\infty$,

$$\begin{cases} |\varphi_{\eta\mathbf{k}t}\rangle \to \exp(i\varepsilon_k^0 t)\,|\varphi_{\eta\mathbf{k}}\rangle \\ |\varphi_{\eta\mathbf{k}}\rangle = U_\eta(0, -\infty)\, a_{\mathbf{k}}^*\,|0\rangle \end{cases} \tag{5–130}$$

$|\varphi_{\eta\mathbf{k}}\rangle$ is a normalized state vector, which we presume to be an eigenvector of the real system, describing a *clothed* particle of wave vector $\mathbf{k}$.

Like $|\varphi_{\eta 0}\rangle$, the wave function (5–129) can be represented with the help of diagrams, which in this case have the structure indicated in Fig. 31. The vertices of these diagrams can vary from $t = -\infty$ to $t = 0$. We see three types of contributions:

(a) The unlinked connected parts: they supply a constant factor $e^{-U_{\eta oc}}$, which we studied in detail in section 2.

(b) The linked parts not containing the line $\mathbf{k}$: they exactly reproduce the ground state $U_{\eta L}(0, -\infty)|0\rangle$.

(c) The linked part containing the line $\mathbf{k}$, which takes account of all the modifications caused by the addition of an extra particle.

This analysis allows us to write $|\varphi_{\eta\mathbf{k}t}\rangle$ in the form

$$|\varphi_{\eta\mathbf{k}t}\rangle = A_{\eta\mathbf{k}t}^*\, U_\eta(0, -\infty)\,|0\rangle \tag{5–131}$$

$A_{\eta\mathbf{k}t}^*$ describes the state of the *excitation* at time $t = 0$. Its matrix elements are given by the *linked connected* diagrams having a line coming from $(\mathbf{k},t)$. In particular, if $t \to -\infty$,

$$A_{\eta\mathbf{k}t}^* \to \exp(i\varepsilon_k^0 t)\, A_{\eta\mathbf{k}}^* \tag{5–132}$$

$A_{\eta\mathbf{k}}^*$ plays the role of a *quasi-particle* creation operator. We shall show

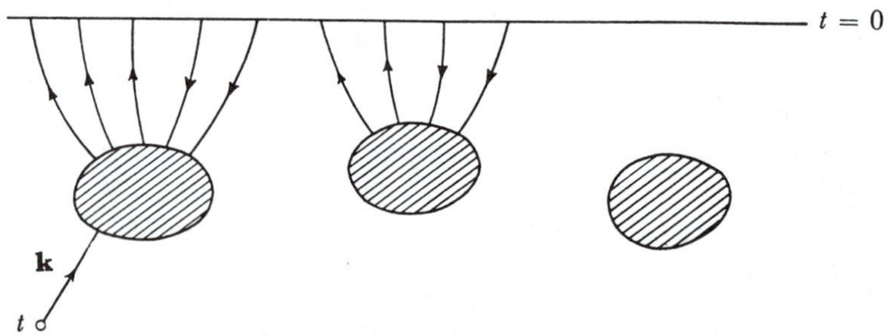

Figure 31

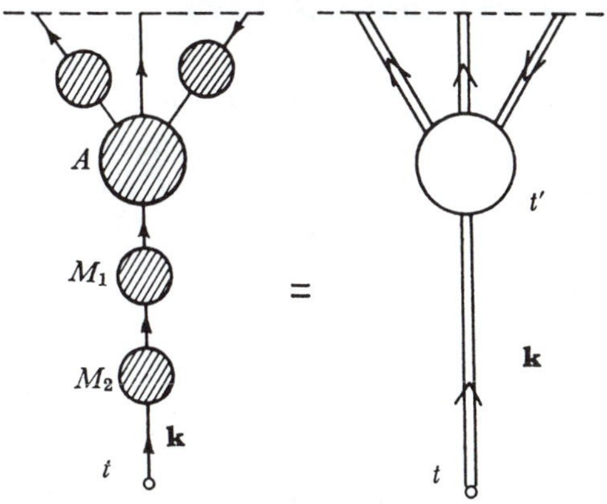

Figure 32

later that this operator is given by (4–36) to within a constant factor, which assures the equivalence of this method with that of Chap. 4.

The problem thus reduces to the study of $A^*_{\eta\mathbf{k}t}$. As in the section on The Green's Functions we must first analyze the extension in time of the corresponding diagrams. The situation is analogous to that which we encountered in studying the Green's functions and is summarized in Fig. 32; each diagram is made up of a "guide" line of wave vector **k**, into which we insert an arbitrary number of self-energy parts M. This line is terminated by an "active" core A (which can be of order zero). From this active core extend the propagators corresponding to the final state of the excitation, which can themselves be subject to diagonal insertions. Each of these elements is localized in time; however, the whole diagram can be extended over as long an interval as desired.

This diagram can be represented symbolically with the help of renormalized propagators (Fig. 32). Let t' be the point of connection of the active core with the lower propagator. We can put $A^*_{\eta\mathbf{k}t}$ into the form

$$A^*_{\eta\mathbf{k}t} = -\mathrm{i}\int_{-\infty}^{0} \mathrm{d}t'\; G_\eta(\mathbf{k}, t', t)\; J^*_{\eta\mathbf{k}t'} + \delta \tag{5–133}$$

where the operator $J^*_{\eta\mathbf{k}t'}$ adds the contributions from the cores A and from the upper lines. G_η is the Green's function in the presence of the interaction $H_1 e^{\eta t}$; δ is a correction arising from the boundary effects on G_η when t is too close to 0. We shall begin our study with the calculation of G_η.

a. Calculation of the Green's function $G_\eta(k,t,t')$

The Green's function is given by (5–97). It still originates from the linked connected diagrams of Fig. 20, which we must evaluate *without taking the limit* $\eta \to 0$. Since the system is no longer invariant with respect to translation in time, G_η is a function of t and t' separately, and no longer of their difference only. If $\eta|t - t'| \ll 1$, we can foresee that

$$\begin{cases} G_\eta(t, t') = G(g, t - t') + O(\eta) \\ g = e^{\eta t} \end{cases} \tag{5–134}$$

(In order to simplify the notation, we omit the coordinate $\mathbf{k}$.) To within a correction of order η, G_η reduces to an ordinary Green's function corresponding to the *instantaneous* value of the coupling constant g. Unfortunately, (5–134) is no longer true when $|t - t'| \gtrsim 1/\eta$; it is thus necessary to calculate G_η from the beginning.

Let $M_\eta(t,t')$ be the contribution of all self-energy parts having their end points at t and t'. We can go from M_η to G_η by a simple rearrangement of Eq. (5–112),

$$G_\eta(t, t') = G_0(t - t') + \int_{-\infty}^{+\infty} dt'' \int_{-\infty}^{+\infty} dt''' \, G_0(t - t'') \, M_\eta(t'', t''') \, G_\eta(t''', t') \tag{5–135}$$

[Recall that G_0 is the unperturbed propagator, given by (3–17).] Let us differentiate (5–135) with respect to t; it is easy to see that, when $t > t'$,

$$\frac{\partial}{\partial t} G_\eta(t, t') = -i\varepsilon_k^0 \, G_\eta(t, t') + i \int_{-\infty}^{+\infty} dt''' \, M_\eta(t, t''') \, G_\eta(t''', t') \tag{5–136}$$

It is convenient to rewrite (5–136) in the form

$$\frac{\partial}{\partial t} [\ln G_\eta(t, t')] = -i\varepsilon_k^0 + i \int_{-\infty}^{+\infty} dt''' \, M_\eta(t, t''') \, \frac{G_\eta(t''', t')}{G_\eta(t, t')} \tag{5–137}$$

We know that $M_\eta(t,t''')$ is a *localized* function, which tends rapidly to 0 for $t - t''' \gg \Delta$. The integration of (5–137) therefore covers only a reduced interval. This is the great advantage of (5–137) over (5–135). (5–137) is a first-order differential equation, which we shall have to integrate over the interval of order $1/\eta$ in which the interaction H_1 is effective. To obtain G_η to order 1, it is thus necessary to know M_η to order η; this will be the first step of our calculation.

Let us consider a diagram contributing to $M_\eta(t,t')$. It contains p

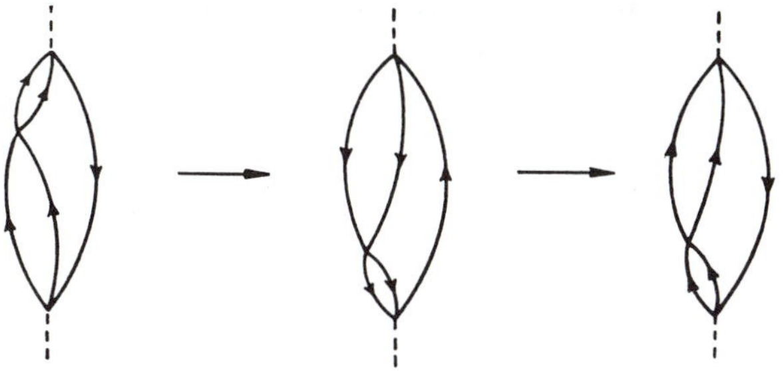

Figure 33

vertices, at times $t_1 \dots t_p$. The adiabatic switching on of the interaction gives rise to a factor

$$\exp[\eta(t_1 + \cdots + t_p)]$$

In fact, all times t_i must be close to t (since the core is localized). To first approximation, we can replace this factor by $e^{p\eta t}$; this amounts to replacing $M_\eta(t,t')$ by $M(g, t-t')$, where $g = e^{\eta t}$ is the value of the coupling constant at time t. In order to evaluate the error thus introduced, we expand the exponential; each diagram gives a contribution proportional to

$$\eta(t_1 + \cdots + t_p - pt) \qquad \textbf{(5–138)}$$

A priori, the calculation of this correction seems difficult; we can, however, find it by an artifice. Let us take the reflection of the diagram with respect to the time $(t+t')/2$, then reverse the sense of all the lines; we thus obtain a new diagram, contributing also to $M_\eta(t,t')$ (see the example in Fig. 33). It is easily verified that the contribution of the lines and the vertices is not changed by this transformation. All that changes is the corrective factor (5–138), which becomes

$$p\eta\,(t' - t) - \eta(t_1 + \cdots + t_p - pt)$$

Let us combine these terms in pairs and get rid of the factor p by the artifice already used to establish Eq. (5–74); we finally obtain

$$M_\eta(t, t') = M(g, t - t') + \eta\,\frac{(t' - t)}{2}\,g\,\frac{\partial}{\partial g}\,M(g, t - t') + O(\eta^2) \qquad \textbf{(5–139)}$$

It should be emphasized that this important relation applies only because M_η is a *localized* function.

We now return to Eq. (5–137). We already know its solution when g is constant; in the limit $t-t' \gg \Delta$, it is given by (5–121). The energy ε_k satisfies Eq. (5–126),

$$\varepsilon_k(g) = \varepsilon_k^0 - [M(g, \varepsilon)]_{\varepsilon=\varepsilon_k(g)} \qquad (5\text{–}140)$$

As for the residue z, it is equal to

$$z(g) = \frac{1}{1 + [\partial M(g, \varepsilon)/\partial\varepsilon]_{\varepsilon=\varepsilon_k(g)}} \qquad (5\text{–}141)$$

We assume that the real solution $G_\eta(t,t')$ will have a similar structure to (5–121), except that the frequency ε_k and the amplitude z will vary slowly in time, following the coupling constant $g = e^{\eta t}$. Therefore we shall try a solution of the form

$$G_\eta(t, t') = f(g, t') \exp\left(-\mathrm{i}\int_{t'}^{t} \varepsilon_k(g'')\, \mathrm{d}t''\right) \qquad (5\text{–}142)$$

(assuming $t-t' \gg \Delta$). Then we must have

$$\frac{\partial}{\partial t} \ln G_\eta(t, t') = -\mathrm{i}\varepsilon_k(g) + \eta b(g, t') \qquad (5\text{–}143)$$

where $b(g,t')$ is a function to be determined, which, in fact, will be shown to be independent of t'.

(5–143) allows us to calculate the last factor of (5–137), which becomes

$$\frac{G_\eta(t''', t')}{G_\eta(t, t')} = \exp[-\mathrm{i}\varepsilon_k(g)\,(t''' - t)] \left[1 + \eta b(g)\,(t''' - t) - \mathrm{i}g\eta\, \frac{\mathrm{d}\varepsilon_k(g)}{\mathrm{d}g}\, \frac{(t''' - t)^2}{2} + O(\eta^2)\right] \qquad (5\text{–}145)$$

Let us put (5–139), (5–143), and (5–145) back into (5–137) and carry out the integration over t'''. In this way we obtain

$$-\mathrm{i}\varepsilon_k(g) + \eta b(g) = -\mathrm{i}\varepsilon_k^0 + \mathrm{i}\left[M(g, \varepsilon) + \frac{\mathrm{i}\eta g}{2}\,\frac{\partial^2}{\partial g\,\partial\varepsilon} M(g, \varepsilon) + \frac{\mathrm{i}\eta g}{2}\,\frac{\mathrm{d}\varepsilon_k(g)}{\mathrm{d}g}\,\frac{\partial^2}{\partial\varepsilon^2} M(g, \varepsilon) + \mathrm{i}\eta b(g)\,\frac{\partial}{\partial\varepsilon} M(g, \varepsilon)\right]_{\varepsilon=\varepsilon_k(g)} + O(\eta^2) \qquad (5\text{–}146)$$

(5–146) can be simplified considerably by using (5–140) and (5–141). We finally obtain

$$b(g) = \frac{g}{2}\frac{\mathrm{d}}{\mathrm{d}g}[\ln z(g)] + O(\eta) \tag{5–147}$$

Equation (5–143) is thus completely determined. In order to integrate it, we write it in the form

$$\frac{\partial}{\partial t}\ln G_\eta(t, t') = -\mathrm{i}\varepsilon_k(g) + \frac{1}{2z}\frac{\mathrm{d}z}{\mathrm{d}g}\frac{\mathrm{d}g}{\mathrm{d}t} + O(\eta^2) \tag{5–148}$$

We adjust the constant of integration in such a way that (5–143) is satisfied when $\eta(t-t') \ll 1$. We thus find

$$G_\eta(t, t') = \mathrm{i}\sqrt{z(g)\,z(g')}\exp\left[+\mathrm{i}\int_t^{t'} \varepsilon_k(g'')\,\mathrm{d}t''\right] + O(\eta) \tag{5–149}$$

(5–149) is *valid* provided that $t-t' \gg \Delta$. [If $t-t' \gtrsim 1/\eta^2$, (5–148) becomes rigorous over the greatest part of the interval of integration; (5–149) remains true to order η.]

Until now we have assumed H_1 to be proportional to $e^{\eta t}$. Actually, (5–149) persists if the interaction *decreases* adiabatically from time 0 to time $+\infty$. We just have to replace g everywhere by $e^{-\eta|t|}$. (The corrections due to self-energy parts close to $t = 0$ involve only a finite time interval, of extent Δ; they are of order η and therefore negligible.)

In general, $\varepsilon_k(g)$ is a complex quantity, having an imaginary part $-i\Gamma_k(g)$. (We saw in Chap. 3 that Γ_k characterizes the damping of the quasi particle.) (5–149) is exact only if the exponential retains a modulus of order 1, which imposes the condition

$$\int_{t'}^{t} \Gamma_k(g'')\,\mathrm{d}t'' \lesssim 1 \tag{5–150}$$

When (5–150) is violated, the exponential becomes very small and the approximations which we have made are no longer valid (in particular, we are led to consider self-energy parts of arbitrarily large extent).

Let us consider the function $G_\eta(t, -\infty)$, which enters into the description of the adiabatic generation of excited states. When t is close to 0, the condition (5–150) can be written

$$\frac{1}{\eta}\int_0^1 \Gamma_k(g'')\,\frac{\mathrm{d}g''}{g''} \lesssim 1$$

Formula (5–149) thus applies only if $1/\eta$ is less than the "average" lifetime of the quasi particle $\langle\Gamma_k\rangle^{-1}$. In other words, it is necessary to turn on the interaction sufficiently quickly so that the excitation is not damped during the process.

On the other hand, the expansion (5–139) is justified only if $1/\eta$ is much greater than the length of a self-energy part Δ. We are thus faced with the dilemma

$$\Delta \ll \frac{1}{\eta} \ll \langle \Gamma_k \rangle^{-1} \tag{5–151}$$

The two conditions (5–151) can be simultaneously realized only in the neighborhood of the Fermi surface. We see that the validity of (5–149) is narrowly limited. In the rest of this section, we shall restrict ourselves to the cases where (5–151) is true; we can then use (5–149) with complete confidence.

b. The S matrix and the problem of reversibility

Let us put the system at $t = -\infty$ in the state $a_{\mathbf{k}}^*|0\rangle$, then switch on the interaction $H_I e^{-\eta|t|}$. The state vector at time $t = +\infty$ is given by

$$S_\eta\, a_{\mathbf{k}}^* \,|\, 0 \rangle = \sum_n |\, n \rangle \langle n \,|\, S_\eta\, a_{\mathbf{k}}^* \,|\, 0 \rangle \tag{5–152}$$

If the time development is reversible, only the term corresponding to $|n\rangle = a_{\mathbf{k}}^*|0\rangle$ can contribute to (5–152). A tendency toward irreversibility will appear in the form of nondiagonal matrix elements. Note that the normalization of the state vector imposes the condition

$$\sum_n \left| \langle n \,|\, S_\eta\, a_{\mathbf{k}}^* \,|\, 0 \rangle \right|^2 = 1 \tag{5–153}$$

The modulus of $\langle 0|a_{\mathbf{k}} S a_{\mathbf{k}}^*|0\rangle$ is thus $\leqslant 1$, the equality corresponding to a reversible development of the system.

Let us first consider the diagonal matrix element. It can be analyzed with the help of diagrams, whose linked part is the same as for G_η [except for the phase factors arising from the two end points: see (5–130)]. It is easily verified that

$$\langle 0 \,|\, a_{\mathbf{k}}\, S_\eta\, a_{\mathbf{k}}^* \,|\, 0 \rangle = \langle 0 \,|\, S_\eta \,|\, 0 \rangle \lim_{\substack{t \to +\infty \\ t' \to -\infty}} \frac{G_\eta(\mathbf{k}, t, t')}{G_0(\mathbf{k}, t, t')} \tag{5–154}$$

Let us replace $G_\eta(t,t')$ by its expression (5–149); the relation (5–154) takes the form

$$< 0 \mid a_{\mathbf{k}}\, S_\eta\, \overset{*}{a}_{\mathbf{k}} \mid 0 > \; = \; < 0 \mid S_\eta \mid 0 > \exp\left[-\mathrm{i}\int_{-\infty}^{+\infty} (\varepsilon_k(g'') - \varepsilon_k^0)\, \mathrm{d}t''\right] \quad \textit{(5–155)}$$

The passage to the limit $(t \to +\infty,\ t' \to -\infty)$ no longer poses any difficulty [note that $z(0) = 1$]. Let us use (5–89) and replace g'' by its value $e^{-\eta|t''|}$. (5–155) is transformed to

$$< 0 \mid a_{\mathbf{k}}\, S_\eta\, \overset{*}{a}_{\mathbf{k}} \mid 0 > \; = \exp\left\{\frac{2a}{\eta} - \frac{2\mathrm{i}}{\eta}\int_0^1 (\varepsilon_k(g'') - \varepsilon_k^0)\frac{\mathrm{d}g''}{g''}\right\} \quad \textit{(5–156)}$$

(5–156) generalizes (5–89) to the case of excited states.

Again, the average value of S_η diverges when $\eta \to 0$. If $\varepsilon_k(g'')$ is real, this divergence is limited to a singular phase factor, which does not affect the structure of the wave function. We saw in section 2 that this singularity actually expresses the energy shift of the system. Unfortunately, $\varepsilon_k(g'')$ also contains an imaginary part $-\ \Gamma_k(g'')$. The singularity does not affect only the phase of (5–156) (as was the case for the ground state), but also its modulus, which becomes

$$\exp\left[-\frac{2}{\eta}\int_0^1 \Gamma_k(g'')\frac{\mathrm{d}g''}{g''}\right] \quad \textit{(5–157)}$$

For the transformation to be *reversible*, it is thus necessary that $\eta \gg \langle \Gamma_k \rangle$. This conclusion is natural; if the quasi particle decays, we shall not recover it in removing the interaction.

According to (5–153), irreversibility implies the appearance of non-diagonal matrix elements of S_η, whose norms make up the deficit left by (5-157). Physically, these terms arise from *real transitions* suffered by the system and correspond to scattering amplitudes. In the perturbation formalism, we obtain contributions of order $1/\eta^{p/2}$, resulting from p successive real transitions. It is necessary to add up all these terms in order to reconstruct the exponential of (5–157).

We shall content ourselves with showing briefly how these real transitions enter. Let us consider a nondiagonal element, represented by the diagram of Fig. 34, and assume that the core B contains no internal line of fixed $\mathbf{k}$—it is localized in time; only its average position t varies from $-\infty$ to $+\infty$. (We ignore the resonances due to collective modes, thereby neglecting the absorption or emission of the latter.) The vectors $\mathbf{k}_1$, $\mathbf{k}_2$, and $\mathbf{k}_3$ are determined by the final state $|n\rangle$ considered. Let us use for

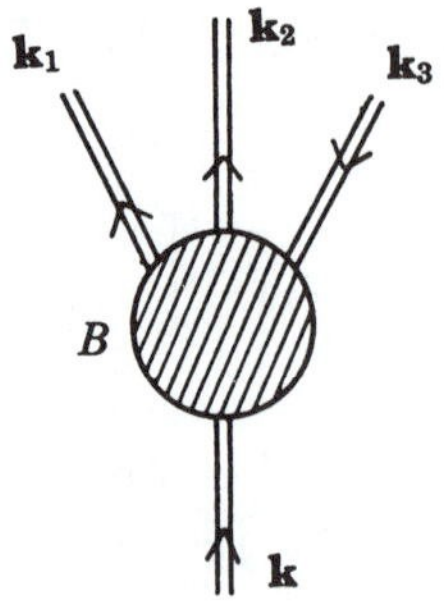

Figure 34

these propagators the value (5–149) (which assumes $\eta \gtrsim \langle \Gamma_k \rangle$). The contribution of the diagram can then be put in the form

$$\int_{-\infty}^{+\infty} f_{\pm}(g) \exp \{ i[\varepsilon_n(g) - \varepsilon_k(g)] \, t \} \, dt \tag{5–158}$$

where $f_{\pm}(g)$ correspond, respectively, to positive and negative values of t. $\varepsilon_n(g)$ is the energy of the quasi particles $\mathbf{k}_1$, $\mathbf{k}_2$, and $\mathbf{k}_3$ for the value g of the coupling constant. The contribution of the core B is incorporated into $f(g)$, which tends to 0 when $t \to \pm \infty$.

Let us first assume that the difference

$$\Delta\varepsilon(g) = \varepsilon_n(g) - \varepsilon_k(g) \tag{5–159}$$

never goes through zero. The exponential of (5–158) oscillates constantly, with a slow drift of the frequency. Equation (5–61) is applicable and shows that (5–158) is of order η. If η is small, this contribution is negligible. On the other hand, if $\Delta\varepsilon(g)$ is zero, we can no longer use (5–61). There is a *resonance* between the initial and final states, and therefore a *real transition*.

These real transitions are produced in the neighborhood of the times $\pm t_0$ for which g takes the value g_0. In principle, the scattered waves at these two moments interfere. Actually, this interference makes no physical sense, since the interval $2t_0$ which separates the two scatterings depends on the choice of η; it is just a mathematical fiction. To avoid this problem, we shall consider only scattering in the neighborhood of $t = t_0$. Let us put

$$\Delta\varepsilon(g) \sim \theta(g - g_0) \sim \theta \, g_0 \, \eta(t + t_0) \tag{5–160}$$

To within corrections of order η, (5–158) can be written

$$f_-(g_0) \int_{-\infty}^{+\infty} dt \exp[\,i\,\theta g_0\,\eta t(t+t_0)] \qquad \textit{(5–161)}$$

(5–161) is easily reduced to a Fresnel integral, equal to

$$f_-(g_0) \exp\left(-\frac{i\theta g_0 \eta t_0^2}{4}\right) \sqrt{\frac{\pi}{\theta g_0 \eta}}$$

As predicted, the scattering amplitude is proportional to $1/\sqrt{\eta}$. The scat-intensity is equal to

$$\left| < n \mid S_\eta\, a_{\mathbf{k}}^* \mid 0 > \right|^2 = \left| f_-(g_0) \right|^2 \frac{\pi}{\theta\, g_0\, \eta} \qquad \textit{(5–162)}$$

It can be represented as the product of a transition probability independent of η by the *duration* of the scattering, $\sim 1/\eta$.

Note: This formulation of the collision problem is unusual. If, however, we assumed $\Delta\epsilon$ to be independent of g, we would obtain a scattering amplitude proportional to $2\eta/(\Delta\epsilon^2+\eta^2)$, that is to say, a scattered intensity $4\eta^2/(\Delta\epsilon^2+\eta^2)^2$. If $\eta \to 0$, this last factor tends to $(2\pi/\eta)\ \delta(\Delta\epsilon)$; we thus recover the usual results of collision theory.

In order to obtain the nondiagonal elements of order $1/\eta$, $1/\eta^{3/2}$, etc., we must appeal to more complicated diagrams, such as that of Fig. 35. A primary transition B_1 is followed by two secondary transitions B_2 and B_3. Together these give a contribution of order $1/\eta^{3/2}$. We shall not explore these questions further; their interest is rather academic.

c. *Wave function and energy of the elementary excitations*

We now investigate the key problem: to what extent does the state

$$\mid \varphi_{\eta\mathbf{k}} > = U_\eta^*(0, -\infty)\, a_{\mathbf{k}} \mid 0 > = A_{\eta\mathbf{k}}^* U_\eta(0, -\infty) \mid 0 > \qquad \textit{(5–163)}$$

defined by Eqs. (5–130) to (5–132) really describe an elementary excitation of the real system? We have seen that $A_{\eta\mathbf{k}}^*$ is given by the diagrams of Fig. 32. These diagrams are analogous to those which we studied in the preceding section, except that the interaction core A can be moved only from $-\infty$ to 0. Again, real transitions can be produced; they give contributions of order $1/\sqrt{\eta}$, $1/\eta$, etc., similar to those calculated for S_η.

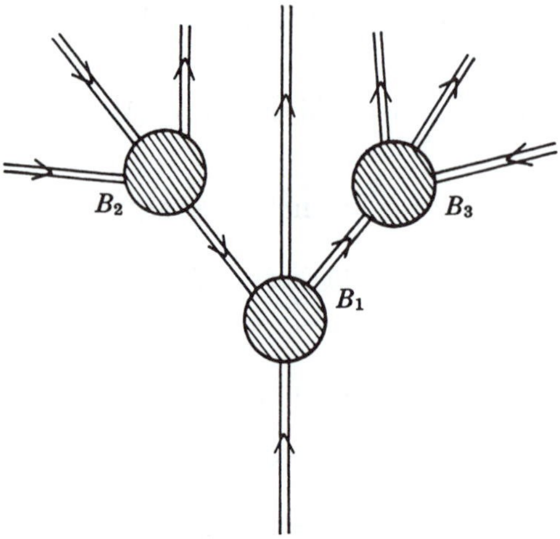

Figure 35

If $\eta \gg \langle \Gamma_k \rangle$, these "singular" terms are of little relative importance as they absorb only a negligible fraction of the total norm [see (5–153)]. Let us set aside the contributions from these real transitions; the rest of the state vector can be expanded in powers of η, as has been done for the ground state [Eqs. (5–59) to (5–62)].

It is essential for the state obtained to have a structure practically independent of the precise value of η, which has no physical meaning. This condition will be satisfied only if η stays within a narrow range:

(a) $1/\eta$ must be greater than the length Δ of a self-energy part; otherwise, corrections of order η become important. This condition amounts to requiring the development to be *adiabatic*: if the interaction is changed too quickly, the system no longer has time to follow and the states obtained are no longer eigenstates.

(b) On the other hand, η must be greater than the average inverse lifetime $\langle \Gamma_k \rangle$. On this condition, the terms of order $1/\eta$ arising from real transitions are of negligible norm. The rest of the wave function is regular, independent of η. This amounts to requiring the development to be *reversible*; if the interaction changes too slowly, the elementary excitation decays before the end of the operation.

η must therefore satisfy conditions (5–151), which limit us to the vicinity of the Fermi surface. With these reservations, the structure of the wave function no longer depends on the precise choice of η, which enters only in a constant phase factor, without physical importance.

Let us point out that our study of the ground state was *exact*, since we could pass to the limit $\eta \to 0$. This is no longer so for the excited states; all our results are *approximate*, subject to an error of order Γ_k, owing to the fact that η is bounded from below by the condition (5–151). This error is not due to a mathematical deficiency; it is of purely physical nature and expresses the uncertainty arising from real transition. This is an important aspect of the many-body problem and more generally of *dissipative* systems. In quantum electrodynamics, conversely, we find that certain excitations are undamped; we can then define their characteristics with as much precision as desired. In the rest of this section, we shall neglect this inaccuracy; it is nevertheless important to keep it in mind.

Let us turn now to more quantitative considerations. By going back to (5–132), (5–133), and (5–149), we can write $A^*_{\eta\mathbf{k}}$ in the form

$$\begin{cases} A^*_{\eta\mathbf{k}} = \mathrm{e}^{\varphi/\eta} \left\{ \int_{-\infty}^{0} \mathrm{d}t' \sqrt{z_k(g')} \exp\left(-\mathrm{i}\int_0^{t'} \varepsilon_k(g'')\,\mathrm{d}t''\right) J^*_{\eta\mathbf{k}t'} + \delta' \right\} \\ \varphi = -\mathrm{i}\eta \int_{-\infty}^{0} \left[\varepsilon_k(g'') - \varepsilon_k^0\right] \mathrm{d}t'' = -\mathrm{i}\int_0^1 \left[\varepsilon_k(g'') - \varepsilon_k^0\right] \dfrac{\mathrm{d}g''}{g''} \end{cases} \tag{5–164}$$

Let us recall that δ' takes account of the boundary effects near $t' = 0$ δ' is regular when $\eta \to 0$. Let us consider the matrix element $\langle n|A^*_{\eta\mathbf{k}}|0\rangle$ We can write it in a form analogous to (5–158),

$$\begin{cases} \langle n \mid A^*_{\eta\mathbf{k}} \mid 0 \rangle = \int_{-\infty}^{0} \mathrm{d}t\, f(g) \exp\left[\mathrm{i}\,\Delta\varepsilon(g)t\right] + \delta'_n \\ \Delta\varepsilon(g) = \varepsilon_n(g) - \varepsilon_k(g) \end{cases} \tag{5–165}$$

where $\varepsilon_n(g)$ is the energy of the elementary excitations contained in $|n\rangle$. If $\Delta\varepsilon(g) \neq 0$, we can apply (5–61). More generally, neglecting real transitions amounts to writing

$$\langle n \mid A^*_{\eta\mathbf{k}} \mid 0 \rangle = -\mathrm{i}\, f(1)\, P\left\{\frac{1}{\Delta\varepsilon(1)}\right\} + \delta'_n + O(\eta) \tag{5–166}$$

η thus enters only indirectly, to ensure the convergence of the integral in (5–165). We can replace (5–164) by

$$\begin{aligned} A^*_{\eta\mathbf{k}} &= \mathrm{e}^{\varphi/\eta} A^*_{\mathbf{k}} \\ A^*_{\mathbf{k}} &= \sqrt{z_k(1)} \int_{-\infty}^{0} \mathrm{d}t' \exp\left[-\mathrm{i}\,\varepsilon_k(1)\,t'\right] J^*_{0\mathbf{k}t'} + \delta' \end{aligned} \tag{5–167}$$

The operator $A^*_{\mathbf{k}}$ thus defined is independent of the precise choice of η

and thus is capable of representing a quasi particle. (5–167) is the generalization to excited states of (5–63).

We must still show that $A^*_{\mathbf{k}}|\varphi_0\rangle$ is an eigenstate of the real system: For the moment we have only a presumption. Furthermore, we do not know the energy of the elementary excitation. In order to make up these deficiencies, we need only repeat the proof given for the ground state [Eqs. (5–65) to (5–76)], replacing $U_\eta(0, -\infty)|0\rangle$ by $U_\eta(0, -\infty)a^*_{\mathbf{k}}|0\rangle$. Equation (5–74) is transformed into

$$(H - E^0_0 - \varepsilon^0_k)\, U_\eta(0, -\infty)\, a^*_{\mathbf{k}}\,|\,0> = i\eta g \frac{\mathrm{d}}{\mathrm{d}g} U_\eta(0, -\infty)\, a^*_{\mathbf{k}}\,|\,0> \Big|_{g=1} \tag{5-168}$$

Using (5–63), (5–131), and (5–167), we obtain

$$U_\eta(0, -\infty)\, a^*_{\mathbf{k}}\,|\,0> = A_{\eta\mathbf{k}}^*\, U_\eta(0, -\infty)\,|\,0> = \exp\left(\frac{a + \varphi}{\eta}\right) A^*_{\mathbf{k}}\,|\,\varphi_0> \tag{5-169}$$

By putting (5–169) back into (5–168), we obtain the relation, valid to within corrections of order η,

$$H\, A^*_{\mathbf{k}}\,|\,\varphi_0> = \left[E^0_0 + \varepsilon^0_k + \mathrm{i}g\frac{\mathrm{d}}{\mathrm{d}g}(a + \varphi)\right]_{g=1} A^*_{\mathbf{k}}\,|\,\varphi_0> \tag{5-170}$$

Let us use (5–76) and (5–164); we finally obtain

$$H\, A^*_{\mathbf{k}}\,|\,\varphi_0> = (E_0 + \varepsilon_k(1))\, A^*_{\mathbf{k}}\,|\,\varphi_0> \tag{5-171}$$

(5–171) shows that the operator $A^*_{\mathbf{k}}$ creates an *elementary excitation* of the physical system, whose energy is $\varepsilon_k(1)$. The adiabatic switching on of the interaction generates *eigenstates.*

The energy of the quasi particle, ε_k is just the pole of the Green's function $G(\mathbf{k},\omega)$. We thus show the *equivalence between the two treatments of the excited states*:

(a) Adiabatic switching on of the interaction
(b) The use of Green's functions

Method (b) is rather abstract and lends itself poorly to physical interpretation. But it is much more general than (a) and remains valid when the perturbation expansion diverges (in particular, for superfluid systems).

In order to complete the comparison of methods (a) and (b), it remains to show that the operator $A^*_{\mathbf{k}}$ defined by (5–167) is identical to that given by the relation (4–36). (4–36) is expressed in the Heisenberg representation. By referring to (5–95), (5–129), and (5–131), we can verify that

$$\mathbf{a}_{\mathbf{k}}^*(t) = U(0, t)\ a_{\mathbf{k}}^*(t)\ U(t, 0) = A_{0\mathbf{k}t}^* \tag{5–172}$$

($A_{0\mathbf{k}t}^*$ is given by the diagrams of Fig. 32, evaluated with $\eta = 0$.) To establish the desired equivalence we must prove that the relation

$$A_{\mathbf{k}}^* = \frac{\alpha}{\sqrt{z_k}} \int_{-\infty}^{0} \exp(-\, i\varepsilon_k t + \alpha t)\ A_{0\mathbf{k}t}^*\ dt \tag{5–173}$$

is satisfied if $\alpha \gg \Delta$.

Let us consider two diagrams contributing, respectively, to $A_{\mathbf{k}}^*$ and $A_{0\mathbf{k}t}^*$, identical after a certain time $t_1 \ll -\Delta$ (Fig. 36). The only difference is between the propagators for times before t_1, which give a factor

$$i \sqrt{z_k} \exp(-\, i\varepsilon_k t_1)$$

for the diagram on the left and $G(k, t_1 - t)$ for the diagram on the right. We can therefore write, for each component, $|n\rangle$,

$$\frac{< n \mid A_{\mathbf{k}}^* \mid 0 >}{< n \mid A_{0\mathbf{k}t}^* \mid 0 >} = \frac{1}{\sqrt{z_k}} \exp(-\, i\varepsilon_k t) + \beta(t) \tag{5–174}$$

the correction $\beta(t)$ arising from diagrams for which t_1 is too close to t. The phase of $\beta(t)$ is fixed by the upper lines and thus varies as $e^{-i\varepsilon_n t}$.

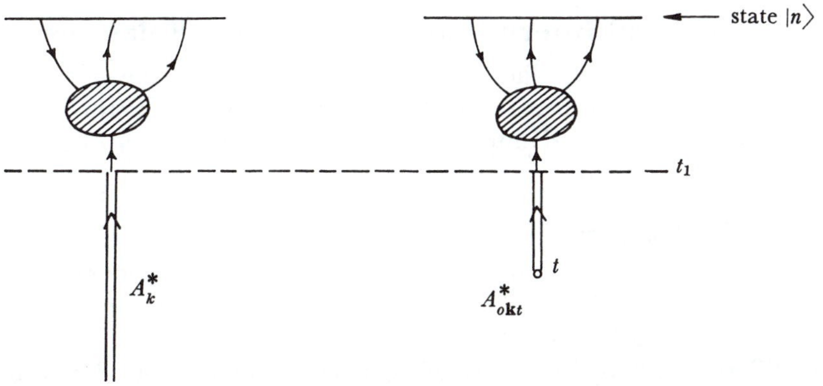

Figure 36

(5–173) is then automatically verified, the *filtering* of frequencies getting rid of the correction $\beta(t)$. The equivalence of our two approximation methods is therefore complete.

The perturbation method is very flexible. By way of example, we calculate the probability of finding a *bare* particle in a quasi particle. The corresponding probability amplitude is equal to

$$\begin{aligned} < \varphi_0 \mid a_{\mathbf{k}} \overset{*}{A_{\eta\mathbf{k}}} \mid \varphi_0 > &= \frac{< 0 \mid U_\eta(+\infty, 0)\, a_{\mathbf{k}}\, U_\eta(0, -\infty)\, \overset{*}{a_{\mathbf{k}}} \mid 0 >}{< 0 \mid S_\eta \mid 0 >} \\ &= \frac{G_\eta(0, -\infty)}{G_0(0, -\infty)} \end{aligned} \tag{5–175}$$

The modulus of this quantity is equal to $\sqrt{z_k}$, and the probability desired is therefore equal to z_k. Similar reasoning allows us to recover the other results of Chap. 4.

We have already mentioned the important theorem of *Van Hove*: the energy of a quasi particle at the Fermi surface is equal to the chemical potential. Within the framework of this section, this theorem becomes obvious. The ground state of $(N + p)$ particles can be written as

$$U_\eta(0, -\infty)\, \overset{*}{a_{\mathbf{k}_1}} \dots \overset{*}{a_{\mathbf{k}_p}} \mid 0 >$$

where the states $\mathbf{k}_1, \ldots, \mathbf{k}_p$ are those closest to the Fermi surface. This state is also obtained by adding to $|\varphi_0\rangle$ the corresponding quasi particles. This simple fact leads immediately to the desired result.

We shall not explore further this aspect of elementary excitations, which does little more than duplicate the results of Chap. 4. Let us simply note that this section gives us a more coherent view of the whole perturbation method, based on the adiabatic switching on of the interaction.

5. Graphical Interpretation of Various Relations

The use of Feynman diagrams considerably reduces the calculations and facilitates the interpretation of the various results. In this section, we shall illustrate these advantages by establishing a number of relations by graphical considerations.

a. Calculation of the ground-state energy

Let ΔE_0 be the variation in the ground-state energy produced by the interaction H_I, given by (5–80). From (5–79), ΔE_0 is obtained from the unlinked connected diagrams having *some vertex fixed at time* $t = 0$. Two diagrams differing only by the choice of the fixed vertex are considered to be identical [we have already emphasized this point when stating Eq. (5–52)].

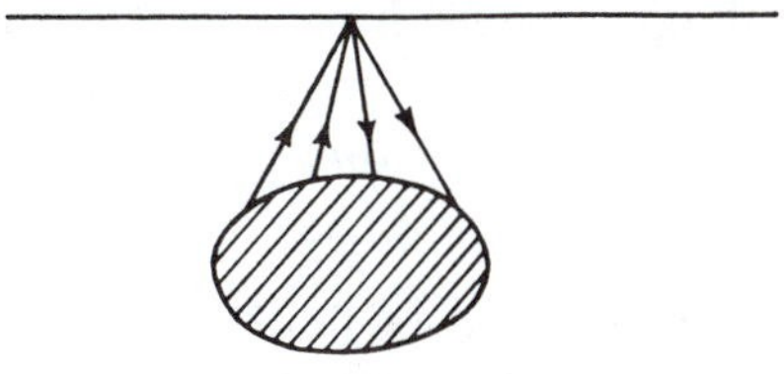

Figure 37

To avoid counting the same diagram more than once, we therefore need an additional rule determining which vertex must be fixed. We can, for example, take the vertex at the maximum time. The diagrams then have the structure indicated in Fig. 37, where all the movable vertices are at negative times. We can decompose this diagram into:

(a) The upper vertex, giving a factor H_I.

(b) A linked diagram having two lines leaving and two lines entering from above, which normally contributes to $U_{\eta L}(0, -\infty)$.

Collecting the constant factors, we can easily verify that

$$\Delta E_0 = \langle 0 | H_{\mathrm{I}} \, U_{\eta L}(0, -\infty) | 0 \rangle \tag{5–176}$$

Using (5–45), we transform (5–176) into

$$\Delta E_0 = \frac{\langle 0 | H_{\mathrm{I}} \, U_\eta(0, -\infty) | 0 \rangle}{\langle 0 | U_\eta(0, -\infty) | 0 \rangle} = \frac{\langle 0 | H_{\mathrm{I}} | \varphi_0 \rangle}{\langle 0 | \varphi_0 \rangle} \tag{5–177}$$

(5–177) is a well-known result, used by Goldstone. We can establish it immediately by noting that

$$\begin{aligned} \langle 0 | H | \varphi_0 \rangle &= E_0 \langle 0 | \varphi_0 \rangle \\ &= \langle 0 | H_0 + H_{\mathrm{I}} | \varphi_0 \rangle = E_0^0 \langle 0 | \varphi_0 \rangle + \langle 0 | H_{\mathrm{I}} | \varphi_0 \rangle \end{aligned} \tag{5–178}$$

Actually, starting from (5–178), we can develop an explicit method of calculation with the help of (5–177).

This analysis of the linked connected diagrams is not the only one possible. On the contrary, we can *label* the fixed vertex to distinguish it from the others. A diagram of order p thus appears p times, since we can select any one of the vertices. The series obtained no longer gives ΔE_0, but

$$g \frac{\mathrm{d}}{\mathrm{d}g} \Delta E_0(g) \bigg|_{g=1} = g \frac{\mathrm{d}E_0}{\mathrm{d}g} \bigg|_{g=1}$$

[see Eq. (5–74)]. These diagrams still have the structure of Fig. 37, but now the movable vertices are free to go anywhere. By removing the fixed vertex, we obtain a diagram contributing to the two-particle Green's function, and no longer to $U_{\eta L}(0, -\infty)$.

Note: In general, each diagram contributing to K appears only once. However, if we start from an unlinked diagram containing two equivalent vertices, we shall obtain the same diagram of K by removing either of these vertices (by "equivalent vertices," we mean that the diagram remains invariant if they are permuted). In fact, the unlinked diagram then contains a factor $\frac{1}{2}$, arising from permutation invariance. The contribution to K is thus changed by a factor $2 \times \frac{1}{2} = 1$. We can avoid all difficulty of this type by returning to labeled diagrams; we are thus assured of counting each diagram only once.

Let us exploit this decomposition by using the sets of rules (5–35), (5–79) and (5–100). After all calculations, we obtain

$$g \frac{\mathrm{d}E_0}{\mathrm{d}g}\bigg|_{g=1} = \frac{1}{4} \sum_{\mathbf{k}_i} V(\mathbf{k}_1, \ldots, \mathbf{k}_4)\, K(\mathbf{k}_1 0, \mathbf{k}_2 0, \mathbf{k}_3 0_+, \mathbf{k}_4 0_+) \tag{5–179}$$

The times 0_+ associated with $\mathbf{k}_3$ and $\mathbf{k}_4$ serve to remove the indeterminacy caused by the closed loops (see the beginning of the chapter). The factor $\frac{1}{4}$ arises from the invariance of the whole diagram when the lines (1,2) or (3,4) are permuted. In proving (5–179), it is necessary to pay great attention to questions of sign, since the decomposition of the diagram "opens" a certain number of closed loops.

Let us go back to the definition (5–17) and take the Fourier transform. Referring to (3–78), we see that

$$g \frac{\mathrm{d}E_0}{\mathrm{d}g}\bigg|_{g=1} = E_{int} = \langle \varphi_0 | H_I | \varphi_0 \rangle \tag{5–180}$$

Here again, relation (5–180) can be established immediately; for this purpose, we differentiate with respect to g the quantity

$$E_0(g) = \langle \varphi_0(g) | H_0 + g H_I | \varphi_0(g) \rangle \tag{5–181}$$

The derivatives of the state vectors give a zero contribution [since the norm of $|\varphi_0(g)\rangle$ is independent of g]. Only the derivative of gH_I remains —hence (5–181). Our formalism is therefore consistent.

Finally, we can obtain E_{int} in a completely different way, by starting from the rules (5–108). Let us represent an unlinked connected diagram of order p in the manner indicated in Fig. 38. Any line can be selected.

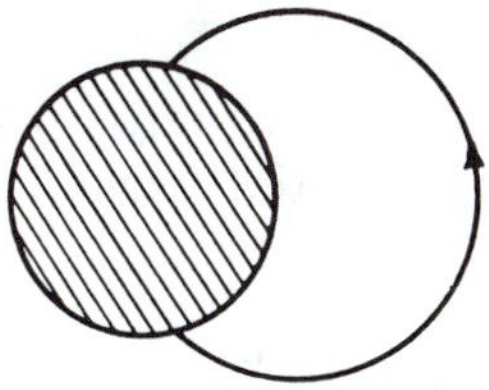

Figure 38

If we sum over all cores A, we find $2p$ times each of the diagrams of ΔE_0. Instead of ΔE_0, we thus calculate

$$2\,g\,\frac{\mathrm{d}}{\mathrm{d}g}\,\Delta\,E_0(g)$$

which, according to (5–180), is just $2E_{\text{int}}$.

Note: If the initial unlinked diagram has two equivalent lines, we obtain it by the above procedure only $2p-1$ times: $2p-2$ times with the factor $\frac{1}{2}$ due to the equivalence, 1 time without this factor $\frac{1}{2}$; this is equivalent to $2p$ times the actual contribution to ΔE_0. Again, we can avoid the difficulty by going back to labeled diagrams.

The core A is composed of some number $\geqslant 1$ of self-energy parts. The corresponding contribution is equal to

$$\frac{M(\mathbf{k},\omega)}{1-G_0(\mathbf{k},\omega)\,M(\mathbf{k},\omega)}=\frac{1}{G_0(\mathbf{k},\omega)}\left\{\frac{G(\mathbf{k},\omega)}{G_0(\mathbf{k},\omega)}-1\right\} \qquad \textit{(5–182)}$$

By going back to the rules (5–105) and (5–108), we obtain

$$2\,E_{int}=-\frac{\mathrm{i}}{2\pi}\sum_{\mathbf{k}}\int_C \mathrm{d}\omega\left[\frac{G(\mathbf{k},\omega)}{G_0(\mathbf{k},\omega)}-1\right] \qquad \textit{(5–183)}$$

[where the contour C is composed of the real axis and the infinite semi-circle in the upper half plane; see the remark which follows (5–105)]. Let us replace G_0 by its explicit form; (5–183) becomes

$$E_{int}=\frac{-\mathrm{i}}{4\pi}\sum_{\mathbf{k}}\int_C \mathrm{d}\omega(\varepsilon_k^0-\omega)\,G(\mathbf{k},\omega) \qquad \textit{(5–184)}$$

We recover the relation (3–33).

b. The dynamical equation

Let us consider the dynamical equation (3–75) and take the Fourier transform with respect to the spatial variables. By using the definition (5–17), we obtain

$$\left(\frac{\partial}{\partial t} + \mathrm{i}\,\varepsilon_k^0\right) G(\mathbf{k}, t, t') + \tfrac{1}{2} \sum_{\mathbf{k}_1\mathbf{k}_2\mathbf{k}_3} V(\mathbf{k}_1, \mathbf{k}_2, \mathbf{k}_3, \mathbf{k})\; K(\mathbf{k}_1 t, \mathbf{k}_2 t, \mathbf{k}_3 t_+, \mathbf{k} t') = \mathrm{i}\delta(t - t') \qquad (5\text{–}185)$$

We propose to interpret (5–185) with the help of diagrams.

The diagrams contributing to $G(\mathbf{k},t,t')$ can be analyzed in the manner indicated in Fig. 39. Every diagram of order $\geqslant 1$ thus appears as the combination of a vertex and a two-particle Green's function. By using the rules for calculation of G and K, we can verify that

$$G(\mathbf{k}, t, t') = G_0(\mathbf{k}, t, t') + \frac{\mathrm{i}}{2} \sum_{\mathbf{k}_1\mathbf{k}_2\mathbf{k}_3} \int \mathrm{d}t_1\, G_0(\mathbf{k}, t, t_1)\, V(\mathbf{k}_1, \mathbf{k}_2, \mathbf{k}_3, \mathbf{k})\; K(\mathbf{k}_1 t_1, \mathbf{k}_2 t_1, \mathbf{k}_3 t_{1+}, \mathbf{k} t') \qquad (5\text{–}186)$$

Figure 39 is the "graphical" version of Eq. (5–186).

Let us apply to (5–186) the operator

$$(\partial/\partial t + \mathrm{i}\,\varepsilon_k^0)$$

According to (3–17), we have

$$\left(\frac{\partial}{\partial t} + \mathrm{i}\,\varepsilon_k^0\right) G_0(\mathbf{k}, t, t') = \mathrm{i}\,\delta(t - t') \qquad (5\text{–}187)$$

From this we immediately deduce the dynamical equation (5–185). The

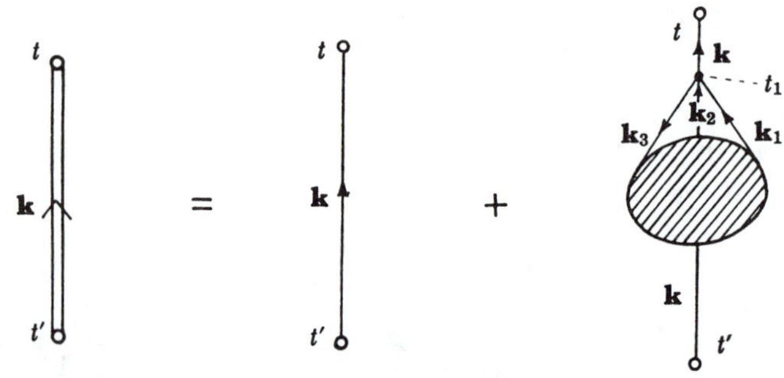

Figure 39

diagrammatic analysis therefore furnishes us with a first integral of this equation.

c. A relation between the self-energy and interaction operators

In conclusion, we prove a new relation between $M(\mathbf{k},\omega)$ and the interaction operator $\Gamma(\mathbf{k}_i,\omega_i)$ defined in Eq. (3–50). Let us consider a diagram of order p contributing to $M(\mathbf{k},\omega)$; it contains $(2p-1)$ lines. Let us select one of these (Fig. 40*a*.) If we sum over all cores A, we shall obtain each M diagram $(2p-1)$ times, thus calculating

$$2g\frac{\partial}{\partial g}M(\mathbf{k},\omega,g)\bigg|_{g=1} - M(\mathbf{k},\omega) \tag{5–188}$$

In order to introduce the interaction operator, we explicitly represent the self-energy parts located on both sides of the chosen line (Fig. 40*b*). The chain going from the point P to the point Q gives a factor

$$\frac{G^2(\mathbf{k}',\omega')}{G_0(\mathbf{k}',\omega')} \tag{5–189}$$

(the line selected contributing at the same time to the propagators above and below). The core B that remains is made up of a single block. In fact, if it were separated into two independent parts B_1 and B_2, we would arrive at one of the two situations indicated by Fig. 41. These two diagrams are to be eliminated, the first because it contains an unlinked connected part, the second because the original diagram of M is reducible. As a consequence, the core B gives a factor

$$\Gamma(\mathbf{k}\,\omega, \mathbf{k}'\omega', \mathbf{k}'\omega', \mathbf{k}\omega)$$

We again find the difficulty mentioned in Chap. 4 [Eqs. (4–85) and

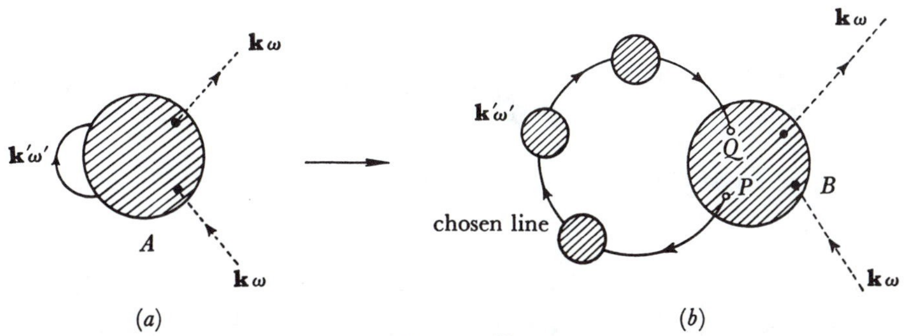

Figure 40

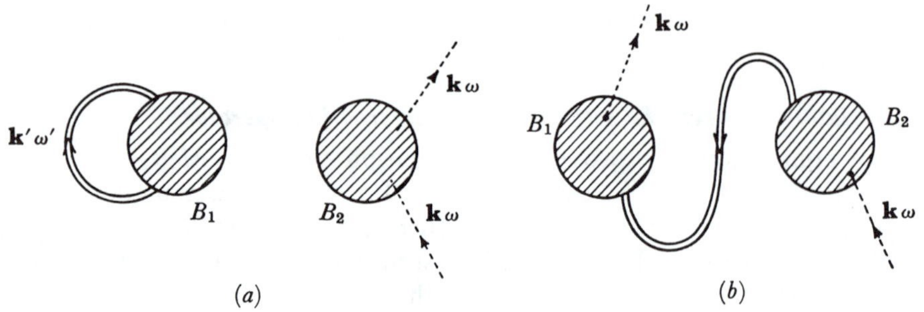

Figure 41

(4–86)]: the "forward" scattering amplitude is defined ambiguously. Its value depends on the order in which we make the transfers of energy and momentum tend to 0. In the present case, this uncertainty is easily removed; whereas conservation of momentum is always rigorous, conservation of energy is only approximate, to within an amount of the order of the velocity η of the adiabatic switching on. We must therefore use the limit Γ°, as in (4–85).

Putting together all these preliminary results, we arrive at the relation

$$\int_C d\omega' \sum_{\mathbf{k}'} (\varepsilon^0_{k'} - \omega')\, \Gamma^0(\mathbf{k}\omega, \mathbf{k}'\omega', \mathbf{k}'\omega', \mathbf{k}\omega)\, G^2(\mathbf{k}'\omega') = 2g \frac{\partial}{\partial g} M(\mathbf{k}, \omega)\Big|_{g=1} - M(\mathbf{k}, \omega) \tag{5-190}$$

(5–190) allows us to calculate the derivative $d\varepsilon_k/dg$ of the quasiparticle energy. In order to do this, we differentiate Eq. (5–140) with respect to g,

$$\frac{d\varepsilon_k}{dg}\left[1 + \frac{\partial M}{\partial \varepsilon}\Big|_{\varepsilon=\varepsilon_k}\right] + \frac{\partial}{\partial g} M(\mathbf{k}, \varepsilon_k) = 0 \tag{5-191}$$

Let us take $\partial M/\partial g$ from (5–190); we obtain

$$\frac{2g}{z_k}\frac{d\varepsilon_k}{dg} = \varepsilon_k - \varepsilon^0_k - \int_C d\omega' \sum_{\mathbf{k}'} (\varepsilon^0_{k'} - \omega')\, \Gamma^0(\mathbf{k}\varepsilon_k, \mathbf{k}'\omega', \mathbf{k}'\omega', \mathbf{k}\varepsilon_k)\, G^2(\mathbf{k}', \omega') \tag{5-192}$$

(5–192) allows us to study the development of the quasi particle as a function of the coupling constant.

6. Renormalization of the Propagators

a. Skeleton diagrams. Functional form of the linked diagrams

Let us take a linked diagram and remove all the self-energy parts it contains. The diagram is thus reduced to a "skeleton". The whole diagram can be obtained by inserting into this skeleton a certain number of self-energy parts at suitable places (see Fig. 42). Instead of considering the complete set of diagrams, associating a factor G_0 with each line, we can limit ourselves to *skeleton diagrams*, provided that we now associate a factor G with each line. This amounts to a partial summation of the perturbation series, which considerably reduces the extent of the calculations.

The result now becomes a *functional* of the function of four variables, $G(\mathbf{k},\omega)$. This functional is extremely complex, since it contains an arbitrarily large number of factors G. We can define it only term by term, by means of the above perturbation expansion. For example, let us consider the self-energy operator $M(\mathbf{k},\omega)$. It is given by the set of skeletons having one entry and one exit point (Fig. 43). The value of M depends on the exact form of the factor G associated with each internal line of the skeleton. We represent this functional dependence in the usual form,

$$M[G]$$

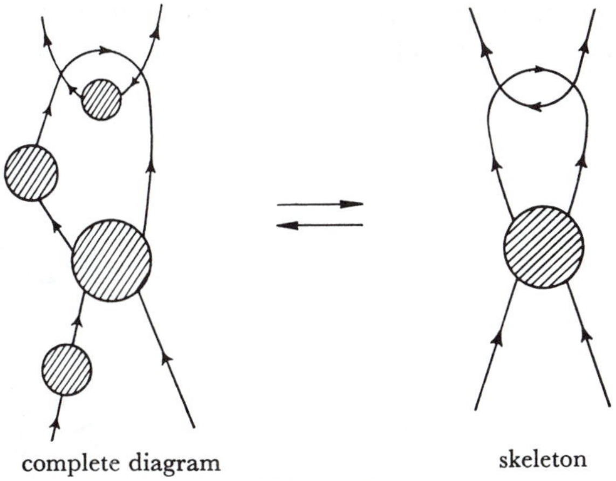

Figure 42

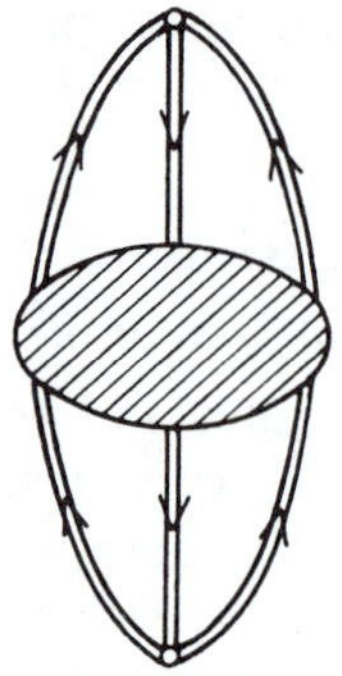

Figure 43

If we change $G(\mathbf{k},\omega)$ by an amount

$$\delta G(\mathbf{k}, \omega) = \varepsilon\, \delta_{\mathbf{kk}'}\, \delta(\omega - \omega') \qquad (5\text{–}193)$$

where ε is an infinitesimal and $\mathbf{k}'$ and ω' given values of the parameters, M varies by an amount δM. By definition

$$\frac{\delta M}{\varepsilon} = \frac{\delta M[G]}{\delta G(\mathbf{k}', \omega')} \qquad (5\text{–}194)$$

is called the functional derivative of M with respect to $G(\mathbf{k}',\omega')$.

b. Functional form of the ground-state energy—variational principle

The preceding discussion applies to all linked diagrams. For example, the interaction operator $\gamma(\mathbf{k}_i,\omega_i)$ is a functional of $G(\mathbf{k}',\omega')$, defined by the expansion in skeleton diagrams. This is no longer so when we turn to unlinked diagrams. For example, let us consider the diagram of Fig. 44.

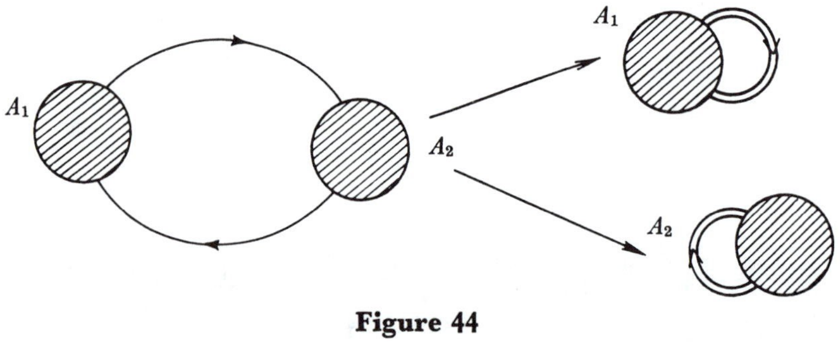

Figure 44

To reduce it to a skeleton, we should remove the cores A_1 and A_2; the result will be of order 0 and will not make sense. We are therefore compelled to leave at least one self-energy part. At this point, there is an ambiguity: must we remove A_1 or A_2? As a consequence, there exists no unlinked skeleton. ΔE_0 cannot be expressed directly as a functional of G.

In order to circumvent this difficulty, we go back to (5–181) and (5–183), which can be written as

$$E_{\text{int}} = g\frac{\mathrm{d}E_0}{\mathrm{d}g} = -\frac{\mathrm{i}}{4\pi}\sum_{\mathbf{k}}\int_C \mathrm{d}\omega\, M(\mathbf{k}, \omega)\, G(\mathbf{k}, \omega) \tag{5–195}$$

Contrary to ΔE_0, E_{int} can be expressed as a functional of G. In order to go back to ΔE_0, we need only integrate with respect to g. This very clever artifice is due to Klein; we follow his treatment very closely. In order to simplify the notation, we represent the set of operators

$$\frac{\mathrm{i}}{2\pi}\sum_{\mathbf{k}}\int_C \mathrm{d}\omega$$

by the single symbol tr. We thus have

$$E_{\text{int}} = -\tfrac{1}{2}\,\mathrm{tr}\, MG \tag{5–196}$$

We have seen that M is a functional of G, defined by skeleton diagrams. Let us designate by $g^n M_n$ the contribution from all skeletons containing n vertices. We have

$$M = \sum_{n=1}^{\infty} g^n M_n \tag{5–197}$$

Actually, each M_n depends indirectly on the coupling constant g, by way of its constituent factors G. M_n is thus *not* the term of order n in the usual expansion but simply the part of the functional M containing $(2n - 1)$ factors G (a skeleton of M with n vertices contains $(2n - 1)$ lines). It is important to note this point carefully.

Let us consider the quantity

$$\mathrm{tr}\, M_n\, G$$

It is natural to represent it by the diagram of Fig. 45 (the line G serves to

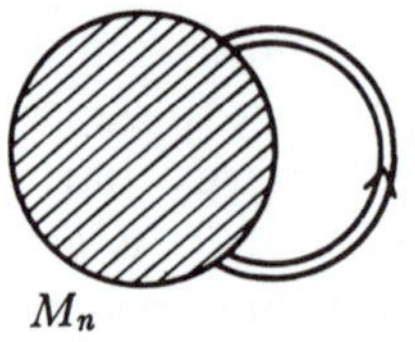

Figure 45

close the skeleton arising from M_n). Let us now calculate

$$\frac{\partial}{\partial g}\{\text{tr } M_n\, G\}$$

Differentiation of the factor G gives a term

$$\text{tr } M_n \frac{\partial G}{\partial g} \tag{5-198}$$

In order to differentiate M_n, we note that this functional depends on g only through the $(2n-1)$ factors G which it contains. $\partial M_n/\partial g$ is therefore the sum of $(2n-1)$ terms, each arising from one line of the skeleton. Let us remove this line, which gives a factor $\partial G(\mathbf{k}',\omega')/\partial g$ Figure 45 then reduces to a diagram contributing to $M_n(\mathbf{k}',\omega')$. Since the operator tr amounts to summing over $\mathbf{k}'$ and ω', each of the terms of the derivative $\partial M_n/\partial g$ is just (5–198). We thus arrive at the important result

$$\frac{\partial}{\partial g}\left\{\text{tr } M_n\, G\right\} = 2n \text{ tr } M_n \frac{\partial G}{\partial g} \tag{5-199}$$

Note: As in the preceding section, we should study the case of equivalent lines carefully; we obtain the same diagram of $M_n(\mathbf{k}',\omega')$ by removing either of the lines. This error is automatically corrected by the factor $\frac{1}{2}$ in the initial expansion of M_n.

We are now in a position to give a functional form for the energy ΔE_0. According to (5–195), we have

$$\Delta E_0 = \int_0^1 \frac{dg'}{g'}\, E_{\text{int}}(g') = -\tfrac{1}{2}\int_0^1 \frac{dg'}{g'}\, \text{tr } M(g')\, G(g') \tag{5-200}$$

Let us replace M by its expansion (5–197) and integrate by parts, using (5–199). We find

$$\begin{aligned}\Delta E_0 &= -\tfrac{1}{2}\sum_{n=1}^{\infty}\frac{g^n}{n}\operatorname{tr} M_n\, G + \sum_{n=1}^{\infty}\int_0^1 g'^n \operatorname{tr}\left[M_n(g')\,\frac{\partial G(g')}{\partial g'}\right] dg' \\ &= -\tfrac{1}{2}\sum_{n=1}^{\infty}\frac{g^n}{n}\operatorname{tr} M_n\, G + \int_0^1 \operatorname{tr}\left[M(g')\,\frac{\partial G}{\partial g'}\right] dg' \end{aligned} \qquad (5\text{–}201)$$

According to (5–110), M is related to G by the equation

$$M = \frac{1}{G_0} - \frac{1}{G} \qquad (5\text{–}202)$$

where G_0 is the propagator in the absence of interactions. We can therefore carry out the integration over g' in (5–201), which leads us to

$$\Delta E_0 = \Phi[G] + \operatorname{tr}\left\{\frac{G}{G_0} - 1\right\} - \operatorname{tr}\left\{\ln\frac{G}{G_0}\right\} \qquad (5\text{–}203)$$

or

$$\Phi[G] = -\tfrac{1}{2}\sum_{n=1}^{\infty}\frac{g^n}{n}\operatorname{tr} M_n[G]\, G \qquad (5\text{–}204)$$

$\Phi[G]$ is a functional of G, defined through the decomposition of M_n into skeleton diagrams. Our objective has thus been attained; we have expressed ΔE_0 *in the form of a functional of the propagator G.*

The expression (5–203) is of great importance, as it is stationary with respect to changes in the value of $G(\mathbf{k},\omega)$. In other words,

$$\frac{\delta(\Delta E_0)}{\delta G(\mathbf{k},\omega)} = 0 \qquad (5\text{–}205)$$

This is a particularly interesting *variational principle*, which explains why numerical calculations of ΔE_0 are insensitive to the choice of the elementary excitation energies.

To establish (5–205), we must calculate the functional derivative of ΔE_0. We see at once that

$$\frac{\delta \Delta E_0}{\delta G(\mathbf{k},\omega)} = \frac{\delta\Phi[G]}{\delta G(\mathbf{k},\omega)} + \frac{1}{G_0(\mathbf{k},\omega)} - \frac{1}{G(\mathbf{k},\omega)} \qquad (5\text{–}206)$$

$\delta\Phi/\delta G$ is evaluated by repeating the arguments which led us to (5–198); we consider the term of order n in (5–204). Its derivative contains $2n$ terms arising from the $2n$ lines of the diagram; these terms are all equal (since we sum over all possible diagrams). Therefore,

$$\frac{\delta\Phi[G]}{\delta G(\mathbf{k},\omega)} = -\frac{1}{2}\sum_{n=1}^{\infty}\frac{g^n}{n}\,2n\,M_n(\mathbf{k},\omega) = -\,M(\mathbf{k},\omega) \tag{5–207}$$

Comparing this result with (5–202), we immediately deduce the variational principle (5–205).

We can consider ΔE_0 as a functional of M, instead of G. We just have to replace G in (5–203) by its explicit value,

$$G = \frac{G_0}{1 - G_0 M} \tag{5–208}$$

We thus have

$$\Delta E_0 = \Phi\left[\frac{G_0}{1 - G_0 M}\right] + \mathrm{tr}\left(\frac{MG_0}{1 - G_0 M}\right) + \mathrm{tr}\left\{\ln(1 - G_0 M)\right\} \tag{5–209}$$

We then can easily verify that

$$\frac{\delta(\Delta E_0)}{\delta M(\mathbf{k},\omega)} = 0 \tag{5–210}$$

Note: We can give a different, more complicated form to $\Delta E_0[G]$. For example, in certain terms of (5–209), we can replace M by its functional expansion (5–197). ΔE_0 thus becomes a *new* functional of G, which is equal to the functional (5–203) when G takes its physical value. This equality is accidental and disappears when G is changed. The variational principle is true only if ΔE_0 is expressed *exactly* in the form (5–203).

This variational principle was proved for the first time by Luttinger and Ward. It is very important. We give some applications of it below.

c. *Application of the variational principle; quasi-particle energy*

We have seen that the state $A_{\mathbf{k}}^*|\varphi_0\rangle$, containing a quasi particle ($k > k_F$), can be adiabatically generated, starting from the unperturbed state $a_{\mathbf{k}}^*|0\rangle$, on condition, however, that the time η^{-1} of switching on of the interaction be much less than the lifetime $\Gamma_k{}^{-1}$. We have described this process, keeping the same propagators G_0 as for the ground state, which leads us to diagrams having a line $\mathbf{k}$ entering from below. We now adopt a different attitude, and we modify $G_0(\mathbf{k},\omega)$ so as to take account of the new bare particle $\mathbf{k}$,

$$G_0 = \frac{1}{k^2/2m - \omega - i\,\delta} \quad \text{becomes} \quad \frac{1}{k^2/2m - \omega + i\,\delta} \tag{5-211}$$

The state $A^*_{\mathbf{k}}|\varphi_0\rangle$ is then obtained from diagrams which have *no* line below. Instead of being manifested by a line below, the quasi particle produces a correction to G_0. Of course, this change has an effect on all physical quantities, in particular on the Green's functions $G(\mathbf{k}',\omega')$.

δ is a positive infinitesimal. We can keep it so on condition that we choose $\eta \gg \Gamma_k$. This will have the effect of ensuring an automatic convergence if one of the vertices of the diagram occurs earlier than $t = -\eta^{-1}$. In practice, we obtain the same result by replacing the infinitesimal δ in G_0 by the finite quantity η. In other words, convergence is assured by acting not on the interaction vertices, but on the propagators.

Let us now consider the Green's function G, given by

$$G(\mathbf{k}', \omega) = \frac{1}{G_0^{-1} - M} = \frac{1}{k'^2/2m - \omega - M(\mathbf{k}', \omega) \pm i\delta}$$

What happens to G when a quasi particle $\mathbf{k}$ is introduced? M experiences only a very small correction, of order $1/\Omega$ (the probability for one of the internal lines of M to have wave vector $\mathbf{k}$ is of order $1/\Omega$). If $\mathbf{k}' \neq \mathbf{k}$, this is the only correction to G; it is unimportant and does not affect the position of the pole of $G(\mathbf{k}',\omega)$. On the other hand, if $\mathbf{k}' = \mathbf{k}$, the term G_0 also changes; G now takes on the new value

$$G(\mathbf{k}, \omega) + \delta G(\mathbf{k}, \omega) = \frac{1}{k^2/2m - \omega - M(\mathbf{k}, \omega) + i\eta} \tag{5-212}$$

The pole $(\varepsilon_k - i\Gamma_k)$ is displaced to

$$\omega = \varepsilon_k - i\,\Gamma_k + i\eta$$

But we know that η must be $\gg \Gamma_k$; the pole of G has therefore crossed the real axis. This is quite natural, since the new elementary excitation of wave vector $\mathbf{k}$ is a hole, and no longer a particle. If Γ_k and η are small, the correction δG can be written as

$$\delta G(\mathbf{k}, \omega) = -2\pi\, i z_k \delta(\omega - \varepsilon_k) \tag{5-213}$$

It is of order unity, instead of $1/\Omega$ when $\mathbf{k}' \neq \mathbf{k}$. We thus recover the results obtained in Chap. 4.

We are now equipped to calculate the excitation energy of the state $A_{\mathbf{k}}^*|\varphi_0\rangle$,

$$\delta E = \langle \varphi_0 | A_{\mathbf{k}} H A_{\mathbf{k}}^* | \varphi_0 \rangle - \langle \varphi_0 | H | \varphi_0 \rangle$$

In the absence of interaction, δE is equal to $k^2/2m$. The correction due to the interaction can be evaluated with the help of (5–203). In the presence of the quasi particle, G is changed by an amount

$$\delta G(\mathbf{k}'\omega) \sim \begin{cases} 1 & \text{if} \quad \mathbf{k}' = \mathbf{k} \\ 1/\Omega & \text{if} \quad \mathbf{k}' \neq \mathbf{k} \end{cases}$$

Actually, the number of values $\mathbf{k}' \neq \mathbf{k}$ is of order Ω; the corresponding contribution to (5–203) is important, and the calculation appears to be difficult. However, we are saved by the variational principle; by virtue of (5–205), the variations δG *have no effect* on ΔE_0; we do not have to concern ourselves with them. Only the explicit variation of G_0 remains. Let us emphasize the fundamental role played by the variational principle (5–205).

The second term of (5–203), proportional to $1/G_0$, does not change appreciably in the transformation (5–211). It remains to consider the last term. Its variation is more subtle, since it is connected with the

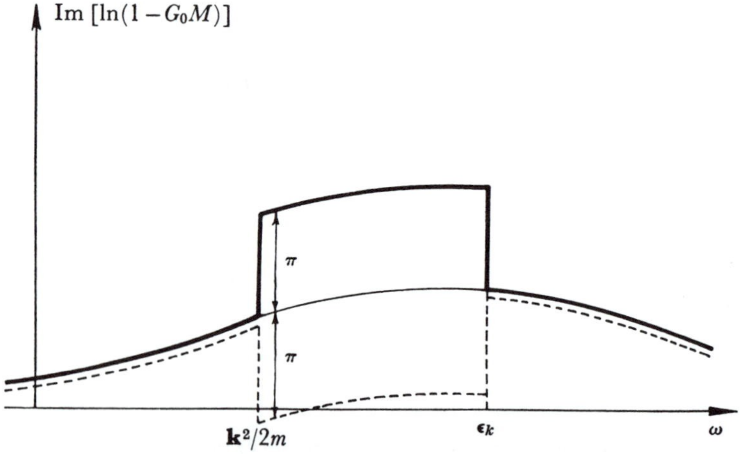

Figure 46

evaluation of the logarithm. If we go back to (5–201), we see that this term enters in the form

$$+ \text{tr} \ln (1 - G_0 M)$$

The real part of this logarithm is well defined and is not modified when the quasi particle is introduced. Let us now consider its imaginary part. In the ground state, it suffers a discontinuity $+i\pi$ at the pole of G_0, $-i\pi$ at the pole of G. In the excited state, these discontinuities are reversed. The situation is illustrated by Fig. 46, where the solid line corresponds to the ground state, the dotted line to the excited state. We see that the introduction of the quasi particle changes the value of the logarithm by $-2i\pi$ in the interval $(k^2/2m, \varepsilon_k)$, where ε_k is the pole of G.

Putting this result back into (5–203), and using the definition of the operator tr, we immediately see that the excitation of the quasi particle $\mathbf{k}$ causes ΔE_0 to change by the quantity

$$(\varepsilon_k - k^2/2m)$$

As a consequence, the excitation energy of the quasi particle is equal to ε_k, pole of the Green's function; we recover the result proved above.

7. Deformations of the Fermi Surface

We saw in section 5 that an elementary excitation of the "particle" type can in no case be transformed into a hole, when the perturbation is switched on adiabatically. The Fermi surface S_F is thus "predetermined," exactly like that of the unperturbed system. In the case studied until now, it is a sphere of radius

$$k_F = (3\pi^2 N)^{1/3}$$

containing N values of the parameter $\mathbf{k}$. This introduces an implicit limitation on the validity of the preceding methods, which we now propose to discuss.

The basic hypothesis of a perturbation expansion is the continuity of all physical quantities as functions of the coupling constant g in the interval $0 \to 1$; we eliminate any possibility of transition, phase change, etc. An equivalent hypothesis consists in assuming the continuity of all quantities as functions of the density $\rho = N/\Omega$ (if ρ is small, it is physically evident that g is small). We saw in Chap. 4 that this simple hypothesis implies

$$V_F = 4\pi^3 N/\Omega \tag{5-214}$$

where V_F is the volume contained by the Fermi surface [see (4–46)]. The fact that S_F contains N values of $\mathbf{k}$ is therefore inherent in any perturbation method. It is a very general result, which can disappear only in pathological cases (if, for example, some of the particles occupy "deep" bound states, as in an impure solid). In the case of *normal* systems, no such complication arises.

On the other hand, we can easily envisage a *continuous* deformation of the Fermi surface as a function of g. For example, particles of mass m can have an anisotropic interaction; S_F then has no reason for being a sphere. Or the particles can have an anisotropic mass; the unperturbed Fermi surface is then an ellipsoid, and S_F can have a very complicated form (while retaining a given volume V_F). Here is a class of phenomena which defy the preceding methods. The situation is further complicated if we take account of spin; the two directions of spin can have different Fermi surfaces (symmetric with respect to the origin because of time-reversal symmetry).

Before describing a formalism applicable to the most general case, let us state exactly why the method developed in the preceding sections will not work. Until now, we have defined the states of the real system by turning on the interaction adiabatically; the Hamiltonian at time t is given by

$$H(t) = H_0 + g\,H_{\mathrm{I}}\,, \qquad g = e^{\eta t} \tag{5-215}$$

To each value of g there corresponds a class of eigenstates $|\varphi(g)\rangle$. In particular, the ground state $|\varphi_0(g)\rangle$ corresponds to a Fermi surface $S_F(g)$. A priori, the only condition imposed on $S_F(g)$ is that it enclose a constant volume [Eq. (5–214)]. For an anisotropic system, $S_F(g)$ will in general depend on g; when g increases from 0 to 1, $S_F(g)$ changes continuously from the unperturbed Fermi surface $S_F^\circ(0)$ to the real surface S_F.

Let us now consider the state $|\varphi(g)\rangle$ adiabatically generated from the ground state $|0\rangle$ of the Hamiltonian H_0. In so far as this state exists, it is characterized by a Fermi surface S_F° identical to that of $|0\rangle$. For all values of $g \neq 0$, $|\varphi(g)\rangle$ is thus an *excited* state of $H_0 + gH_1$; we go from $|\varphi_0(g)\rangle$ to $|\varphi(g)\rangle$ by creating an equal number of quasi particles and quasi holes, in such a way as to distort the Fermi surface from $S_F(g)$ to S_F°. This situation is illustrated in Fig. 47. $|\varphi(g)\rangle$ is obtained by adding to $|\varphi_0(g)\rangle$ the holes A and A' and the particles B and B'.

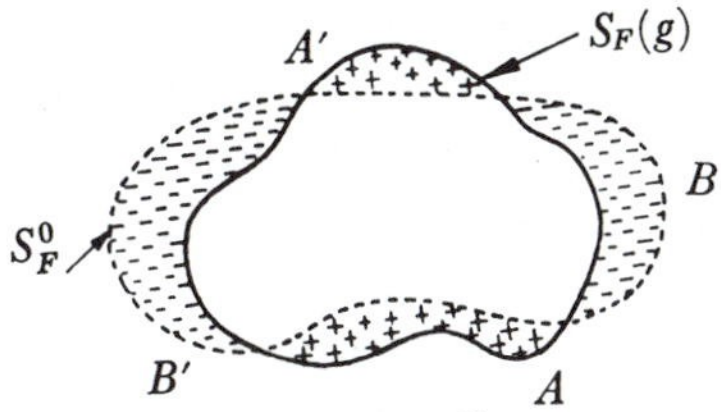

Figure 47

We made some reservations as to the existence of $|\varphi(g)\rangle$. In fact, $|\varphi(g)\rangle$ is unstable. Each of the elementary excitations present has a finite lifetime; furthermore, collisions between elementary excitations can occur. The corresponding lifetime Γ^{-1} is very short; it is determined by the elementary excitations which are farthest from $S_F(g)$. The situation is analogous to that studied for a single quasi particle in the section on Adiabatic Generation of the Excited States. If $S_F(g)$ is very close to S_F°, we can adiabatically generate $|\varphi(g)\rangle$ on condition that we choose $\eta \gg \Gamma$. When $S_F(g)$ departs too much from S_F°, $|\varphi(g)\rangle$ becomes too ambiguous to be calculated.

Let us assume that $\eta \to 0$; in the course of the adiabatic switching on, the quasi particles are in perpetual transition; the state obtained tends to be "thermalized." We have seen that these real transitions introduce into $U_{\eta L}(0, -\infty)$ singular terms of order $1/\sqrt{\eta}$, $1/\eta$, etc. The expansion of $U_{\eta L}$ therefore has no limit for $\eta \to 0$. Rigorously, this does not exclude the existence of a limit not expandable in a perturbation series. Physically, we have the feeling that, when $\eta \to 0$, the quasi particles created by changing g are immediately thermalized, distributed over the whole Fermi surface $S_F(g)$; if this hypothesis is exact, the adiabatic switching on generates at each instant the ground state $|\varphi_0(g)\rangle$. This does not further our study, since the corresponding time development operator cannot be expressed as the limit for $\eta \to 0$ of a perturbation expansion. In conclusion, the method developed above is incapable of giving us an explicit expression for the ground state. At the very most, it can describe an approximate excited state.

We could try to avoid the difficulty by starting from the unperturbed state having the same Fermi surface S_F as the real ground state. Actually, we again find the same problems, since, for each value of $g \neq 1$, the state thus obtained is not the ground state, and so there can be real transitions and divergent terms $\sim 1/\sqrt{\eta}$, etc. By keeping η finite, we can obtain a rough expression for $|\varphi_0\rangle$; on the other hand, we cannot have an exact result, since $U_{\eta L}$ has no limit when $\eta \to 0$.

To treat the problem rigorously, we must modify the process of adiabatic switching and introduce a Hamiltonian $H(g)$, different from $H_0 + gH_1$. The choice of $H(g)$ will be dictated by the following conditions:

(1) $H(1)$ is equal to the physical Hamiltonian $H_0 + H_1$.

(2) $H(0)$ is simple and diagonal.

(3) For each value $0 \leqslant g \leqslant 1$, the ground state $|\varphi_0(g)\rangle$ is a *normal* state, having the same Fermi surface S_F as the real ground state $|\varphi_0(1)\rangle$. (We assume for the moment that S_F is known.)

We put $g = e^{\eta t}$, which amounts to going adiabatically from $H(0)$ to $H(1)$. The time development of the system is then characterized by an operator $\tilde{U}_\eta(t,t')$. $\tilde{U}_\eta$ is different from the operator U_η considered above, but it is nevertheless calculated in the same way. Most of the properties of U_η are found again in $\tilde{U}_\eta$; in particular, the quasi-particle distribution function remains fixed while g varies, except in the case of a real transition (in other words, S_F is a constant of the motion). If at time $t = -\infty$ we start from $|\varphi_0(0)\rangle$, we find at time t a state having the same Fermi surface, which, according to our hypothesis, is just $|\varphi_0(g)\rangle$. The adiabatic development thus leaves the system in its ground state at each instant. There can be no real transitions; we can make $\eta \to 0$, and thus define *exactly* the ground state $|\varphi_0(1)\rangle$; our objective has been attained.

For $H(g)$ we take an expression of the form

$$\begin{cases} H(g) = \tilde{H}_0 + \tilde{H}_I \\ \tilde{H}_0 = H_0 + \sum_{\mathbf{k}} \lambda_{\mathbf{k}}\, a_{\mathbf{k}}^* a_{\mathbf{k}} \\ \tilde{H}_I = gH_I - \sum_{\mathbf{k}} \xi_{\mathbf{k}}(g)\, a_{\mathbf{k}}^* a_{\mathbf{k}} \end{cases} \tag{5–216}$$

where $\xi_{\mathbf{k}}(g)$ is a continuous function of g, satisfying the two conditions

$$\begin{cases} \xi_{\mathbf{k}}(0) = 0 \\ \xi_{\mathbf{k}}(1) = \lambda_{\mathbf{k}} \end{cases} \tag{5–217}$$

We choose the parameter $\lambda_{\mathbf{k}}$ so that the ground state of H_0 has the same Fermi surface S_F as the real ground state. Note that this condition allows us great freedom in the choice of $\lambda_{\mathbf{k}}$. We choose a simple form, increasing regularly with the modulus of k.

In principle, it would now be necessary to determine $\xi_{\mathbf{k}}(g)$. Actually, this is useless, since we shall never have to make use of it explicitly. The unperturbed state corresponds to $g = 0$, and the various physical quantities enter only for $g = 1$; the conditions (5–217) give us all necessary information. It is therefore sufficient to show that there *exists* a function $\xi_{\mathbf{k}}(g)$ having the desired properties. This is almost obvious. Let us consider the ground state of the Hamiltonian $\tilde{H}_0 + gH_1$. Whatever the corresponding Fermi surface $\tilde{S}_F$ may be, it is always possible to deform it into S_F with the help of a perturbation of the type

$$\sum_{\mathbf{k}} \xi_{\mathbf{k}}(g)\, a^*_{\mathbf{k}}\, a_{\mathbf{k}}$$

Since $S_F(g)$ is a continuous function of g, $\xi_{\mathbf{k}}(g)$ will also be continuous.

Note: Several authors have suggested taking

$$\xi_{\mathbf{k}}(g) = +\, g\, \lambda_{\mathbf{k}}$$

In general, this choice is bad, since the corresponding ground state will in all probability have a Fermi surface depending on g (although for $g = 0$ and $g = 1$ we obtain S_F). To keep S_F constant, we must choose a more complicated expression for $\xi_{\mathbf{k}}(g)$.

We now calculate the time-development operator $\tilde{U}_\eta$ by treating $\tilde{H}_1$ as a perturbation. The algebra developed in the section can be transposed without difficulty. The modifications are the following:

(a) The "unperturbed" propagator associated with each line of a diagram is now

$$\tilde{G}_0(\mathbf{k}, \omega) = \frac{1}{k^2/2m + \lambda_{\mathbf{k}} - \omega \pm i\delta} \tag{5–218}$$

The sign + corresponds to the interior of the Fermi surface S_F, the sign − to the exterior. $\tilde{G}_0$ differs from G_0 in two ways: first, the energy $k^2/2m$ is increased by $\lambda_{\mathbf{k}}$; second, the Fermi surface is modified.

(b) In addition to the usual vertices from H_1, there appears a new type of vertex, corresponding to the interaction

$$-\sum_{\mathbf{k}} \xi_{\mathbf{k}}(g)\, a^*_{\mathbf{k}}\, a_{\mathbf{k}}$$

These "binary" vertices have only one line entering and one leaving, both having the same wave vector $\mathbf{k}$; we represent them graphically by Fig. 48.

To calculate the contribution from a diagram, it is simplest first to ignore the binary vertices; the rules established in sections 2 and 3 are then valid. It remains simply to study the correction arising from the

Figure 48

insertion of a binary vertex into the line **k**. The contraction

$$-\mathrm{i}\,\tilde{G}_0(\mathbf{k},\omega)$$

is then replaced by

$$(-\mathrm{i}\,\tilde{G}_0(\mathbf{k},\omega))^2\,(-\xi_{\mathbf{k}}(g))\,(-\mathrm{i})$$

The first factor arises from the two lines associated with the binary vertex; the second is just the interaction matrix element; finally, the factor $(-i)$ results from (5–14) (the addition of a binary vertex increasing the order by 1). In short, each binary vertex contributes a factor $\xi_{\mathbf{k}}(g)$.

With this corrected formalism, all the results previously established remain exact. The Green's functions G and K are given by the same diagrams as before. Let us consider G. We can again decompose it into a series of irreducible self-energy parts. Let $\tilde{M}(\mathbf{k},\omega)$ be the "corrected" self-energy operator. We obtain all the diagrams of G by lining up an arbitrary number of self-energy parts, connected by lines $\tilde{G}_0$. As a consequence,

$$G(\mathbf{k},\omega)=\frac{\tilde{G}_0(\mathbf{k},\omega)}{1-\tilde{G}_0(\mathbf{k},\omega)\,\tilde{M}(\mathbf{k},\omega)} \qquad (5\text{–}219)$$

With the use of (5–218), this expression can be written as

$$G(\mathbf{k},\omega)=\frac{1}{k^2/2m+\lambda_{\mathbf{k}}-\omega-\tilde{M}(\mathbf{k},\omega)\pm\mathrm{i}\delta} \qquad (5\text{–}220)$$

The self-energy operator $\tilde{M}$ involves two types of diagrams, indicated

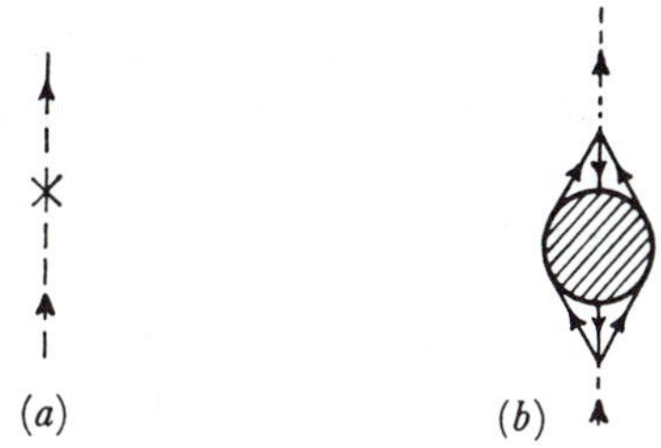

Figure 49

in Fig. 49. The diagram (a) is unique; its contribution is equal to $\xi_{\mathbf{k}}(1) = \lambda_{\mathbf{k}}$. The diagrams (b) are characterized by vertices of the usual type at both ends; let $M(\mathbf{k},\omega)$ be the corresponding contribution. We have

$$\tilde{M}(\mathbf{k}, \omega) = \lambda_{\mathbf{k}} + M(\mathbf{k}, \omega) \tag{5–221}$$

Putting (5–221) back into (5–220), we see that

$$G(\mathbf{k}, \omega) = \frac{1}{k^2/2m - \omega - M(\mathbf{k}, \omega) \pm i\,\delta} \tag{5–222}$$

The parameter $\lambda_{\mathbf{k}}$ has disappeared, at least explicitly.

Again, we can reduce the diagrams (b) contributing to M to skeleton diagrams. M thus becomes a functional of G. Note that a skeleton diagram can contain no binary vertices (these always occur as a diagonal insertion). As a consequence, the parameter $\lambda_{\mathbf{k}}$ does not appear in the functional $M[G]$.

Knowing the functional $M[G]$, we can, with the help of (5–222), obtain G. Since $\lambda_{\mathbf{k}}$ does not appear at any stage of the calculation, the solution does not depend on it. This conclusion justifies our method a posteriori; the observable physical result cannot depend on an arbitrary parameter. (5–222) resembles our old result (5–111), with, however, a basic difference: the sign of $\pm i\delta$ refers to the true Fermi surface S_F, and no longer to the Fermi surface of the unperturbed system. This is the only effect that $\lambda_{\mathbf{k}}$ has had, but it is essential.

In practice, we choose at the start a Fermi surface S_F (enclosing, of course, a given volume). For each value of $\mathbf{k}$, this fixes the term $\pm i\delta$ appearing in (5–222). We then can determine the functional $M[G]$. In general, the result depends on S_F. The pole $\varepsilon_{\mathbf{k}}$ of $G(\mathbf{k},\omega)$ will have a negative imaginary part if k is outside of S_F (particle-type excitations), positive if k is inside of S_F (hole-type excitations). If S_F is well chosen, $\varepsilon_{\mathbf{k}}$ is constant over the whole Fermi surface, equal to the chemical potential

μ. This assures us of being in the ground state. The solution of the problem thus requires a "self-consistent" determination of S_F.

The ground-state energy is calculated as for an isotropic system. Its unperturbed value is equal to

$$\widetilde{E}_0^0 = \sum_{\mathbf{k}\in S_F} \left(\frac{k^2}{2m} + \lambda_{\mathbf{k}}\right) \tag{5-223}$$

The correction $\Delta\widetilde{E}_0$, arising from the perturbation

$$H_{\mathrm{I}} - \sum_{\mathbf{k}} \lambda_{\mathbf{k}}\, a_{\mathbf{k}}^* a_{\mathbf{k}}$$

is given by the unlinked connected diagrams. We can reproduce the analysis given in the section on Renormalization of the Propagators and express the ground-state energy as a functional of the propagator G. We shall just quote the result. Let us define

$$\Delta E_0 = \Delta\widetilde{E}_0 + \sum_{\mathbf{k}\in S_F} \lambda_{\mathbf{k}} \tag{5-224}$$

(5–224) represents the correction to the ground-state energy arising from the physical interaction H_1; it is the analog of the ΔE_0 defined by (5–77). ΔE_0 is still given by the functional expansion (5–203). The only difference comes from

$$G_0 = \frac{1}{k^2/2m - \omega \pm i\delta} \tag{5-225}$$

the sign $\pm i\delta$ referring to the *true* Fermi surface.

Note: To prove this result directly, it is best to carry out a partial renormalization of the propagators by eliminating all binary vertices. $\widetilde{G}_0$ is then replaced by G_0. In this operation, we lose a diagram of $\Delta\widetilde{E}_0$, namely the one having only a single vertex, which is binary. The contribution of this diagram is

$$-\sum_{\mathbf{k}\in S_F} \lambda_{\mathbf{k}}$$

As a consequence, ΔE_0 is given by the same diagrams as for the isotropic system, evaluated with the same propagators G_0. (5–203) is thus preserved, since it is a purely algebraic relation, resulting from the structure of the diagrams.

If we start with the wrong Fermi surface, the energy E_0 thus obtained is complex. It then corresponds to an excited state, of finite lifetime. It seems that we could determine S_F by minimizing the real part of E_0 (this is only a conjecture). The only condition imposed on S_F is that of having a constant volume. It is convenient to eliminate this condition by introducing a Lagrangian multiplier; we shall try to minimize the "free energy"

$$F = H - \mu N \tag{5-226}$$

for an arbitrary value of μ, without any restriction on S_F. Then we shall fix μ so that S_F encloses the desired volume. This amounts to adjusting the number of particles

$$N = - \partial F_0 / \partial \mu \tag{5-227}$$

In practice, we shall never try to *calculate* the Fermi surface, which is much too difficult. What is most important to us is to know that it exists (that is to say, that the system is normal). The preceding discussion shows that we can then use a perturbation formalism to study all physical properties of our choice. This formalism is defined unambiguously with the help of skeleton diagrams, involving the correct Fermi surface. There is no difficulty in extending the method developed for isotropic systems.

Chapter 6

Interaction of Two Elementary Excitations

In the preceding chapter, we have considered only a single elementary excitation, quasi-particle or quasi-hole. We now extend the field of this study by considering the interaction between two elementary excitations. It is no longer the single-particle Green's function G which interests us, but the two-particle function K, defined by (3–39). We immediately discard the free part of K, which yields no new information, and we consider only the *interaction* operator $\Gamma(\mathbf{p}_i)$, given by (3–50). We shall use indiscriminately the *complete* interaction operator γ defined by (3–43) or (3–49) and the *reduced* operator Γ, with the δ functions removed. The choice of one or the other of these quantities is just a question of convenience.

The study of γ will give us information about a number of problems:

(a) Scattering of two elementary excitations
(b) Bound states of two elementary excitations, collective modes
(c) Correlation function, response to an external field
(d) Justification of the Landau theory.

In the course of this discussion, we shall encounter certain important phenomena, such as multiple scattering or screening, and we shall indicate how to treat them in an approximate way.

We shall study γ by relying on its perturbation expansion, established in Chap. 5. The function $\gamma(\mathbf{k}_i, \omega_i)$ is defined by (5–114); it is given by the diagrams of Fig. 26 stripped of the four external propagators $G(k_i, \omega_i)$, evaluated with the same rules as $K(\mathbf{k}_i, \omega_i)$. We propose to analyze the structure of these diagrams.

1. Multiple Scattering of Two Elementary Excitations

a. Multiple scattering of two particles

Let us assume our system to be reduced to only two particles. These can interact only with each other. The interaction core then reduces to the diagrams of Fig. 1 (with an arbitrary number of bubbles). (We have represented by dotted lines the emerging propagators, which are not included in γ.) γ thus consists of an *iteration* of an elementary act of interaction; there is *multiple scattering* of the two particles. This important phenomenon affects the physical "observables" (for example, the scattering cross section).

Let us now return to our N-body system, and let us consider the diagram of γ indicated in Fig. 2; the double lines represent *renormalized* propagators G. There are two *successive* acts of interaction, between which we again find two well-individualized particles. Note that each of these can interact "on its own" with the medium and carry along a self-energy cloud (it is characterized by a propagator G and not by G_0); however, these two self-energy clouds are *independent* of each other. On the interior of A, we have, conversely, interaction between the self-energy clouds; the two particles have lost their individuality. We shall say that there is *multiple scattering* of the two incident particles.

We shall call an *irreducible* diagram of $\gamma(\mathbf{k}_i,\omega_i)$ any diagram that cannot be decomposed as in Fig. 2. Let $J(\mathbf{k}_i,\omega_i)$ be the contribution of these irreducible diagrams, called the *irreducible interaction of two particles* (not to be confused with the irreducible interaction of a particle and a hole, which we shall study later). J is evaluated using the same rules as for γ or for K. To obtain the complete interaction γ, we must iterate the irreducible interaction J an arbitrary number of times. This process is

Figure 1

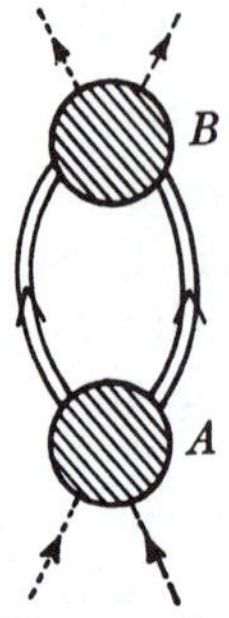

Figure 2

illustrated in Fig. 3. This "graphical" equation is easily expressed in mathematical form,

$$\gamma(1\ 2\ 3\ 4) = J(1\ 2\ 3\ 4) + \tfrac{1}{2}\sum_{5,6} J(1\ 2\ 5\ 6)\, G(5)\, G(6)\, \gamma\,(5\ 6\ 3\ 4) \qquad (6\text{–}1)$$

where we have made the abbreviations

$$\begin{cases} \gamma(\mathbf{k}_1\omega_1 \ldots\ldots \mathbf{k}_4\omega_4) = \gamma(1 \ldots\ldots 4), \text{ etc...} \\ G(\mathbf{k}_5\omega_5) = G(5), \text{ etc...} \\ \sum\limits_{\mathbf{k}_5} \int d\omega_5 = \sum\limits_5, \text{ etc...} \end{cases} \qquad (6\text{–}2)$$

Note the factor $\frac{1}{2}$ in (6–1), arising because the two intermediate lines of Fig. 3 are equivalent (replacing J by the diagram symmetric with respect to a vertical axis, we obtain the same total diagram of γ).

Equation (6–1), relating γ to J, is the *Bethe-Salpeter equation.* It is an integral equation for γ, whose solution is in general difficult. It is sometimes convenient to Fourier-transform it and to rewrite it as a function of the space and time variables. We thus obtain

$$\begin{aligned} \gamma(\mathbf{x}_1, \mathbf{x}_2, \mathbf{x}_3, \mathbf{x}_4) &= J(\mathbf{x}_1, \mathbf{x}_2, \mathbf{x}_3, \mathbf{x}_4) \\ &+ \tfrac{1}{2}\iint d\mathbf{x}_5\, d\mathbf{x}_5'\, d\mathbf{x}_6\, d\mathbf{x}_6'\; J(\mathbf{x}_1, \mathbf{x}_2, \mathbf{x}_5, \mathbf{x}_6)\; G(\mathbf{x}_5, \mathbf{x}_5')\; G(\mathbf{x}_6,\mathbf{x}_6')\; \gamma\; (\mathbf{x}_5', \mathbf{x}_6', \mathbf{x}_3, \mathbf{x}_4) \end{aligned} \qquad (6\text{–}3)$$

The Bethe-Salpeter equation is extremely important, as it contains all the properties of a two-particle system: bound states, scattering amplitude, etc.

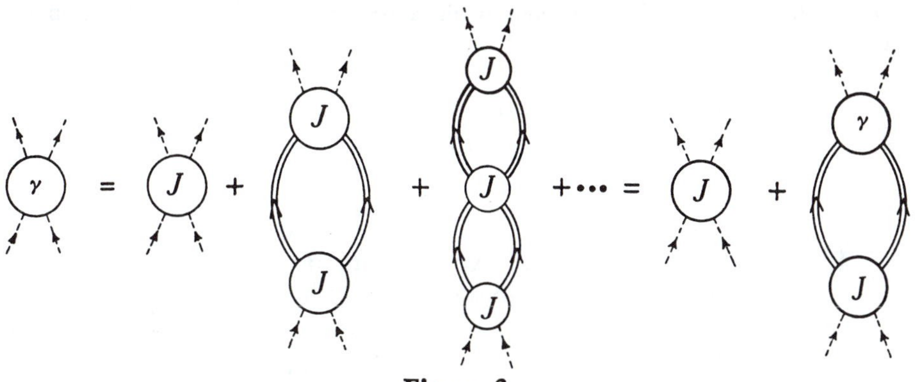

Figure 3

We can rewrite it by introducing the two-particle Green's function $K(1,2,3,4)$, instead of the interaction γ. For this purpose, we decompose K in the way indicated in Fig. 4.

This graphical equation can be written in the form

$$K(1\ 2\ 3\ 4) = K_L(1\ 2\ 3\ 4) + \tfrac{1}{2}\sum_{5,6} G(1)\ G(2)\ J(1\ 2\ 5\ 6)\ K(5\ 6\ 3\ 4) \qquad \textit{(6–4)}$$

where K_L is the free part of the Green's function (again the factor $\frac{1}{2}$ arises from the equivalence of the two intermediate lines).

Note: We can obtain (6–4) directly from (6–1). We know that

$$\begin{cases} K_L(1\ 2\ 3\ 4) = G(3)\ G(4)\ [\delta_{13}\ \delta_{24} - \delta_{14}\ \delta_{23}] \\ \delta K(1\ 2\ 3\ 4) = G(1)\ G(2)\ G(3)\ G(4)\ \gamma(1\ 2\ 3\ 4) \end{cases}$$

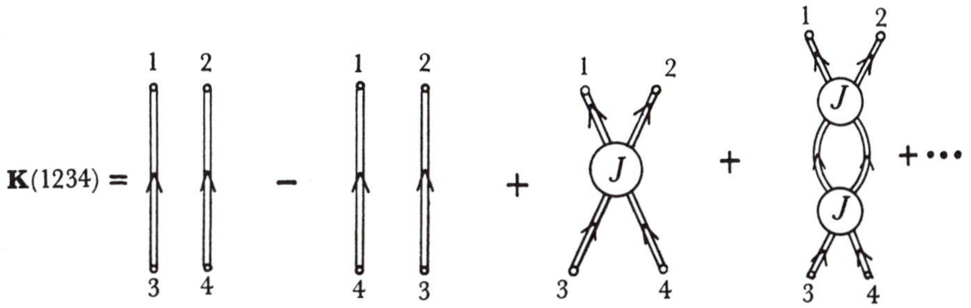

Figure 4

Let us multiply (6–1) by $G_1G_2G_3G_4$ and use the antisymmetry of J (and of γ); we obtain at once

$$\delta K(1\ 2\ 3\ 4) = \tfrac{1}{2}\, G_1\, G_2 \sum_{5,6} J(1\ 2\ 5\ 6)\, K(5\ 6\ 3\ 4)$$

Hence the relation (6–4).

Equation (6–4) recalls the expansion made for the single-particle Green's function G. Let us compare Fig. 4 with Fig. 24 of Chap. 5. In (5–110), we decompose G into a series of propagators G_0, separated by self-energy parts. In (6–4), we do the same thing, but for two particles instead of one; the irreducible interactions J separate regions where the two particles propagate independently (each of them being renormalized). Unfortunately, we thus obtain an integral equation, instead of the algebraic equation (5–110).

By way of example, let us return to the case of two particles. The only irreducible diagram is then made up of a single vertex (Fig. 5). By applying the rules (5–107), we obtain

$$J(1\ 2\ 3\ 4) = -\frac{i}{2\pi}\, V(\mathbf{k}_1, \ldots, \mathbf{k}_4)\, \delta(\omega_1 + \omega_2 - \omega_3 - \omega_4) \qquad \textit{(6–5)}$$

(the minus sign arises from the sign rule proved for K). The interaction γ is obtained from (6–1), the propagators G reducing to G_0 in this simple case,

$$G(k, \omega) = \frac{1}{k^2/2m - \omega - i\eta} \qquad \textit{(6–6)}$$

Note that γ depends only on

$$\omega_1 + \omega_2 = \omega_3 + \omega_4 = E \qquad \textit{(6–7)}$$

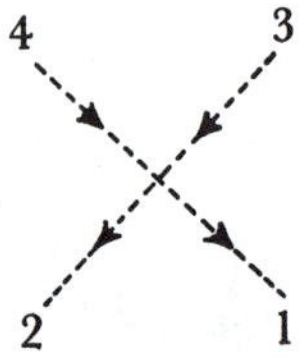

Figure 5

It is convenient to set

$$\gamma(\mathbf{k}_1\,\omega_1, \ldots, \mathbf{k}_4\,\omega_4) = -\frac{i}{2\pi}\, T(\mathbf{k}_1, \ldots, \mathbf{k}_4\,;\, E)\, \delta(\omega_1 + \omega_2 - \omega_3 - \omega_4) \qquad (6\text{–}8)$$

(6–1) can then be put in the form

$$T(\mathbf{k}_1, \ldots, \mathbf{k}_4\,;\, E) =$$

$$V(\mathbf{k}_1, \ldots, \mathbf{k}_4) - \frac{i}{4\pi} \sum_{\mathbf{k}_5\mathbf{k}_6} \int d\omega_5 \frac{1}{(\varepsilon_{k_5} - \omega_5 - i\eta)\,(\varepsilon_{k_6} - E + \omega_5 - i\eta)} \qquad (6\text{–}9)$$

$$\times\; V(\mathbf{k}_1\,\mathbf{k}_2\,\mathbf{k}_5\,\mathbf{k}_6)\; T(\mathbf{k}_5\,\mathbf{k}_6\,\mathbf{k}_3\,\mathbf{k}_4\,;\, E)$$

(where $\varepsilon_k = k^2/2m$). The integration over the energies is trivial and gives

$$T(\mathbf{k}_1, \ldots, \mathbf{k}_4\,;\, E) = V(\mathbf{k}_1, \ldots, \mathbf{k}_4) + \tfrac{1}{2} \sum_{\mathbf{k}_5\mathbf{k}_6} \frac{V(\mathbf{k}_1\,\mathbf{k}_2\,\mathbf{k}_5\,\mathbf{k}_6)\; T(\mathbf{k}_5\,\mathbf{k}_6\,\mathbf{k}_3\,\mathbf{k}_4\,;\, E)}{\varepsilon_{k_5} + \varepsilon_{k_6} - E - i\eta} \qquad (6\text{–}10)$$

We can, if we wish, give a more explicit form to (6–10) by replacing $V(\mathbf{k}_1 \ldots \mathbf{k}_4)$ by its expression (5–17).

The quantity $T(\mathbf{k}_1 \ldots \mathbf{k}_4;E)$ is called the *reaction matrix* of the two particles. It governs the scattering properties. We shall not stress this point, which forms the basis of the formal theory of scattering, and which is treated in detail in many other works. Let us point out that, for a hard-core interaction, the reaction matrix is essential; in fact, the components $V(\mathbf{k}_1 \ldots \mathbf{k}_4)$ are then infinite. However, we can calculate T by returning to configuration space, then going back to $T(\mathbf{k}_1 \ldots \mathbf{k}_4;E)$; we thus obtain a well-defined result, physically reasonable. Each term of the perturbation expansion diverges; nevertheless, if we make the partial summation of the diagrams of Fig. 1, we obtain a convergent result. Such an assertion is mathematical heresy; actually, it simply indicates that the matrix T must not be expanded in a series.

b. Multiple scattering of a hole and a particle

This is a much more interesting problem than that studied in the preceding paragraph, for two reasons:

(a) Hole-particle bound states are numerous (collective modes, excitons, whereas those of two particles are not found in normal gases.

(b) An external field tends to *excite* the system, that is to say, to create a hole and a particle near each other whose interaction is essential.

Let us consider a diagram of $\gamma(1,2,3,4)$, and let us look for multiple scattering of a particle and a hole. There are two types of reduction

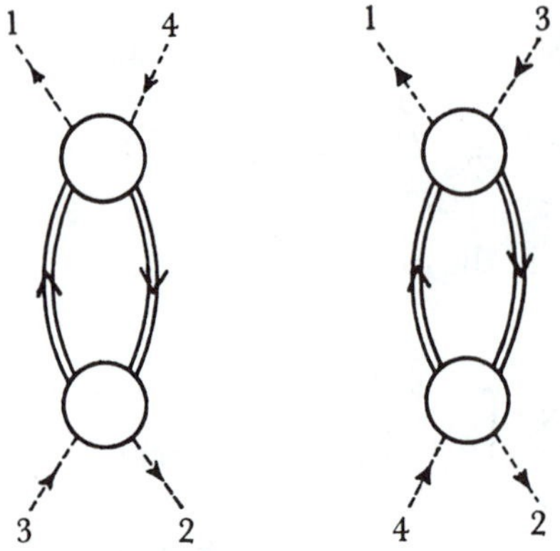

Figure 6

possible, indicated in Fig. 6. We can associate hole 1 with particle 3 *or* with particle 4 (this ambiguity did not appear for two particles). We shall only study the multiple scattering of pair 1–4, ignoring that of pair 1–3. Thus, diagram 6*a* is reducible, although diagram 6*b* is irreducible. Because of the antisymmetry of γ, this choice is of no importance, and retains complete generality in its statement.

We shall call $I(1,2,3,4)$ the contribution to γ from all irreducible diagrams, not capable of decomposition according to the scheme of Fig. 6*a*. I is the *irreducible hole-particle interaction* corresponding to the interaction $1, 4 \to 2, 3$. To obtain γ, we must iterate I an arbitrary number of times, as indicated in Fig. 7.

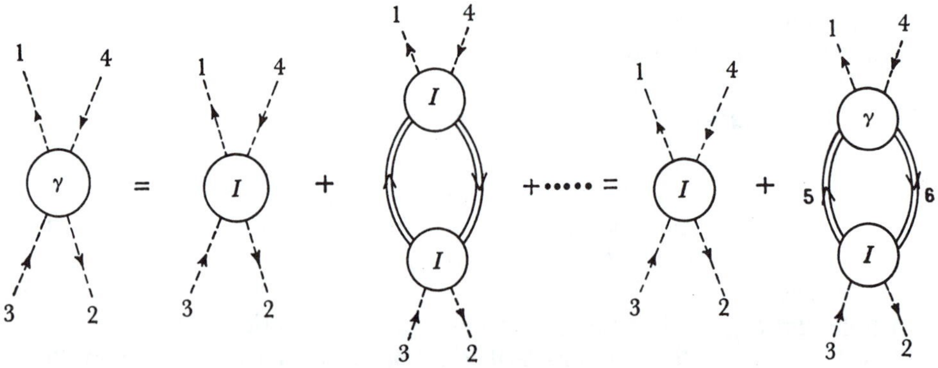

Figure 7

This graphical equation is equivalent to the relation

$$\gamma(1\ 2\ 3\ 4) = I(1\ 2\ 3\ 4) + \sum_{5,6} I(5\ 2\ 3\ 6)\, G(5)\, G(6)\, \gamma(1\ 6\ 5\ 4) \qquad \textit{(6–11)}$$

where we again use the condensed notation of (6–2). In order to prove (6–11), we must pay careful attention to questions of sign. It is best to draw all diagrams so that the open line coming from 3 terminates at 1; each factor I then carries a plus sign. (6–11) is the Bethe-Salpeter equation for particle-hole scattering; it is equivalent to (6–1). Note the absence of the factor $\frac{1}{2}$, the intermediate lines no longer being equivalent. As in the preceding paragraph, we can take the Fourier transform and write (6–11) in a form analogous to (6–3).

The notation used up to now is concise but fairly obscure. We are going to make it more explicit and simpler. First let us consider the problem of spin. Two cases are possible, according to whether $\sigma_3 = \pm\sigma_2$. Let us assume $\sigma_3 = -\sigma_2 = \pm\frac{1}{2}$. The initial particle-hole pair has a total z component of spin equal to ± 1 (triplet state). This spin must be conserved; as a consequence, in Fig. 6 we must have

$$\sigma_3 = -\sigma_2 = \sigma_5 = -\sigma_6 = \sigma_1 = -\sigma_4 = \pm 1/2 \qquad \textit{(6–12)}$$

The spins of the intermediate lines are completely determined. Let us now turn to the second case $\sigma_3 = \sigma_2 = \pm\frac{1}{2}$; the total z component of spin is zero (which can correspond to either a singlet or a triplet state). Conservation of spin imposes the conditions

$$\sigma_5 = \sigma_6\,, \qquad \sigma_1 = \sigma_4 \qquad \textit{(6–13)}$$

which allow complete freedom concerning the sign of σ_5 and of σ_1. It is thus necessary to sum over the spin of the intermediate lines.

We now introduce the reduced notation $(k,\omega) = p$, and we set

$$\begin{cases} p_3 = p - \varpi/2 \\ p_2 = p + \varpi/2 \end{cases} \qquad \begin{cases} p_1 = p' - \varpi/2 \\ p_4 = p' + \varpi/2 \end{cases} \qquad \textit{(6–14)}$$

ϖ measures the total wave vector and energy which are conserved. p and p' are the relative wave vectors and energies before and after the interaction. We define

$$\begin{cases} I(p_1\,\sigma, p_2\,{-\sigma}, p_3\,\sigma, p_4\,{-\sigma}) = {}^1I(p, p'\,;\varpi)\,\delta(p_1 + p_2 - p_3 - p_4) \\ I(p_1\,\sigma', p_2\,\sigma, p_3\,\sigma, p_4\,\sigma') = {}^0I(\mathbf{p}, \mathbf{p}'\,;\varpi)\,\delta(p_1 + p_2 - p_3 - p_4) \end{cases} \qquad \textit{(6–15)}$$

and similarly

$$\begin{cases} \gamma(p_1\,\sigma, p_2\,{-\sigma}, p_3\,\sigma, p_4\,{-\sigma}) = {}^1\Gamma(p, p'\,;\varpi)\,\delta(p_1+p_2-p_3-p_4) \\ \gamma(p_1\,\sigma', p_2\,\sigma, p_3\,\sigma, p_4\sigma') = {}^0\Gamma(\mathbf{p}, \mathbf{p}'\,;\varpi)\,\delta(p_1+p_2-p_3-p_4) \end{cases} \tag{6–16}$$

We recall the conventions of notation

$$\begin{cases} \mathbf{p} = (p, \sigma) \\ \delta(p-p') = \delta(\omega-\omega')\,\delta_{k,k'} \\ \sum\limits_{p} = \sum\limits_{k} \int d\omega \end{cases} \tag{6–17}$$

(6–15) and (6–16) assume that $|\varphi_0\rangle$ is invariant with respect to simultaneous rotation of all the spins.

With all these preliminaries, we can rewrite (6–11) in the forms

$$\begin{aligned} {}^1\Gamma(p, p'\,;\varpi) &= {}^1I(p, p'\,;\varpi) \\ &\quad + \sum_{p''} {}^1I(p, p''\,;\varpi)\,G(p''-\varpi/2)\,G(p''+\varpi/2)\,{}^1\Gamma(p'', p'\,;\varpi) \end{aligned} \tag{6–18}$$

$$\begin{aligned} {}^0\Gamma(\mathbf{p}, \mathbf{p}';\varpi) &= {}^0I(\mathbf{p}, \mathbf{p}'\,;\varpi) \\ &\quad + \sum_{\mathbf{p}''} {}^0I(\mathbf{p}, \mathbf{p}''\,;\varpi)\,G(p''-\varpi/2)\,G(p''+\varpi/2)\,{}^0\Gamma(\mathbf{p}'', \mathbf{p}'\,;\varpi) \end{aligned} \tag{6–19}$$

The Bethe-Salpeter equation is thus decoupled into two independent integral equations, one for the total spin σ_z equal to 1, the other for $\sigma_z = 0$. We can simplify once more and decouple (6–19) by putting

$${}^0\Gamma(p\sigma, p'\sigma\,;\varpi) \pm {}^0\Gamma(p\sigma, p'\,{-\sigma}\,;\varpi) = {}^0\Gamma_\pm(p, p';\varpi) \tag{6–20}$$

as well as an analogous equation defining ${}^\circ I_\pm(p,p';\varpi)$. (6–19) is equivalent to the pair of equations

$$\begin{aligned} {}^0\Gamma_\pm(p, p';\varpi) &= {}^0I_\pm(p, p'\,;\varpi) \\ &\quad + \sum_{p''} {}^0I_\pm(p, p''\,;\varpi)\,G(p''-\varpi/2)\,G(p''+\varpi/2)\,{}^0\Gamma_\pm(p'', p'\,;\varpi) \end{aligned} \tag{6–21}$$

${}^\circ\Gamma_+$ and ${}^\circ\Gamma_-$ are thus completely decoupled.

It can be shown that ${}^\circ\Gamma_+$ describes the interaction of a particle-hole pair in the singlet state, whereas ${}^\circ\Gamma_-$ refers to the component of the triplet state of zero σ_z. The direction of the total spin cannot be involved (at least in the absence of spin-orbit coupling). We must therefore have

$$^1I = {^0I_-}, \qquad {^1\Gamma} = {^0\Gamma_-} \tag{6–22}$$

We shall prove this relation in Appendix F.

Let us summarize this discussion. We started from the Bethe-Salpeter equation (6–11), depending on the orbital and spin variables. We have decomposed it into two independent equations, depending only on the orbital variables concerned, respectively, with the triplet and singlet states of the spins of the particle and the hole.

Later in this chapter we shall be interested primarily in $^\circ\Gamma$, defined by (6–16). Actually, according to (6–22), all the results proved for $^\circ\Gamma$ remain true for $^1\Gamma$.

2. Bound States of Two Excitations. Collective Modes

Let q be the wave vector and ε the energy associated with the four-vector ϖ. These are the *total* wave vector and energy of the hole-particle pair. We propose to study $^\circ\Gamma(\mathbf{p},\mathbf{p}';\varpi)$ as a function of ε, q being fixed once for all.

We saw in Chap. 3 that the particle-hole bound states lead to singularities in the Green's function $K(\mathbf{k}_i,\omega_i)$. In order to explore this question, it is convenient to use the function $K(\mathbf{p},\mathbf{p}';\varpi)$, defined by analogy with (6–14) and (6–16). Let us return to Eq. (D–4); we see that $^\circ K(\mathbf{p},\mathbf{p}';\varpi)$ contains a term (among many others)

$$\frac{i}{2\pi}\sum_{mns}\frac{[a_{\mathbf{k}+q/2}]_{0m}\,[a^*_{\mathbf{k}-q/2}]_{ms}\,[a_{\mathbf{k}'-q/2}]_{sn}\,[a^*_{\mathbf{k}'+q/2}]_{n0}}{(\omega_{n0}-\omega'-\varepsilon/2-i\eta)\,(\omega+\varepsilon/2-\omega_{m0}+i\eta)\,(\varepsilon-\omega_{s0}+i\eta)} \tag{6–23}$$

Let $|\varphi_{q\lambda}\rangle$ be a particle-hole bound state, of total wave vector q and energy $\omega_{q\lambda}$ (the index λ distinguishes the various bound states of the same q). For a given q, the states $|\varphi_{q\lambda}\rangle$ are discrete (whereas the independent particle-hole pairs form a continuum). The state $|\varphi_{q\lambda}\rangle$ appears among the intermediate states $|\varphi_s\rangle$ of (6–23); its contribution contains a factor

$$\frac{1}{\varepsilon-\omega_{q\lambda}+i\eta} \tag{6–24}$$

$|\varphi_{q\lambda}\rangle$ thus gives rise to a discrete *pole* with respect to the variable ε; this confirms the results of Chap. 3 [see Eq. (3–56)]. Note that this pole is present whatever the values of k, ω, k' and ω' are: this expresses the spreading out of the relative wave function in p space.

The preceding discussion assumes that $|\varphi_{q\lambda}\rangle$ is a *rigorous* eigenstate of the system, which can appear in (6–23); $|\varphi_{q\lambda}\rangle$ must not be damped, and $\omega_{q\lambda}$ is real. This case occurs for the actual bound states, such as excitons in an insulator (these are lying in the gap and as a consequence are *stable*). Actually, there also exist *unstable* bound states of the phonon type (or any other collective mode), which are immersed in the continuum; they also give rise to poles with respect to ε, but now the pole $\omega_{q\lambda}$ will be complex, its imaginary part characterizing the damping.

The investigation of bound states is thus reduced, for a given q, to the investigation of the poles of $^{\circ}K(\mathbf{p},\mathbf{p}';\varpi)$. These poles are also present in the interaction $^{\circ}\Gamma(\mathbf{p},\mathbf{p}';\varpi)$. (The singlet bound states will appear as poles of $^{\circ}\Gamma_{+}$, the triplet states as poles of $^{\circ}\Gamma_{-}$.)

Let us consider Eq. (6–19). If $\varepsilon \to \omega_{q\lambda}$, $^{\circ}\Gamma$ tends to infinity. The inhomogeneous term $^{\circ}I$ (which remains finite) becomes negligible. In this limit we have

$$^{0}\Gamma(\mathbf{p},\mathbf{p}';\varpi) = \sum_{\mathbf{p}''} {}^{0}I(\mathbf{p},\mathbf{p}'';\varpi)\, G(p''-\varpi/2)\, G(p''+\varpi/2)\, {}^{0}\Gamma(\mathbf{p}'',\mathbf{p}';\varpi) \qquad (6\text{–}25)$$

(6–25) is a *homogeneous* integral equation, which has a solution only for certain *eigenvalues* of the parameter ε. These eigenvalues determine precisely the poles $\omega_{q\lambda}$ of Γ. Let us emphasize this very important result:

(a) The interaction $^{\circ}\Gamma(\mathbf{p},\mathbf{p}';\varpi)$ is given unambiguously by the *inhomogeneous* Bethe-Salpeter equation (6–19).

(b) In order to determine the poles of $^{\circ}\Gamma$ (that is to say, the bound states), we look for the eigenvalues of the *homogeneous* Bethe-Salpeter equation (6–25), with the term $^{\circ}I$ removed.

We are thus led to the solution of the eigenvalue equation (6–25). Note that the variable $\mathbf{p}'$ no longer comes in explicitly; it is simply a parameter, without any effect on the solution. Let $u(\mathbf{p},\varpi_{\lambda})$ be the eigenfunction of (6–25) associated with the eigenvalue $\omega_{q\lambda}$. The general solution of (6–25) is the product of $u(\mathbf{p},\varpi_{\lambda})$ with any function of $\mathbf{p}'$ (by virtue of the general properties of linear homogeneous equations). In order to remove this ambiguity, we use the antisymmetry of $\Gamma(\mathbf{p}_1,\mathbf{p}_2,\mathbf{p}_3,\mathbf{p}_4)$,

$$\Gamma(\mathbf{p}_1\mathbf{p}_2\mathbf{p}_3\mathbf{p}_4) = \Gamma(\mathbf{p}_2\mathbf{p}_1\mathbf{p}_4\mathbf{p}_3) \qquad (6\text{–}26)$$

By comparing (6–26) with (6–14), we see that

$$^{0}\Gamma(\mathbf{p},\mathbf{p}';\varpi) = {}^{0}\Gamma(\mathbf{p}',\mathbf{p};-\varpi) \qquad (6\text{–}27)$$

The solution of (6–25) must therefore be written, up to a constant multiplicative factor, in the form

$$^{0}\Gamma(\mathbf{p}, \mathbf{p}'\,;\varpi) = u(\mathbf{p}, \varpi_\lambda)\, u(\mathbf{p}', -\varpi_\lambda).\ \text{const} \tag{6–28}$$

Actually, (6–25) claims to determine only the singular part of $^{\circ}\Gamma$. We shall therefore write

$$^{0}\Gamma(\mathbf{p}, \mathbf{p}'\,;\varpi) \sim \frac{u(\mathbf{p}, \varpi_\lambda)\, u(\mathbf{p}', -\varpi_\lambda)}{\varepsilon - \omega_{q\lambda} + i\eta}.\ \text{const} \qquad \text{when } \varepsilon \to \omega_{q\lambda} \tag{6–29}$$

(the constant depends on the norm chosen for the eigenfunction u).

We know how to interpret the eigenvalues $\omega_{q\lambda}$; they give the energies of the bound states. But how should we interpret the eigenvectors u_λ and the solution (6–29)? To answer this question, we must first go back to the Green's function $^{\circ}K(\mathbf{p},\mathbf{p}';\varpi)$. When $\varpi \to \varpi_\lambda$, we can neglect the free part K_L, which is regular, with respect to the bound part δK, which is singular. As a consequence, (6–29) can be written

$$^{0}K(\mathbf{p}, \mathbf{p}'\,;\varpi) \sim \frac{f(\mathbf{p}, \varpi_\lambda)\, f(\mathbf{p}', -\varpi_\lambda)}{\varepsilon - \omega_{q\lambda} + i\eta}.\ \text{const} \qquad \text{when } \varpi \to \varpi_\lambda \tag{6–30}$$

where the function f is defined by

$$f(\mathbf{p}, \varpi_\lambda) = u(\mathbf{p}, \varpi_\lambda)\, G(p + \varpi_\lambda/2)\, G(p - \varpi_\lambda/2) \tag{6–31}$$

We propose to compare (6–30) with the direct expansion of the function $^{\circ}K$.

Let us return to (D–4) and look for *all* the terms of $^{\circ}K(\mathbf{p},\mathbf{p}';\varpi)$ that contain in the denominator a factor $(\varepsilon - \omega_{q\lambda} + i\eta)$. We have already found the term (6–23). We must add to it the terms obtained by permuting $a_{\mathbf{k}'-q/2}$ and $a^*_{\mathbf{k}'+q/2}$, or $a_{\mathbf{k}+q/2}$ and $a^*_{\mathbf{k}-q/2}$, or both. There are no others. [If in (6–23) we permute the first two operators with the last two, the state s has a wave vector $-q$ instead of q and the denominator contains a factor $(\varepsilon + \omega_{s0} - i\eta)$.] By collecting all these singular terms and comparing with (6–30), we easily verify that

$$\text{cons.}\ f(\mathbf{p}, \varpi_\lambda) = \sum_m \left\{ \frac{[a_{\mathbf{k}+q/2}]_{0m}\, [a^*_{\mathbf{k}-q/2}]_{m\lambda}}{\omega + \omega_{q\lambda}/2 - \omega_{m0} + i\eta} - \frac{[a^*_{\mathbf{k}-q/2}]_{0m}\, [a_{\mathbf{k}+q/2}]_{m\lambda}}{\omega_{q\lambda}/2 - \omega - \omega_{m0} + i\eta} \right\} \tag{6–32}$$

(Again f is defined up to a multiplicative constant.)

To interpret (6–31), we take its Fourier transform with respect to the energy,

$$f(\mathbf{k}, t\,;\,\varpi_\lambda) = \frac{1}{2\pi}\int d\omega\,\exp(-\,i\omega t)\,f(\mathbf{k}, \omega\,;\,\varpi_\lambda) \qquad (6\text{–}33)$$

Let us replace f by its expression (6–32). The integrations are easy; the result can be put into the form

$$f(\mathbf{k}, t\,;\,\varpi_\lambda) = \text{cons}\cdot\exp\left(\frac{i\omega_{q\lambda}t}{2}\right) < \varphi_0 \mid T\{\,\mathbf{a}_{\mathbf{k}+q/2}(t)\ \mathbf{a}^*_{\mathbf{k}-q/2}(0)\,\} \mid \varphi_{q\lambda} > \qquad (6\text{–}34)$$

(the operators **a** and **a*** are here expressed in the Heisenberg representation). Comparing (6–34) with (3–52), we see that, up to a phase factor and a constant multiplicative factor, f is just the Fourier transform of $\tilde{\chi}_\lambda(\mathbf{x},0)$ with respect to r. In particular, for $t = 0$, f is the wave function for the relative motion of the hole and the particle in the bound state, expressed in k space. Rigorously, this discussion is valid only for a stable state. For an unstable state, the notion of wave function becomes a little obscure. We shall then *take* $f(\mathbf{k},0,\varpi_\lambda)$ to represent the relative wave function of the bound state.

Let us summarize our results. The hole-particle bound states are completely determined by the homogeneous Bethe-Salpeter equation (6–25). The eigenvalues give the energies $\omega_{q\lambda}$, and the eigenfunctions are the Fourier transforms of the wave functions describing the relative motion of the bound state.

Let us emphasize the generality of these methods. They apply as well to the collective modes of a dilute gas or to plasmons as to excitons in an insulating crystal. They thus give a certain unity to areas which until now have been well differentiated.

3. Correlation Functions and Vertex Operators

Let us place our system in an external field, and let $\mathscr{H}_1$ be the Hamiltonian coupling the system to the field. The total Hamiltonian becomes

$$H = H_0 + H_I + \mathcal{H}_1 \qquad (6\text{–}35)$$

We shall consider two types of external fields: scalar and vector fields. For a scalar field, $\mathscr{H}_1$ involves only the particle density $\rho(r)$,

$$\mathcal{H}_1 = \int d^3r\,\rho(r)\,\Phi(r, t) \qquad (6\text{–}36)$$

$\Phi(r,t)$ is some scalar potential. For a vector field, on the other hand, $\mathscr{H}_1$ involves the current density at the point r, $J(r)$,

$$\mathcal{H}_1 = \int d^3r\, \boldsymbol{J}(r) \cdot \boldsymbol{A}(r, t) \qquad (6\text{–}37)$$

$A(r,t)$ is a vector potential. In the particular case of an electron gas, (6–36) corresponds to the coupling with an external test charge, whereas (6–37) is involved in the coupling to an electromagnetic wave.

The density and current density operators are defined by

$$\begin{cases} \rho(r) = \sum_i \delta(r - r_i) \\ \boldsymbol{J}(r) = \frac{1}{2} \sum_i \left[(\boldsymbol{P}_i/m)\, \delta(r - r_i) + \delta(r - r_i)\, (\boldsymbol{P}_i/m) \right] \end{cases} \qquad (6\text{–}38)$$

It is convenient to introduce their Fourier transforms,

$$\begin{cases} \rho_q = \sum_i \exp(-iqr_i) = \sum_{\mathbf{k}} a^*_{\mathbf{k}-q/2}\, a_{\mathbf{k}+q/2} \\ \boldsymbol{J}_q = \frac{1}{2} \sum_i \left[\exp(-iqr_i), \frac{\boldsymbol{P}_i}{m} \right]_+ = \sum_{\mathbf{k}} \frac{\boldsymbol{k}}{m}\, a^*_{\mathbf{k}-q/2}\, a_{\mathbf{k}+q/2} \end{cases} \qquad (6\text{–}39)$$

With this notation, we can rewrite $\mathscr{H}_1$ in the form

$$\mathcal{H}_1 = \begin{cases} \sum_q \Phi_q(t)\, \rho_{-q} & \text{scalar coupling} \\ \sum_q \boldsymbol{A}_q(t) \cdot \boldsymbol{J}_{-q} & \text{vector coupling} \end{cases} \qquad (6\text{–}40)$$

In order to simplify, we shall assume that $\mathscr{H}_1$ is periodic in space and time, having the structure of a traveling wave. For scalar coupling, for example, we set

$$\Phi_{q'}(t) = \begin{cases} \Phi_0 \exp(-i\varepsilon t) & \text{if} \quad q' = q \\ \Phi_0^* \exp(+i\varepsilon t) & \text{if} \quad q' = -q \\ 0 \text{ otherwise} \end{cases} \qquad (6\text{–}41)$$

We are thus in the same position as in Chap. 2.

We shall see in sections 7 and 8 that the response of the system to these external fields involves the two functions

$$\Delta_4(\mathbf{k}, t, t' \,;\, \varpi) = \int_{-\infty}^{+\infty} dt'' \exp(-i\varepsilon t'') < \varphi_0 \mid T\left\{ a^*_{\mathbf{k}-q/2}(t')\, a_{\mathbf{k}+q/2}(t)\, \rho_{-q}(t'') \right\} \mid \varphi_0 >$$

$$\Delta_\alpha(\mathbf{k}, t, t' \,;\, \varpi) = \int_{-\infty}^{+\infty} dt'' \exp(-i\varepsilon t'') < \varphi_0 \mid T\left\{ a^*_{\mathbf{k}-q/2}(t')\, a_{\mathbf{k}+q/2}(t)\, J_{-q\alpha}(t'') \right\} \mid \varphi_0 > \qquad (6\text{–}42)$$

Again, ϖ is an abbreviated notation for (q,ε); $\alpha = (1,2,3)$ refers to one of the three coordinates of the current. For reasons of symmetry, Δ_4 and Δ_α are independent of the spin σ.

Returning to (6–38), we see that the expressions (6–42) involve only the two-particle Green's function K. More precisely,

$$\Delta_4(k, t, t' \,;\, \varpi) =$$

$$\sum_{\mathbf{k}''} \int_{-\infty}^{+\infty} dt'' \exp(-i\varepsilon t'')\, K\left[(\mathbf{k}'' - q/2)\, t'', (\mathbf{k} + q/2)\, t\,,\, (\mathbf{k} - q/2)\, t', (\mathbf{k}'' + q/2)\, t''\right]$$

$$\Delta_\alpha(k, t, t' \,;\, \varpi) = \qquad (6\text{–}43)$$

$$\sum_{\mathbf{k}''} \frac{k''_\alpha}{m} \int_{-\infty}^{+\infty} dt'' \exp(-i\varepsilon t'')\, K\left[(\mathbf{k}'' - q/2)\, t'', (\mathbf{k} + q/2)\, t\,,\, (\mathbf{k} - q/2)\, t', (\mathbf{k}'' + q/2)\, t''\right]$$

This form is quite cumbersome. In order to simplify it, we replace K by its Fourier expansion (3–47), and we use the reduced notation ${}^0K(\mathbf{p},\mathbf{p}';\varpi)$ defined by (6–16). We thus obtain

$$\left\{\begin{aligned} &\Delta_4(k, t, t' \,;\, \varpi) = \\ &\quad \frac{1}{2\pi} \sum_{\mathbf{k}''} \int_{-\infty}^{+\infty} d\omega\, d\omega'' \exp[-i\omega(t - t')] \exp\left[-i\varepsilon \left(\frac{t + t'}{2}\right)\right] {}^0K(\mathbf{p}, \mathbf{p}'' \,;\, \varpi) \\ &\Delta_\alpha(k, t, t' \,;\, \varpi) = \\ &\quad \frac{1}{2\pi} \sum_{\mathbf{k}''} \frac{k''_\alpha}{m} \int_{-\infty}^{+\infty} d\omega\, d\omega'' \exp[-i\omega(t - t')] \exp\left[-i\varepsilon \left(\frac{t + t'}{2}\right)\right] {}^0K(\mathbf{p}, \mathbf{p}'' \,;\, \varpi) \end{aligned}\right. \qquad (6\text{–}44)$$

We see that

$$\exp\left[\mathrm{i}\varepsilon\left(\frac{t+t'}{2}\right)\right]\Delta_4 \qquad (\text{or } \Delta_\alpha)$$

is a function only of $t - t'$. Let us define the Fourier transforms,

$$\Delta_{4(\alpha)}(k, \omega\,;\varpi) = \int_{-\infty}^{+\infty}\Delta_{4(\alpha)}(k, t, t'\,;\varpi)\exp[\mathrm{i}\omega(t-t')]\exp\left[\mathrm{i}\varepsilon\left(\frac{t+t'}{2}\right)\right]\mathrm{d}(t-t') \tag{6–45}$$

The preceding definitions then take the very simple form

$$\begin{cases} \Delta_4(p\,;\varpi) = \sum_{\mathbf{p}'} {}^0K(\mathbf{p}, \mathbf{p}'\,;\varpi) \\ \Delta_\alpha(p\,;\varpi) = \sum_{\mathbf{p}'} \frac{k'_\alpha}{m}\, {}^0K(\mathbf{p}, \mathbf{p}'\,;\varpi) \end{cases} \tag{6–46}$$

[We use the convention (6–17).]

Just as for $^\circ K$, the quantities Δ_4 and Δ_α can be represented by diagrams, indicated in Fig. 8. The central core is an ordinary γ, surrounded by four propagators G. The two internal propagators come together again at the vertex M, with which we associate a factor 1 if it corresponds to Δ_4, a factor k'_α/m if it corresponds to Δ_α. It is convenient to draw the diagram so that the line coming from $(\mathbf{p} - \varpi/2)$ terminates on $(\mathbf{p}' - \varpi/2)$. We can then apply the usual rules of calculation.

This graphical representation emphasizes the physical meaning of the quantities Δ. At the point M, there is interaction with the external field and creation of a pair $(\mathbf{k}' + q/2)$, $(\mathbf{k}' - q/2)$. Actually, these two excita-

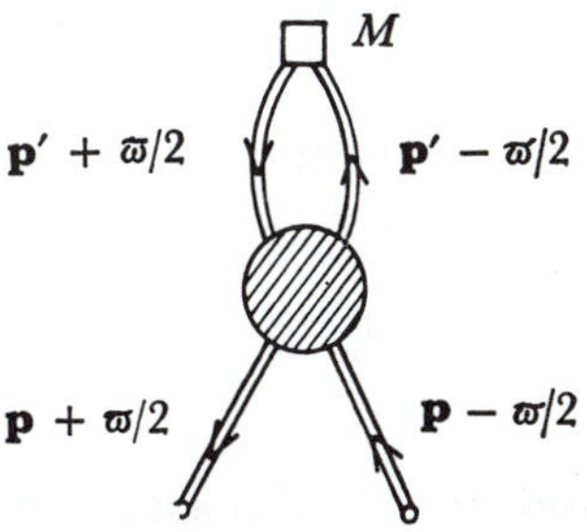

Figure 8

tions are created near one another; they have a tendency to interact. Δ_4 and Δ_α measure, respectively, for a scalar and a vector field, the probability of recovering, at the end of a certain time, these two excitations in the state $(\mathbf{k} + q/2)$, $(\mathbf{k} - q/2)$.

It is convenient to remove the two external lines from the diagram of Fig. 8, putting

$$\begin{cases} \Delta_4(p\,;\varpi) = \Lambda_4(p\,;\varpi)\, G(p + \varpi/2)\, G(p - \varpi/2) \\ \Delta_\alpha(p\,;\varpi) = \Lambda_\alpha(p\,;\varpi)\, G(p + \varpi/2)\, G(p - \varpi/2) \end{cases} \tag{6-47}$$

We shall designate the quantities Λ_4 and Λ_α by the names of scalar and vector "vertex parts." These vertex parts are frequently used in quantum electrodynamics. We shall see later in this chapter that they also play a very important role in the many-body problem.

The first term of the vertex parts is of order 0; it arises from the free part of K in (6–46) and is equal to 1 for Λ_4, k_α/m for Λ_α. The other terms arise from δK. Comparing (6–46) with (6–47), we see that

$$\begin{aligned} \Lambda_4(p\,;\varpi) &= 1 + \sum_{\mathbf{p}'} {}^0\Gamma(\mathbf{p}, \mathbf{p}'\,;\varpi)\, G(p' - \varpi/2)\, G(p' + \varpi/2) \\ \Lambda_\alpha(p\,;\varpi) &= \frac{k_\alpha}{m} + \sum_{\mathbf{p}'} \frac{k'_\alpha}{m}\, {}^0\Gamma(\mathbf{p}, \mathbf{p}'\,;\varpi)\, G(p' - \varpi/2)\, G(p' + \varpi/2) \end{aligned} \tag{6-48}$$

As a consequence the singularities of Γ occur also in Λ.

Many properties are common to Λ_4 and Λ_α. To shorten the writing of the equations, we introduce the notation μ to designate all four indices (1,2,3,4), and we put

$$\lambda_\mu(k) = \begin{cases} 1 & \text{if} \quad \mu = 4 \\ k_\alpha/m & \text{if} \quad \mu = \alpha \end{cases} \tag{6-49}$$

(6–48) can thus be written in the unique form

$$\Lambda_\mu(p\,;\varpi) = \lambda_\mu(k) + \sum_{\mathbf{p}'} {}^0\Gamma(\mathbf{p}, \mathbf{p}'\,;\varpi)\, G(p' - \varpi/2)\, G(p' + \varpi/2)\, \lambda_\mu(k') \tag{6-50}$$

We can separate out in Λ_μ the effect of hole-particle multiple scattering, as we have already done for Γ. Let us compare (6–19) and

(6–50). A simple calculation shows that

$$\Lambda_\mu(p\,;\varpi) = \lambda_\mu(k) + \sum_{\mathbf{p}'} {}^0I(\mathbf{p}, \mathbf{p}'\,;\varpi)\, G(p' - \varpi/2)\, G(p' + \varpi/2)\, \Lambda_\mu(p'\,;\varpi) \qquad (6\text{–}51)$$

(6–51) is the Bethe-Salpeter equation for the vertex parts. We can obtain it directly by looking at Fig. 8. Let us start from the bottom. If the diagram is of order zero, we obtain λ_μ; if not, there is at least one irreducible interaction I, plus a complete vertex part Λ_μ, hence (6–51).

We shall use these results in the following sections. For the moment, we leave the vertex parts and study the correlation functions. Let us look again at Eq. (6–42); using (6–38), we easily verify that

$$\sum_{\mathbf{k}} \Delta_4(k, t, t\,;\varpi) = \int_{-\infty}^{+\infty} dt'' \exp(-i\varepsilon t'') < \varphi_0 \mid P\{\rho_q(t)\,\rho_{-q}(t'')\} \mid \varphi_0 > \qquad (6\text{–}52)$$

The matrix element entering into (6–52) is just the correlation function $\Omega\bar{S}(q, t'' - t)$, defined by (3–62). As a consequence, we find, after integration over t'',

$$\exp(i\varepsilon t) \sum_{\mathbf{k}} \Delta_4(k, t, t\,;\varpi) = \Omega\,\bar{S}(q, \varepsilon) \qquad (6\text{–}53)$$

Using the definition (6–45), we can put this relation into the form

$$2\pi\,\Omega\,\bar{S}(q, \varepsilon) = \sum_{\mathbf{p}} \Delta_4(p, \varpi) \qquad (6\text{–}54)$$

Thus, knowledge about Δ_4 includes that about $\bar{S}$; the correlation function is a by-product of the preceding study.

$\bar{S}$ is the *density* correlation function. It is easy to extend the preceding results to other correlation functions. By analogy with (3–62), let us define

$$\left\{\begin{aligned} S_{\alpha\beta}(q, t) &= 2\pi < \varphi_0 \mid P\{\mathbf{J}_{q\alpha}(t)\,\mathbf{J}_{-q\beta}(0)\} \mid \varphi_0 > \\ S_{\alpha 4}(q, t) &= 2\pi < \varphi_0 \mid P\{\mathbf{J}_{q\alpha}(t)\,\rho_{-q}(0)\} \mid \varphi_0 > \\ S_{4\alpha}(q, t) &= 2\pi < \varphi_0 \mid P\{\rho_q(t)\,\mathbf{J}_{-q\alpha}(0)\} \mid \varphi_0 > \\ S_{44}(q, t) &= 2\pi < \varphi_0 \mid P\{\rho_q(t)\,\rho_{-q}(0)\} \mid \varphi_0 > = 2\pi\,\Omega\,\bar{S}(q, t) \end{aligned}\right. \qquad (6\text{–}55)$$

Up to a factor $2\pi\Omega$, $S_{\alpha\beta}$ is the current correlation function, from which we can derive the conductivity. $S_{\alpha 4}$ and $S_{4\alpha}$ are mixed functions, giving, for example, the density fluctuation induced by a vector field (see Chap. 2).

We define the Fourier transforms of these various functions by the unique relation

$$S_{\mu\nu}(q, \varepsilon) = \int_{-\infty}^{+\infty} S_{\mu\nu}(q, t) \exp(i\varepsilon t)\, dt \qquad (6\text{–}56)$$

By repeating the proof which led us to (6–54), we can easily verify that

$$S_{\mu\nu}(\varpi) = \sum_{\mathbf{p}} \lambda_\mu(k)\, \Delta_\nu(p\,;\varpi) \qquad (6\text{–}57)$$

[see the definition (6–49)]. This very simple relation is in fact very powerful.

In the same way as for Δ_ν, $S_{\mu\nu}$ can be represented by diagrams. It suffices in Fig. 8 to close the two lower lines at a vertex N analogous to M (Fig. 9) and to sum over the variable $\mathbf{p}$. With the vertices M and N are associated, respectively, the factors $\lambda_\nu(k')$ and $\lambda_\mu(k)$. We obtain $\Delta_\nu(p;\varpi)$ from $°K(\mathbf{p},\mathbf{p}';\varpi)$ by closing one end of the diagram, then $S_{\mu\nu}(\varpi)$ by closing the second. This process of formation is very natural.

In conclusion, we shall express the correlation functions with the help of the vertex operators. A glance at Fig. 9 or, alternatively, the comparison of (6–47) and (6–57) shows that

$$S_{\mu\nu}(\varpi) = \sum_{\mathbf{p}} \lambda_\mu(k)\, G(p + \varpi/2)\, G(p - \varpi/2)\, \Lambda_\nu(p\,;\varpi) \qquad (6\text{–}58)$$

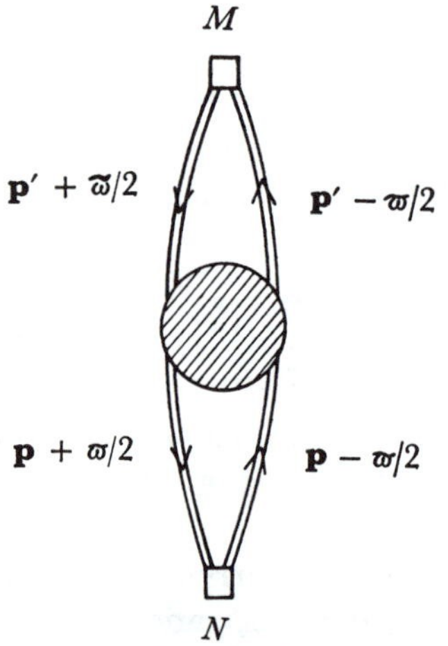

Figure 9

Note that the diagram of Fig. 9 is symmetric. Inverting the roles of M and N amounts to changing ϖ to $-\varpi$. As a consequence, $S_{\mu\nu}$ satisfies the symmetry relation

$$S_{\mu\nu}(\varpi) = S_{\nu\mu}(-\varpi) \tag{6-59}$$

This can be established directly by the techniques of Chap. 2.

With the help of these correlation functions, we can calculate the response of the system to an external field; we shall exploit these results in the sections on Response to a Scalar Field and Response of an Electron Gas to an Electromagnetic Field.

4. The Limit $\varpi \to 0$ (Short-Range Forces)

We propose to study the limit for $\varpi \to 0$ of the various quantities previously defined. We shall see that these limits are not unique.

a. The limits of $G(p - \varpi/2)\, G(p + \varpi/2)$

In Eq. (6–19), there appears the product

$$G(p - \varpi/2)\, G(p + \varpi/2) \tag{6-60}$$

It is tempting to assert that, when $\varpi \to 0$, (6–60) tends to $G^2(p)$; this is, unfortunately, false! According to (4–15), the Green's function G can be written as

$$G(k, \omega) = G^{\text{inc}}(k, \omega) + \frac{z_k}{\varepsilon_k - \omega + i\eta} \tag{6-61}$$

η is positive or negative according to whether k is inside or outside of the Fermi surface S_F. The incoherent part G^{inc} arises from the continuous background in the spectral densities $A\pm$; it is a regular function of ω. The product (6–60) can thus be decomposed into four terms. All the difficulties come from

$$\frac{z_{k-q/2}}{(\varepsilon_{k-q/2} - \omega + \varepsilon/2 + i\eta')} \cdot \frac{z_{k+q/2}}{(\varepsilon_{k+q/2} - \omega - \varepsilon/2 + i\eta'')} \tag{6-62}$$

(6–62) is highly singular, since the two poles with respect to ω merge when $\varpi \to 0$. The other terms, containing G^{inc}, have at most *one* pole; their limit for $\varpi \to 0$ is well defined.

Since $q \to 0$, we can put

$$\begin{cases} z_{k \pm q/2} = z_k \\ \varepsilon_{k \pm q/2} = \varepsilon_k \pm q \cdot v_k/2 \end{cases} \tag{6–63}$$

where v_k is the velocity of the quasi particle **k**. In order to remove the ambiguity associated with (6–62), let us compare each of its factors with its limit for $q = 0$,

$$\frac{z_k}{\varepsilon_k - \omega + i\eta} \tag{6–64}$$

The relative signs of the three quantities η, η', and η'' will determine the limit of (6–62).

Let us draw the surfaces S_+ and S_- obtained by displacing the Fermi surface S_F by $\pm q/2$ (Fig. 10). We thus divide k space into six regions, for which the signs of η, η', and η'' are given in Table 1. In regions I and II, η, η', and η'' have the same sign. The two poles of (6–62) come together on the same side of the real axis. When $\varpi \to 0$, (6–62) then tends to

$$\left(\frac{z_k}{\varepsilon_k - \omega + i\eta}\right)^2 \tag{6–65}$$

We can verify this assertion by studying the limit for $\varpi \to 0$ of the Fourier transform of (6–60) with respect to ω. (6–65) is not true in regions III to VI, where the poles approach each other while remaining on opposite sides of the real axis.

Let us take, for example, region III: η and η' are positive, and η'' is negative. Since we are infinitely near the Fermi surface, we can neglect

Table 1

Region	I	II	III	IV	V	VI
η	+	−	+	−	+	−
η'	+	−	+	+	−	−
η''	+	−	−	−	+	+

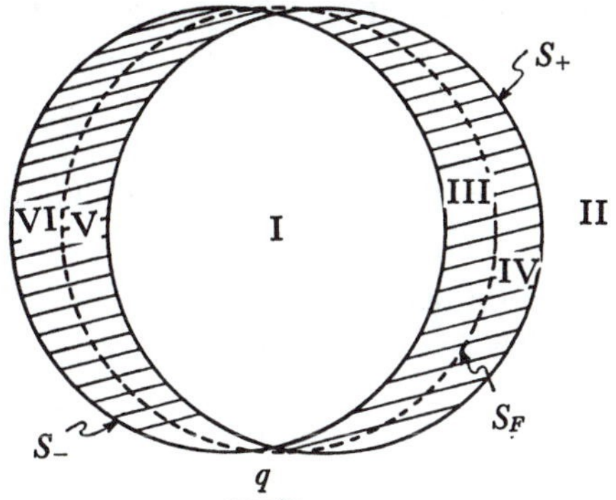

Figure 10

damping of the quasi particles and assume ε_k to be real. As a consequence,

$$\frac{z_k}{\varepsilon_k + q\cdot v_k/2 - \omega - \varepsilon/2 + i\eta''} = \frac{z_k}{\varepsilon_k + q\cdot v_k/2 - \omega - \varepsilon/2 + i\eta} + 2i\pi\, z_k\, \delta(\varepsilon_k + q\cdot v_k/2 - \omega - \varepsilon/2) \tag{6-66}$$

Putting this expression back into (6–62), we can put the latter into the form

$$\frac{z_k^2}{(\varepsilon_k + q\cdot v_k/2 - \omega - \varepsilon/2 + i\eta)(\varepsilon_k - q\cdot v_k/2 - \omega + \varepsilon/2 + i\eta)} + \frac{2i\pi\, z_k^2\, \delta(\varepsilon_k + q\cdot v_k/2 - \omega - \varepsilon/2)}{\varepsilon - q\cdot v_k + i\eta} \tag{6-67}$$

In the first term of (6–67), the two poles are above the real axis. When $\varpi \to 0$, this term tends to (6–65). The second term is singular, tending to

$$\frac{2i\pi\, z_k^2\, \delta(\varepsilon_k - \omega)}{\varepsilon - q\cdot v_k + i\eta} \tag{6-68}$$

Applying the same analysis to regions IV, V, and VI, we verify that in all cases the limit of (6–62) can be written as

$$\left(\frac{z_k}{\varepsilon_k - \omega + i\eta}\right)^2 \pm \frac{2i\pi\, z_k^2\, \delta(\varepsilon_k - \omega)}{\varepsilon - q\cdot v_k \pm i\,|\eta|} \tag{6-69}$$

the plus sign corresponding to regions III and IV and the minus sign to regions V and VI.

It is desirable to combine the results of (6–65) (regions I and II) and (6–69) (regions III to VI) in a single equation. For this purpose, let us consider a surface element $d\sigma$ of the Fermi surface, whose normal n_k is directed toward the exterior (Fig. 11). Let θ be the angle between q and n_k; $\cos\theta$ is positive in regions III and IV, negative in regions V and VI. The correction term in (6–69) appears in the shaded region, of volume

$$d\sigma\, q\,|\cos\theta| = q\,|\cos\theta|\,\delta(k_n)\,d^3k \tag{6–70}$$

(where k_n is the minimum distance from the point k to the surface S_F). Combining (6–69) with (6–70), we see that for every value of k we can write

$$\begin{cases} \lim\limits_{\varpi\to 0}(6.62) = \left(\dfrac{z_k}{\varepsilon_k - \omega + i\eta}\right)^2 + 2i\pi\, z_k^2\, \dfrac{q\cdot n_k}{\varepsilon - q\cdot v_k + i\alpha}\,\delta(\varepsilon_k - \omega)\,\delta(k_n) \\ \text{Sign}\ (\alpha) = \text{sign}\ (q\cdot n_k) \end{cases} \tag{6–71}$$

(To convince oneself of this, it is sufficient to integrate over a small volume d^3q).

The factor $\delta(k_n)$ restricts k to the Fermi surface S_F. As a consequence, ε_k is equal to the chemical potential μ. $\delta(k_n)$ can thus be written in the form

$$\delta(k_n) = |\,v_k\,|\,\delta(\varepsilon_k - \mu) \tag{6–72}$$

Combining (6–71) with the regular part of the quantity (6–60) we find

$$\begin{cases} \lim\limits_{\varpi\to 0} G(p - \varpi/2)\,G(p + \varpi/2) = G^2(p) + R(p\,;\varpi) \\ R(p\,;\varpi) = 2i\pi\, z_k^2\, \dfrac{q\cdot v_k}{\varepsilon - q\cdot v_k + i\alpha}\,\delta(\mu - \omega)\,\delta(\mu - \varepsilon_k) \end{cases} \tag{6–73}$$

This very important result summarizes the preceding discussion.

Note: In proving (6–73), we have not assumed the system to be isotropic; this equation is therefore very general. Although in practice we are mainly interested in isotropic systems, we shall carry out the calculations of this section without making this supplementary hypothesis, contenting ourselves with occasionally pointing out the simplifications which it makes.

We see from (6–73) that the limit of $G(p-\varpi/2)\,G(p+\varpi/2)$ when $\varpi \to 0$ is not unique. It depends on the limiting value of the ratio

$$\boldsymbol{r} = \boldsymbol{q}/\varepsilon \tag{6–74}$$

We shall designate by $R^r(p)$ the limit of $R(p;\varpi)$ when q and ε tend

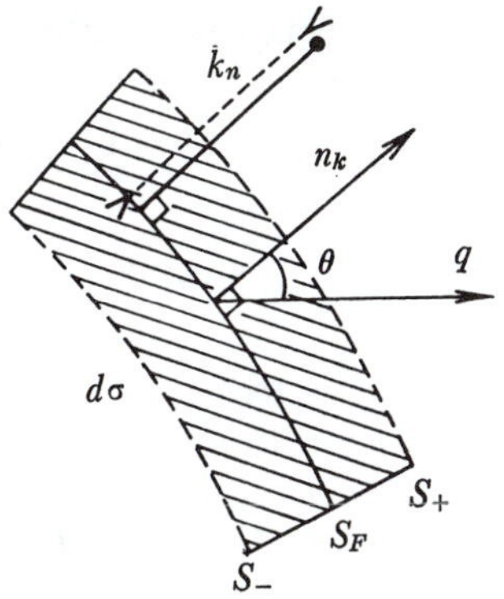

Figure 11

simultaneously to 0, the ratio r remaining constant. According to (6–73), we have

$$R^r(p) = 2i\pi\, z_k^2 \frac{r \cdot v_k}{1 - r \cdot v_k + i\alpha}\, \delta(\mu - \omega)\, \delta(\mu - \varepsilon_k) \qquad \textit{(6–75)}$$

$R = 0$ when $r = 0$; in this case, the product (6–60) tends simply to $G^2(p)$.

Note that the singular term R enters only on the Fermi surface, the energy ω being equal to μ. This characteristic is very important; it will allow us to justify the phenomenological theory of Landau.

b. The limit of the interaction operator (short-range forces)

Let us consider $°\Gamma(\mathbf{p},\mathbf{p}';\varpi)$. We can represent it with the help of skeleton diagrams, obtained by suppressing all self-energy parts in the complete diagrams (see Chap. 5, section 6). With each line of the skeleton is associated a factor G, instead of G_0.

Let us take a skeleton diagram of order p, having $(2p - 2)$ lines. Its contribution contains p factors $V(\mathbf{k}_1, \ldots, \mathbf{k}_4)$ and $(2p - 2)$ factors G. We limit ourselves to interactions *of finite range*, such as the interaction between two neutrons or between two atoms of He^3. In this case, the Fourier transform $V(q)$ of the binary interaction remains finite for $q = 0$; the matrix elements $V(\mathbf{k}_1, \ldots, \mathbf{k}_4)$ are therefore regular, whatever the values of $\mathbf{k}_1, \ldots \mathbf{k}_4$. This conclusion is false for Coulomb interactions, whose range is infinite; we shall treat this special case in section 6.

A singularity of $°\Gamma$ when $\varpi \to 0$ thus can arise only from the G

factors. In fact, the only trouble is that studied in the preceding paragraph, occurring when the poles of two factors G coincide. Let p_1 and p_2 be the variables characterizing these two propagators. There will be a singularity if by construction

$$p_2 = p_1 \pm \varpi$$

This requires that the diagram be *reducible* (see Fig. 6). In an irreducible diagram contributing to $°I$, the difference $(p_2 - p_1)$ is *never* fixed by construction; it can vary from $-\infty$ to $+\infty$. As a consequence, the function $°I(\mathbf{p},\mathbf{p}';\varpi)$ is regular when $\varpi \to 0$. Its limit $°I(\mathbf{p},\mathbf{p}')$ does not depend on the ratio r defined by (6–74).

Let $°\Gamma^r(\mathbf{p},\mathbf{p}')$ be the limit of $°\Gamma(\mathbf{p},\mathbf{p}';\varpi)$ when $\varpi \to 0$, r remaining constant. The symmetry relation (6–27) allows us to write

$$^0\Gamma^r(\mathbf{p}, \mathbf{p}') = {}^0\Gamma^r(\mathbf{p}', \mathbf{p}) \qquad \textit{(6–76)}$$

To calculate $°\Gamma^r$, we take the limit of Eq. (6–19). Using (6–73), we obtain

$$^0\Gamma^r(\mathbf{p}, \mathbf{p}') = {}^0I(\mathbf{p}, \mathbf{p}') + \sum_{\mathbf{p}''} {}^0I(\mathbf{p}, \mathbf{p}'') \{ G^2(p'') + R^r(p'') \} \, {}^0\Gamma^r(\mathbf{p}'', \mathbf{p}') \qquad \textit{(6–77)}$$

Let us consider the particular case $r = 0$:

$$^0\Gamma^0(\mathbf{p}, \mathbf{p}') = {}^0I(\mathbf{p}, \mathbf{p}') + \sum_{\mathbf{p}''} {}^0I(\mathbf{p}, \mathbf{p}'') \, G^2(p'') \, {}^0\Gamma^0(\mathbf{p}'', \mathbf{p}') \qquad \textit{(6–78)}$$

By manipulating (6–78) and (6–77), we can easily verify that

$$^0\Gamma^r(\mathbf{p}, \mathbf{p}') = {}^0\Gamma^0(\mathbf{p}, \mathbf{p}') + \sum_{\mathbf{p}''} {}^0\Gamma^0(\mathbf{p}, \mathbf{p}'') \, R^r(p'') \, {}^0\Gamma^r(\mathbf{p}'', \mathbf{p}') \qquad \textit{(6–79)}$$

Actually, it is simpler to prove (6–79) graphically. Let us consider a reducible diagram contributing to $°\Gamma^r$ (Fig. 12). With each pair of intermediate lines is associated a factor $(G^2 + R^r)$. Let us expand the product so obtained. We first have "regular" terms not containing *any* factors R^r, giving a total contribution $°\Gamma^0(\mathbf{p},\mathbf{p}')$. In the other terms, let us isolate the *first* singular factor $R^r(p'')$ starting from the bottom; the lower part of the diagram contains only regular terms, hence a factor $°\Gamma^0(\mathbf{p},\mathbf{p}'')$; the upper part can contain anything and therefore gives a contribution $°\Gamma^r(\mathbf{p}'',\mathbf{p}')$. (6–79) is thus proved.

Equation (6–79) allows us to relate the different limits of $°\Gamma$ to each other. It is equivalent to the Bethe-Salpeter equation. We can write it in

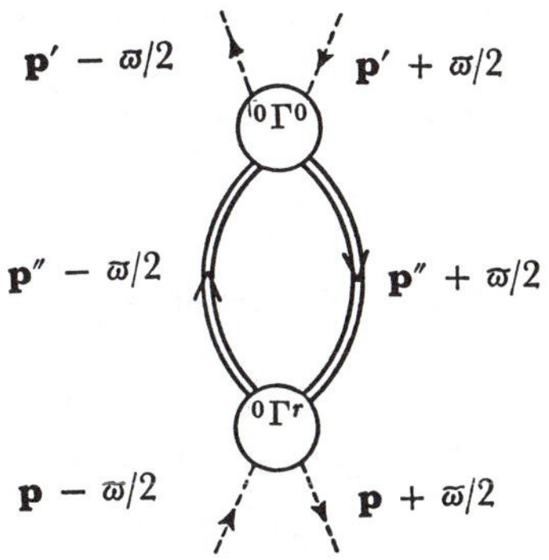

Figure 12

a different way, referring not to the limit $r = 0$, but to the case $r = \infty$. Let us set

$$\begin{cases} R^r(p) = R^\infty(p) + \widetilde{R}^r(p) \\ R^\infty(p) = -2i\pi\, z_k^2\, \delta(\mu - \omega)\, \delta(\mu - \varepsilon_k) \\ \widetilde{R}^r(p) = \dfrac{2i\pi z_k^2}{1 - r\cdot v_k + i\alpha}\, \delta(\mu - \omega)\, \delta(\mu - \varepsilon_k) \end{cases} \tag{6-80}$$

and assume the limit ${}^0\Gamma^\infty(\mathbf{p},\mathbf{p}')$ to be known, given by (6–77). Repeating the proof which led us to (6–79), we easily verify that

$${}^0\Gamma^r(\mathbf{p}, \mathbf{p}') = {}^0\Gamma^\infty(\mathbf{p}, \mathbf{p}') + \sum_{\mathbf{p}''} {}^0\Gamma^\infty(\mathbf{p}, \mathbf{p}'')\, \widetilde{R}^r(p'')\, {}^0\Gamma^r(\mathbf{p}'', \mathbf{p}') \tag{6-81}$$

According to the value of r, it may be preferable to choose (6–79) or (6–81).

These two equations are actually integral equations. Because of the two δ functions which R^r contains, the integrals extend only over the Fermi *surface* S_F. The situation is therefore much simpler than for (6–19). Note that R^0 and R^∞ are both independent of the direction of q. As a consequence, ${}^0\Gamma^0(\mathbf{p},\mathbf{p}')$ and ${}^0\Gamma^\infty(\mathbf{p},\mathbf{p}')$ also have this property. (For an isotropic system, these two functions depend only on the angle between k and k').

In practice, we shall never use the complete function ${}^0\Gamma^r(\mathbf{p},\mathbf{p}')$, much too rich in information, but only the *reduced* function

$$\begin{cases} {}^0\Gamma^r(\mathbf{k}, \mathbf{k}') = {}^0\Gamma^r(\mathbf{k}\mu, \mathbf{k}'\mu) \\ k \text{ et } k' \text{ on the Fermi surface} \end{cases} \qquad (6\text{–}82)$$

This reduced function satisfies the integral equations (6–79) and (6–81). Let us introduce the two quantities

$$\begin{aligned} f(\mathbf{k}, \mathbf{k}') &= 2\pi i z_k z_{k'} {}^0\Gamma^0(\mathbf{k}, \mathbf{k}') \\ A(\mathbf{k}, \mathbf{k}') &= 2\pi i z_k z_{k'} {}^0\Gamma^\infty(\mathbf{k}, \mathbf{k}') \end{aligned} \qquad (6\text{–}83)$$

We saw in Chap. 4 that $f(\mathbf{k},\mathbf{k}')$ was just the interaction between two quasi-particles in Landau's sense [Eq. (4–85)]. $A(\mathbf{k},\mathbf{k}')$ is related to the *forward* scattering amplitude of the two quasi particles $\mathbf{k}$ and $\mathbf{k}'$. We shall re-establish these results in section 7 by a more rigorous method than that used in Chap. 4.

f and A are related by Eq. (6–79), which we can write [see (6–75)] as

$$A(\mathbf{k}, \mathbf{k}') = f(\mathbf{k}, \mathbf{k}') - \sum_{\mathbf{k}''} f(\mathbf{k}, \mathbf{k}'')\, \delta(\varepsilon_{k''} - \mu)\, A(\mathbf{k}'', \mathbf{k}') \qquad (6\text{–}84)$$

We can decouple (6–84) by introducing the even and odd components of A and f, by analogy to (6–20),

$$f^\pm(k, k') = f(k\sigma, k'\sigma) \pm f(k\sigma, k'-\sigma) \qquad (6\text{–}85)$$

For an *isotropic* system, A and f depend only on the angle θ between k and k'. It is then sufficient to decompose A and f into Legendre polynomial series,

$$\begin{cases} f^\pm(\theta) = \sum_{l=0}^{\infty} f_l^\pm\, P_l(\cos\theta) \\ A^\pm(\theta) = \sum_{l=0}^{\infty} A_l^\pm\, P_l(\cos\theta) \end{cases} \qquad (6\text{–}86)$$

to obtain the solution to Eq. (6–84),

$$A_l^\pm = \frac{f_l^\pm}{1 + (m^* k_F/2\pi^2)\,[f_l^\pm/(2l+1)]} \qquad (6\text{–}87)$$

(m^* is the quasi particle effective mass). According to Chap. 1, we know that f_0^+ can be deduced from the velocity of sound, f_1^+ from the effective mass, and f_0^- from the spin susceptibility. We can find from these the corresponding coefficients of A, thus obtaining valuable information for the study of transport phenomena.

It is sometimes convenient to define, by analogy with (6–83), the function

$$f^r(\mathbf{k}, \mathbf{k}') = 2\pi\, \mathrm{i}\, z_k\, z_{k'}\, {}^0\Gamma^r(\mathbf{k}, \mathbf{k}') \tag{6–88}$$

Equations (6–79) and (6–81) then allow us to write

$$\left\{\begin{aligned} f^r(\mathbf{k}, \mathbf{k}') &= f(\mathbf{k}, \mathbf{k}') + \sum_{\mathbf{k}''} f(\mathbf{k}, \mathbf{k}'') \frac{r\cdot v_{k''}}{1 - r\cdot v_{k''} + \mathrm{i}\alpha}\, \delta(\varepsilon_{k''} - \mu)\, f^r(\mathbf{k}'', \mathbf{k}') \\ &= A(\mathbf{k}, \mathbf{k}') + \sum_{\mathbf{k}''} A(\mathbf{k}, \mathbf{k}'') \frac{\delta(\varepsilon_{k''} - \mu)}{1 - r\cdot v_{k''} + \mathrm{i}\alpha}\, f^r(\mathbf{k}'', \mathbf{k}') \end{aligned}\right. \tag{6–89}$$

In contrast to A and f, f^r depends on the direction of the vector r.

Let us turn now to the limits of the vertex operators, $\Lambda_\mu^r(p)$. These limits can be obtained from the definitions (6–48) and (6–58) and from the preceding results on ${}^\circ\Gamma$. Actually, it is simpler to analyze directly the diagrams of Fig. 8. The method is the same as that used to prove (6–79); as for ${}^\circ\Gamma$, there are several variants, according to whether the singular terms are investigated starting from the top or the bottom of the diagram and whether we refer to $r = 0$ or to $r = \infty$. We shall only state the results, leaving to the reader the task of proving them.

$$\Lambda_\mu^r(p) = \Lambda_\mu^0(p) + \sum_{\mathbf{p}'} {}^0\Gamma^0(\mathbf{p}, \mathbf{p}')\, R^r(p')\, \Lambda_\mu^r(p') \tag{6–90a}$$

$$= \Lambda_\mu^0(p) + \sum_{\mathbf{p}'} {}^0\Gamma^r(\mathbf{p}, \mathbf{p}')\, R^r(p')\, \Lambda_\mu^0(p') \tag{6–90b}$$

$$= \Lambda_\mu^\infty(p) + \sum_{\mathbf{p}'} {}^0\Gamma^\infty(\mathbf{p}, \mathbf{p}')\, \tilde{R}^r(p')\, \Lambda_\mu^r(p') \tag{6–90c}$$

$$= \Lambda_\mu^\infty(p) + \sum_{\mathbf{p}'} {}^0\Gamma^r(\mathbf{p}, \mathbf{p}')\, \tilde{R}^r(p')\, \Lambda_\mu^\infty(p') \tag{6–90d}$$

The choice of one of these equations depends on the problem studied; they are all equivalent, Eq. (6–90*b*) and (6–90*d*) giving a formal solution of Eqs. (6–90*a*) and (6–90*c*). We shall see in the following section that the limits Λ_μ° and Λ_μ^{∞} can be calculated very easily. Equations (6–90) are thus very helpful.

On the Fermi surface ($\omega = \varepsilon_k = \mu$), the relations (6–90) take a particularly simple form. Let us take

$$\begin{cases} \Lambda_\mu^r(k, \varepsilon_k) = \Lambda_\mu^r(k) \\ \varepsilon_k = \mu \end{cases} \tag{6–91}$$

Using the definitions (6–75), (6–80), (6–83), and (6–88), we can write

$$z_k \Lambda_\mu^r(k) = \begin{cases} z_k \Lambda_\mu^0(k) + \sum_{\mathbf{k}'} f(\mathbf{k}, \mathbf{k}') \dfrac{r \cdot v_{k'}}{1 - r \cdot v_{k'} + i\alpha} \delta(\mu - \varepsilon_{k'}) \, z_{k'} \Lambda_\mu^r(k') & (a) \\ z_k \Lambda_\mu^0(k) + \sum_{\mathbf{k}'} f^r(\mathbf{k}, \mathbf{k}') \dfrac{r \cdot v_{k'}}{1 - r \cdot v_{k'} + i\alpha} \delta(\mu - \varepsilon_{k'}) \, z_{k'} \Lambda_\mu^0(k') & (b) \\ z_k \Lambda_\mu^\infty(k) + \sum_{\mathbf{k}'} A(\mathbf{k}, \mathbf{k}') \dfrac{1}{1 - r \cdot v_{k'} + i\alpha} \delta(\mu - \varepsilon_{k'}) \, z_{k'} \Lambda_\mu^r(k') & (c) \\ z_k \Lambda_\mu^\infty(k) + \sum_{\mathbf{k}'} f^r(\mathbf{k}, \mathbf{k}') \dfrac{1}{1 - r \cdot v_{k'} + i\alpha} \delta(\mu - \varepsilon_{k'}) \, z_{k'} \Lambda_\mu^\infty(k') & (d) \end{cases} \tag{6–92}$$

We have gone to the trouble of writing down these equations in detail because each of them will prove useful at some stage of our theory.

It now remains to study the limit of the correlation functions $S_{\mu\nu}^r$. These are given by the diagrams of Fig. 9. These diagrams contain an arbitrary number ≥ 1 of pairs of lines ($p - \varpi/2$, $p + \varpi/2$), each giving a factor $[G^2(p) + R^r(p)]$. Let us expand the product thus obtained. We first have terms which are completely regular (without factors R^r), whose contribution is equal to $S_{\mu\nu}^{\circ}$. In all other cases, we isolate the first factor $R^r(p)$ found above the bottom of the diagram (Fig. 13). Below these lines, there are only regular terms, whose contribution is $\Lambda^{\circ}(p)$ (see Fig. 8). Above, we can have any combination of factors G^2 and R^r, and hence a contribution $\Lambda_\nu^r(p)$. In summary, we have

$$S_{\mu\nu}^r = S_{\mu\nu}^0 + \sum_{\mathbf{p}} \Lambda_\mu^0(p) \, R^r(p) \, \Lambda_\nu^r(p) \tag{6–93}$$

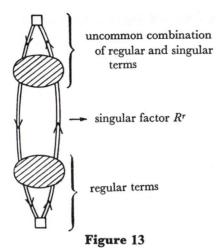

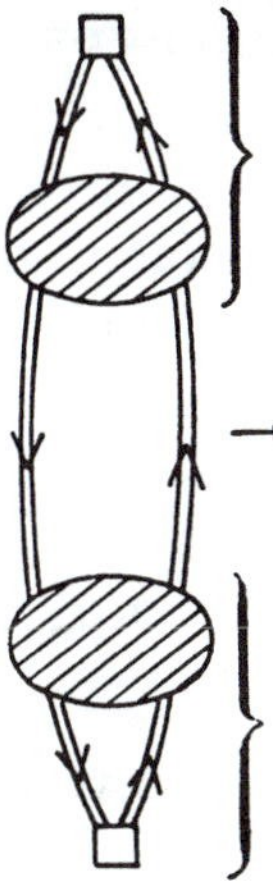

Figure 13

We can invert the roles of μ and ν since, according to (6–59), $S^r_{\mu\nu}$ is a symmetric tensor,

$$S^r_{\mu\nu} = S^r_{\nu\mu} \tag{6–94}$$

Finally, we can refer to the limit $r = \infty$ instead of $r = 0$. We thus obtain the relation

$$S^r_{\mu\nu} = S^\infty_{\mu\nu} + \sum_{\mathbf{p}} \Lambda^\infty_\mu(p)\, \widetilde{R}^r(p)\, \Lambda^r_\nu(p) \tag{6–95}$$

[see (6–80)]. The relations (6–93) and (6–95) complement each other.

Thanks to the two δ functions contained in R^r and $\widetilde{R}^r$, we can put these results into the simpler form

$$\begin{cases} S^r_{\mu\nu} = S^0_{\mu\nu} + 2i\pi \sum_{\mathbf{k}} z_k \Lambda^0_\mu(k) \dfrac{r \cdot v_k}{1 - r \cdot v_k + i\alpha}\, \delta(\mu - \varepsilon_k)\, z_k \Lambda^r_\nu(k) & \text{(6–96a)} \\ \quad\;\; = S^\infty_{\mu\nu} + 2i\pi \sum_{\mathbf{k}} z_k \Lambda^\infty_\mu(k) \dfrac{1}{1 - r \cdot v_k + i\alpha}\, \delta(\mu - \varepsilon_k)\, z_k \Lambda^r_\nu(k) & \text{(6–96b)} \end{cases}$$

We see that only intermediate states *on the Fermi surface* are involved—that is, only those for which the notion of quasi particle is valid. It therefore seems possible to calculate the correlation functions by studying the dynamics of these quasi particles.

5. The Ward Pitaerskii Identities

In order to complete this formalism, we must calculate the limiting values $\Lambda^{\circ}_{\mu}(p)$ and $\Lambda^{\infty}_{\mu}(p)$, $S^{\circ}_{\mu\nu}$ and $S^{\infty}_{\mu\nu}$. This can be done with the help of the Ward identities, relating the vertex operators to the derivatives of the self-energy operator $M(p)$. (The application of these identities to the many body problem is due to Pitaerskii).

Let us first consider the derivative

$$\frac{\partial}{\partial\omega} M(k, \omega) = \lim_{\varepsilon\to 0} \frac{1}{\varepsilon} \left[M(k, \omega + \varepsilon) - M(k, \omega)\right] \tag{6–97}$$

As a consequence of energy conservation at each vertex, the passage from ω to $\omega + \varepsilon$ amounts to increasing by ε the energy of *all the internal lines* of the diagrams contributing to M. Let us consider a skeleton diagram of order p, containing $(2p - 1)$ lines. The $(2p - 1)$ factors $G(k',\omega')$ are replaced by $G(k', \omega' + \varepsilon)$. To first order in ε, the variation in M is the sum of $(2p - 1)$ terms, each arising from an internal line of the original skeleton diagram.

Let us choose one of these terms, represented in Fig. 14. The core A has the structure of an interaction operator. This core is necessarily irreducible; if not, the original diagram would contain a diagonal insertion and would not be a skeleton diagram. (We see in Fig. 15 that the core A_2 is a diagonal insertion in the intermediate lines.) We can thus write

$$\frac{M(k, \omega + \varepsilon) - M(k, \omega)}{\varepsilon} = \sum_{\mathbf{p}'} \mathcal{J}(\mathbf{p}, \mathbf{p}') \left\{ \frac{G(k', \omega' + \varepsilon) - G(k', \omega')}{\varepsilon} \right\} \tag{6–98}$$

The kernel $\mathcal{J}(\mathbf{p},\mathbf{p}')$ is made up of all *irreducible* diagrams obtained by suppressing *any* line of *any* skeleton diagram contributing to M. Let us consider an irreducible diagram D; it appears at least once in $\mathcal{J}$ and arises from a diagram of M obtained by reclosing the ends $\mathbf{p}'$ of D with a line G. If D appears twice in $\mathcal{J}$, this can be done only by opening two

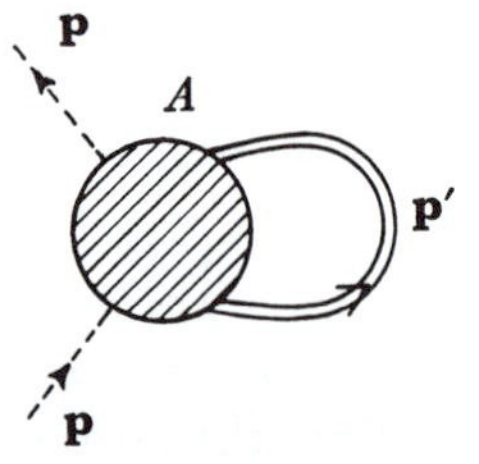

Figure 14

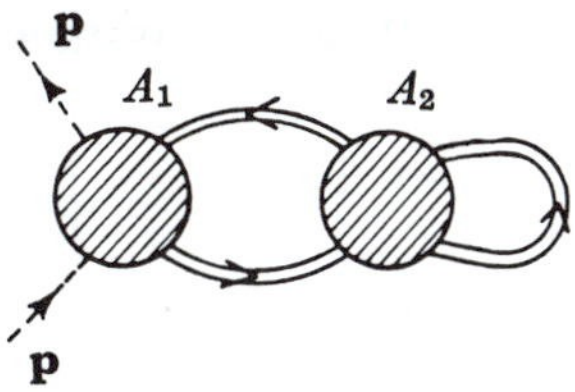

Figure 15

lines of the same diagram of M; these two lines are then equivalent, and the diagram of M contains a factor $\frac{1}{2}$ from the start, with the result that in the end D is multiplied by a factor 1. This conclusion persists if the diagram D appears more than twice in $\mathcal{J}$ (this can be shown easily by using labeled instead of unlabeled diagrams). We thus prove term by term the identity

$$\mathfrak{I}(\mathbf{p}, \mathbf{p}') = {}^0I(\mathbf{p}, \mathbf{p}') \tag{6–99}$$

(the limit of $°I$ when $\varpi \to 0$ is defined without any ambiguity).

Let us turn now to the second factor of (6–98). According to (5–111), we have (even for finite ε)

$$\frac{G(k', \omega' + \varepsilon) - G(k', \omega')}{\varepsilon} = \frac{[\varepsilon + M(k', \omega' + \varepsilon) - M(k', \omega')]}{\varepsilon} G(k', \omega')\, G(k', \omega' + \varepsilon) \tag{6–100}$$

When $\varepsilon \to 0$, the first factor of (6–100) tends to $[1 + \partial M(p')/\partial\omega']$. As for the second factor, its limit is given by (6–73). Here we have $r = 0$; as a consequence, the limit is simply $G^2(p')$. Collecting these results, we see that $\partial M/\partial\omega$ satisfies the integral equation

$$\frac{\partial M(p)}{\partial\omega} = \sum_{\mathbf{p}'} {}^0I(\mathbf{p}, \mathbf{p}')\, G^2(p') \left[1 + \frac{\partial M(p')}{\partial\omega'}\right] \tag{6–101}$$

Let us return to Eq. (6–51), in the case $\mu = 4$, and take its limit for $\varpi \to 0$, $r = 0$. We find

$$\Lambda_4^0(p) = 1 + \sum_{\mathbf{p}'} {}^0I(\mathbf{p}, \mathbf{p}')\, G^2(p')\, \Lambda_4^0(p') \tag{6–102}$$

Comparing (6–101) with (6–102), we see immediately that

$$\Lambda_4^0(p) = 1 + \frac{\partial M(p)}{\partial \omega} \tag{6–103}$$

This relation is known as a *Ward identity.* It is not unique; there exist three others, giving, respectively, Λ_4^∞, Λ_α^0, and Λ_α^∞. An analogous equation exists in quantum electrodynamics; it then expresses invariance with respect to a gauge transformation.

Let us now consider $\partial M(p)/\partial k_\alpha$. Repeating the analysis made for (6–98), we obtain

$$\frac{\partial M(p)}{\partial k_\alpha} = \operatorname*{Lim}_{q_\alpha \to 0} \sum_{\mathbf{p}'} {}^0I(\mathbf{p}, \mathbf{p}') \left\{ \frac{G(k' + q_\alpha, \omega') - G(k', \omega')}{q_\alpha} \right\} \tag{6–104}$$

Again using (5–111), we can write

$$\frac{G(k' + q_\alpha, \omega') - G(k', \omega')}{q_\alpha} = \left[\frac{-q_\alpha k'_\alpha/m + M(k' + q_\alpha, \omega') - M(k', \omega')}{q_\alpha}\right] G(k', \omega')\, G(k' + q_\alpha, \omega') \tag{6–105}$$

The last factor in (6–105) corresponds to the case $r = \infty$ (by definition). When $q_\alpha \to 0$, we therefore obtain

$$\frac{\partial M(p)}{\partial k_\alpha} = \sum_{\mathbf{p}'} {}^0I(\mathbf{p}, \mathbf{p}')\, [G^2(p') + R^\infty(p')] \left[\frac{\partial M(p')}{\partial k'_\alpha} - \frac{k'_\alpha}{m}\right] \tag{6–106}$$

Let us take (6–51) for the case $\mu = \alpha$, $r = \infty$ and make ϖ tend to 0. We find

$$\Lambda_\alpha^\infty(p) = \frac{k_\alpha}{m} + \sum_{\mathbf{p}'} {}^0I(\mathbf{p}, \mathbf{p}')\, [G^2(p') + R^\infty(p')]\, \Lambda_\alpha^\infty(p') \tag{6–107}$$

A comparison of (6–106) and (6–107) shows that

$$\Lambda_\alpha^\infty(p) = \frac{k_\alpha}{m} - \frac{\partial M(p)}{\partial k_\alpha} \tag{6–108}$$

(6–108) is our second Ward identity.

M is a function, not only of k and ω, but also, implicitly, of the Fermi surface S_F. Let us designate by $M(k + q_\alpha, \omega; q_\alpha)$ the value of the self-energy operator *when we simultaneously translate the wave vector k and the Fermi surface by an amount* q_α. In other words, $M(k + q_\alpha, \omega; q_\alpha)$ is the value of $M(k,\omega)$ seen from a reference frame moving with respect to the system at a uniform velocity $(-q_\alpha/m)$. We define the derivative,

$$\frac{\mathrm{d}M(p)}{\mathrm{d}k_\alpha} = \lim_{q_\alpha \to 0} \frac{1}{q_\alpha} [M(k + q_\alpha, \omega \,; q_\alpha) - M(k, \omega)] \qquad (6\text{–}109)$$

It is important not to confuse dM/dk_α with $\partial M/\partial k_\alpha$. In the first case, the Fermi surface follows the motion of k, whereas, in the second case it does not. We shall see later that $\partial M/\partial k_\alpha$ is related to the *velocity* of the quasi particle k, whereas dM/dk_α is related to the *current* carried.

Again, we analyze the variation of M diagram by diagram. $dM(p)/dk_\alpha$ is obtained by replacing in (6–104) $G(k' + q_\alpha, \omega')$ by $G(k' + q_\alpha, \omega'; q_\alpha)$. We are thus led to calculate [see (6–105)].

$$\lim_{q_\alpha \to 0} \left[\frac{G(k' + q_\alpha, \omega' \,; q_\alpha) - G(k', \omega')}{q_\alpha}\right] = \left[\frac{-k'_\alpha}{m} + \frac{\mathrm{d}M(p')}{\mathrm{d}k'_\alpha}\right] \lim_{q_\alpha \to 0} G(k', \omega')\, G(k' + q_\alpha, \omega' \,; q_\alpha) \qquad (6\text{–}110)$$

When we translate k and the Fermi surface *simultaneously*, the *relative* position of k and S_F does not change. The poles of $G(k',\omega')$ and of $G(k' + q_\alpha, \omega'; q_\alpha)$ are therefore always on the *same side* of the real axis, for *any* k'. As a consequence,

$$G(k', \omega')\, G(k' + q_\alpha, \omega' \,; q_\alpha) \xrightarrow[q_\alpha \to 0]{} G^2(p') \qquad (6\text{–}111)$$

Everything goes through as if we were considering the case $r = 0$.

Collecting these results, we see that $dM(p)/dk_\alpha$ satisfies the integral equation,

$$\frac{\mathrm{d}M(p)}{\mathrm{d}k_\alpha} = \sum_{\mathbf{p}'} {}^0I(\mathbf{p}, \mathbf{p}')\, G^2(p') \left[\frac{\mathrm{d}M(p')}{\mathrm{d}k'_\alpha} - \frac{k'_\alpha}{m}\right] \qquad (6\text{–}112)$$

This should be compared with the relation, obtained from (6–51),

$$\Lambda^0_\alpha(p) = \frac{k_\alpha}{m} + \sum_{\mathbf{p}'} {}^0I(\mathbf{p}, \mathbf{p}')\, G^2(p')\, \Lambda^0_\alpha(p') \qquad (6\text{–}113)$$

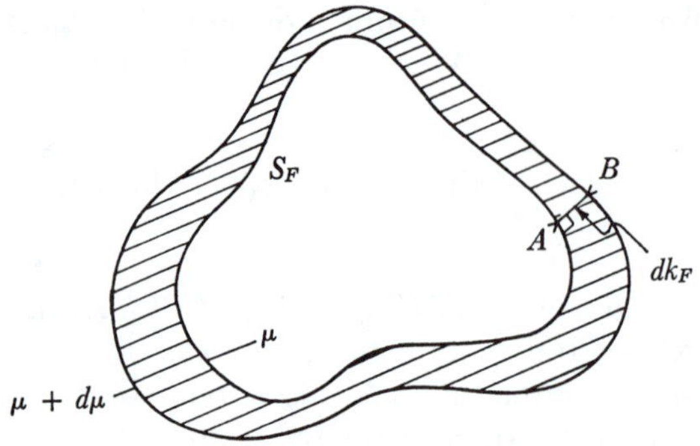

Figure 16

We see at once that

$$\boxed{\Lambda_\alpha^0(p) = \frac{k_\alpha}{m} - \frac{\mathrm{d}M(p)}{\mathrm{d}k_\alpha}} \tag{6–114}$$

It remains to establish a Ward identity for Λ_μ^∞; this is a little more difficult. Let us assume that the chemical potential μ varies by $d\mu$ ($d\mu > 0$). The Fermi surface expands. Let dk_F be the "normal" displacement of S_F at any point (see Fig. 16). For an anisotropic system dk_F can depend on direction. We shall put

$$\mathrm{d}k_F = \frac{\mathrm{d}k_F}{\mathrm{d}\mu}\,\mathrm{d}\mu \tag{6–115}$$

The energy ε_k of a quasi particle depends on μ. We can relate $\partial\varepsilon_k/\partial\mu$ to $dk_F/d\mu$ by noting that the energy of a quasi particle at the Fermi surface must remain equal to the chemical potential. Thus (see Fig. 16)

$$\varepsilon(k_B,\, \mu + \mathrm{d}\mu) - \varepsilon(k_A,\, \mu) = \frac{\partial\varepsilon_k}{\partial\mu}\,\mathrm{d}\mu + v_k\,\mathrm{d}k_F = \mathrm{d}\mu$$

from which we obtain the relation

$$\frac{\partial\varepsilon_k}{\partial\mu} = 1 - v_k\,\frac{\mathrm{d}k_F}{\mathrm{d}\mu} \tag{6–116}$$

The self-energy operator M depends implicitly on μ. Let us define

$$\frac{\mathrm{d}M(p)}{\mathrm{d}\mu} = \lim_{\mathrm{d}\mu\to 0} \frac{M(k, \omega + \mathrm{d}\mu\,;\,\mu + \mathrm{d}\mu) - M(k, \omega\,;\,\mu)}{\mathrm{d}\mu} \tag{6–117}$$

We have manifestly

$$\frac{\mathrm{d}M(p)}{\mathrm{d}\mu} = \frac{\partial M(p)}{\partial \mu} + \frac{\partial M(p)}{\partial \omega} \tag{6–118}$$

The decomposition made for the three other Ward identities applies again. Transposing (6–104) we obtain

$$\frac{\mathrm{d}M(p)}{\mathrm{d}\mu} = \lim_{\mathrm{d}\mu\to 0} \sum_{\mathbf{p}'} {}^{0}I(\mathbf{p}, \mathbf{p}') \left[\frac{G(k', \omega' + \mathrm{d}\mu\,;\,\mu + \mathrm{d}\mu) - G(k', \omega'\,;\,\mu)}{\mathrm{d}\mu}\right] \tag{6–119}$$

Using (5–111), we put the bracket in (6–119) into the form

$$\left[\frac{\mathrm{d}M(p')}{\mathrm{d}\mu} + 1\right] \lim_{\mathrm{d}\mu\to 0} G(k', \omega' + \mathrm{d}\mu\,;\,\mu + \mathrm{d}\mu)\, G(k', \omega'\,;\,\mu) \tag{6–120}$$

The difficulty lies in the evaluation of the limit of

$$G(k, \omega + \mathrm{d}\mu\,;\,\mu + \mathrm{d}\mu)\, G(k, \omega\,;\,\mu) \tag{6–121}$$

when $d\mu \to 0$.

When k is outside the shaded region in Fig. 16, the two factors of (6–121) have their poles on the same side of the real axis; the limit of the product is then equal to $G^2(p;\mu)$. This is false in the shaded region, where the singular part of the product is written as

$$\frac{z_k^2}{[\varepsilon_k + (\partial\varepsilon_k/\partial\mu)\,\mathrm{d}\mu - \omega - \mathrm{d}\mu + i\eta]\,[\varepsilon_k - \omega - i\eta]}, \quad \eta > 0 \tag{6–122}$$

(k passes from the exterior to the interior of S_F when μ increases by $d\mu$). By analogy with (6–67) and (6–69), we can put this expression into the form

$$\left(\frac{z_k}{\varepsilon_k - \omega - i\eta}\right)^2 - \frac{2\, i\pi z_k^2 \delta(\varepsilon_k - \omega)}{\mathrm{d}\mu(1 - \partial\varepsilon_k/\partial\mu)} \tag{6–123}$$

The second term of (6–123) appears only in the shaded region; this fact we shall express by the factor $v_k\, dk_F\, \delta(\varepsilon_k - \mu)$ [see (6–70) and (6–72)]. Using the identity (6–116), we finally obtain

$$\lim_{d\mu\to 0} G(k, \omega + d\mu\,;\, \mu + d\mu)\, G(k, \omega\,;\, \mu) = G^2(p\,;\, \mu) - 2i\pi\, z_k^2 \delta(\varepsilon_k - \mu)\, \delta(\varepsilon_k - \omega) \tag{6–124}$$

We thus obtain the same limit as in the case $r = \infty$.

Let us put the result (6–124) back into (6–119); we obtain the integral equation satisfied by $dM(p)/d\mu$,

$$dM(p)/d\mu = \sum_{\mathbf{p}'} {}^0I(\mathbf{p}, \mathbf{p}')\, [G^2(p') + R^\infty(p')]\, [1 + dM(p')/d\mu] \tag{6–125}$$

From (6–51) we deduce the relation

$$\Lambda_4^\infty(p) = 1 + \sum_{\mathbf{p}'} {}^0I(\mathbf{p}, \mathbf{p}')\, [G^2(p') + R^\infty(p')]\, \Lambda_4^\infty(p') \tag{6–126}$$

By comparison, we find

$$\boxed{\Lambda_4^\infty(p) = 1 + dM(p)/d\mu} \tag{6–127}$$

This relation concludes the set of four Ward identities.

These four identities are particularly interesting on the Fermi surface ($\omega = \varepsilon_k = \mu$). Note first that, according to (5–111), the residue of $G(k, \omega)$ at the pole ε_k is equal to

$$z_k = \frac{1}{1 + \partial M(p)/\partial\omega}\bigg|_{\omega=\varepsilon_k} \tag{6–128}$$

Comparing this with (6–103), we deduce that

$$\Lambda_4^0(k) = 1/z_k \tag{6–129}$$

Furthermore, the quasi-particle energy ε_k is given by the equation

$$k^2/2m - \varepsilon_k - M(k, \varepsilon_k) = 0 \tag{6–130}$$

Let us differentiate this equation with respect to k_α. We find

$$0 = \frac{k_\alpha}{m} - \frac{\partial M(k, \varepsilon_k)}{\partial k_\alpha} - \frac{\partial \varepsilon_k}{\partial k_\alpha}\left[1 + \frac{\partial M}{\partial\omega}\bigg|_{\omega=\varepsilon_k}\right] \tag{6–131}$$

Let us compare this with (6–108) and use (6–128). We obtain

$$\Lambda_\alpha^\infty(k) = \frac{1}{z_k}\frac{\partial \varepsilon_k}{\partial k_\alpha} = \frac{1}{z_k} v_{k\alpha} \tag{6–132}$$

where v_k is the velocity of the quasi particle k. Similarly, taking the total derivative d/dk_α of (6–130) and using (6–114), we find

$$\Lambda_\alpha^0(k) = \frac{1}{z_k}\frac{d\varepsilon_k}{dk_\alpha} \tag{6–133}$$

Finally, let us take the derivative of (6–130) with respect to the chemical potential μ,

$$\frac{\partial \varepsilon_k}{\partial \mu}\left(1 + \frac{\partial M}{\partial \omega}\right)_{\omega=\varepsilon_k} + \frac{\partial M}{\partial \mu} = 0 \tag{6–134}$$

By making use of (6–118), (6–127), and (6–128), we can put this relation into the form

$$\frac{1}{z_k}\frac{\partial \varepsilon_k}{\partial \mu} = \frac{1}{z_k} - \Lambda_4^\infty(k) \tag{6–135}$$

Let us refer now to (6–116); we see that

$$\Lambda_4^\infty(k) = \frac{1}{z_k} v_k \frac{dk_F}{d\mu} \tag{6–136}$$

It remains to clarify the physical meaning of $d\varepsilon_k/dk_\alpha$. $G(k + q_\alpha, \omega;\, q_\alpha)$ is just the Green's function $G(k,\omega)$ seen from a reference system moving with the velocity $-q_\alpha/m$. In this moving system, the energy of the quasi particle k is changed by an amount $q_\alpha(d\varepsilon_k/dk_\alpha)$. This point has already been treated in Chap. 1, Eqs. (1–25) to (1–28). According to (1–28), $d\varepsilon_k/dk_\alpha$ is just the component $J_{k\alpha}$ of the current carried by the quasi particle k [the definitions of q_α in Chaps. 1 and 6 are opposite in sign, which makes the minus sign disappear from (1–28)]. For a translationally invariant system, the total current operator J commutes with the Hamiltonian H. The total current thus remains constant during the adiabatic switching on of H_1; in particular, J_k is equal to its value in the absence of interactions,

$$J_k = k/m \tag{6–137}$$

We summarize below the expressions for the four Ward identities on the Fermi surface:

$$\begin{cases} z_k \Lambda_4^0(k) = 1 \\ z_k \Lambda_4^\infty(k) = v_k \, (\mathrm{d}k_F/\mathrm{d}\mu) \\ z_k \Lambda_\alpha^\infty(k) = v_{k\alpha} \\ z_k \Lambda_\alpha^0(k) = J_{k\alpha} \end{cases} \tag{6–138}$$

These are fundamental relations, since they involve observable quantities. They constitute a starting point from which we can calculate any limit $\Lambda_\mu^r(k)$, with the help of the integral equations (6–92).

Using (6–92), we can relate $\Lambda_\mu^\circ(k)$ and $\Lambda_\mu^\infty(k)$. Replacing these quantities by their expressions (6–138), we obtain the relations

$$J_{k\alpha} = v_{k\alpha} + \sum_{\mathbf{k}'} f(\mathbf{k}, \mathbf{k}') \, \delta(\mu - \varepsilon_{k'}) \, v_{k'\alpha} \tag{6–139a}$$

$$J_{k\alpha} = v_{k\alpha} + \sum_{\mathbf{k}'} A(\mathbf{k}, \mathbf{k}') \, \delta(\mu - \varepsilon_{k'}) \, J_{k'\alpha} \tag{6–139b}$$

$$v_k \frac{\mathrm{d}k_F}{\mathrm{d}\mu} = 1 - \sum_{\mathbf{k}'} f(\mathbf{k}, \mathbf{k}') \, \delta(\mu - \varepsilon_{k'}) \, v_{k'} \frac{\mathrm{d}k_F'}{\mathrm{d}\mu} \tag{6–140a}$$

$$v_k \frac{\mathrm{d}k_F}{\mathrm{d}\mu} = 1 - \sum_{\mathbf{k}'} A(\mathbf{k}, \mathbf{k}') \, \delta(\mu - \varepsilon_{k'}) \tag{6–140b}$$

Equation (6–139*a*) confirms the result (1–33), established by Landau's method. Moreover, (6–140*a*) is equivalent to Eq. (1–20). This formalism is thus very close to that of Chap. 1.

Before leaving the vertex operators, we shall mention one last relation, known as a "continuity" equation, relating $\Lambda_4^r(k)$ to $\Lambda_\alpha^r(k)$,

$$\sum_\alpha \{ r_\alpha(\Lambda_\alpha^r - \Lambda_\alpha^\infty) \} = \Lambda_4^r - \Lambda_4^0 \tag{6–141}$$

To prove (6–141), we express Λ_α^r with the help of (6–92*d*), Λ_4^r with the help of (6–92*b*), and we use the identities (6–138).

Let us now apply these results to the study of the correlation functions $S_{\mu\nu}^r$. Let us first consider the limiting case $S_{\mu 4}^\circ$. According to (6–58), and with the use of (6–103), we can write

$$S^0_{\mu 4} = \sum_{\mathbf{p}} \lambda_\mu(k) \left[1 + \frac{\partial M(p)}{\partial \omega}\right] \lim_{\substack{\varpi \to 0 \\ r=0}} \{ G(p + \varpi/2)\, G(p - \varpi/2) \} \qquad \textit{(6–142)}$$

Referring to Eq. (6–100), we see that

$$\left[1 + \frac{\partial M(p)}{\partial \omega}\right] \lim_{\substack{\varpi \to 0 \\ r=0}} \{ G(p + \varpi/2)\, G(p - \varpi/2) \} = \frac{\partial G(p)}{\partial \omega} \qquad \textit{(6–143)}$$

This results directly from the hypothesis $r = 0$. Inserting (6–143) into (6–142), and performing the integration over ω, we verify that

$$S^0_{44} = S^0_{\alpha 4} = 0 \qquad \textit{(6–144)}$$

Let us consider $S^\infty_{\mu\alpha}$. With the help of (6–58) and (6–108), we can write it as

$$S^\infty_{\mu\alpha} = \sum_{\mathbf{p}} \lambda_\mu(k) \left[\frac{k_\alpha}{m} - \frac{\partial M(p)}{\partial k_\alpha}\right] \lim_{\substack{\varpi \to 0 \\ r=\infty}} G(p + \varpi/2)\, G(p - \varpi/2) \qquad \textit{(6–145)}$$

This limit is the same as that involved in (6–105). As a consequence, the preceding relation becomes

$$S^\infty_{\mu\alpha} = - \sum_{\mathbf{p}} \lambda_\mu(k) \frac{\partial G(p)}{\partial k_\alpha} \qquad \textit{(6–146)}$$

Let us first perform the integration over ω. We know that in doubtful cases we must close the contour in the upper half plane. But, according to (3–24),

$$\int_C G(p)\, d\omega = -2i\pi\, m_k \qquad \textit{(6–147)}$$

where m_k is the *bare* particle distribution function in the state $|\varphi_0\rangle$. (6–146) can thus be written as

$$S^\infty_{\mu\alpha} = 2i\pi \sum_{\mathbf{k}} \lambda_\mu(k) \frac{\partial m_k}{\partial k_\alpha} \qquad \textit{(6–148)}$$

Integrating by parts, we easily verify that

$$\begin{cases} S^{\infty}_{4\alpha} = 0 \\ S^{\infty}_{\beta\alpha} = -2i\pi(N/m)\,\delta_{\alpha\beta} \end{cases} \tag{6–149}$$

We shall see in section 8 that (6–149) ensures the gauge invariance of the theory.

(6–144) and (6–149), supplemented by the symmetry relation (6–94), give us special limits of $S_{\mu\nu}$. The most general limit is then obtained from (6–96), which we can simplify with the aid of the Ward identities (6–138),

$$S^{r}_{4\mu} = S^{0}_{4\mu} + 2i\pi \sum_{\mathbf{k}} \frac{r \cdot v_k}{1 - r \cdot v_k + i\alpha}\,\delta(\mu - \varepsilon_k)\, z_k \Lambda^{r}_{\mu}(k) \tag{6–150}$$

$$S^{r}_{\alpha\mu} = S^{\infty}_{\alpha\mu} + 2i\pi \sum_{\mathbf{k}} \frac{v_{k\alpha}}{1 - r \cdot v_k + i\alpha}\,\delta(\mu - \varepsilon_k)\, z_k \Lambda^{r}_{\mu}(k) \tag{6–151}$$

Comparing (6–150) with (6–151), we see at once that

$$\sum_{\alpha} \{ r_\alpha [S^{r}_{\alpha\mu} - S^{\infty}_{\alpha\mu}] \} = S^{r}_{4\mu} - S^{0}_{4\mu} \tag{6–152}$$

These "continuity" equations relate the density to the current correlation functions; they express the conservation of charge.

To complete our collection of limiting cases, we shall calculate S^{∞}_{44} and $S^{0}_{\alpha\beta}$. We find from (6–150) and (6–136)

$$S^{\infty}_{44} = -2i\pi \sum_{\mathbf{k}} \delta(\mu - \varepsilon_k)\, v_k \frac{dk_F}{d\mu} \tag{6–153}$$

Let us refer to Fig. 16. The factor $v_k \cdot dk_F\, \delta(\mu - \varepsilon_k)$ amounts to limiting the summation over $\mathbf{k}$ to the shaded region. In this region we have

$$\sum_{\mathbf{k}} 1 = dN$$

(the total number of particles N is equal to the number of values of k *inside* S_F). As a consequence, (6–153) can be written

$$S^{\infty}_{44} = -2i\pi\, dN/d\mu \tag{6–154}$$

$d\mu/dN$ is related to the compressibility by the relation (1–16); we shall interpret this result in section 7.

Finally, (6–138) and (6–151) allow us to write

$$S^0_{\alpha\beta} + \frac{2i\pi N}{m}\,\delta_{\alpha\beta} = 2i\pi \sum_{\mathbf{k}} v_{k\alpha}\, J_{k\beta}\,\delta(\mu - \varepsilon_k) \qquad (6\text{–}155)$$

Let us replace the sum over **k** by an integral and note that

$$v_{k\alpha}\,\delta(\mu - \varepsilon_k)\,\mathrm{d}^3k = \frac{v_{k\alpha}}{|v_k|}\,\mathrm{d}\sigma \qquad (6\text{–}156)$$

where $d\sigma$ is the surface element of S_F [see (6–70)]. The integral (6–155) thus appears as a flux across S_F, to which we can apply Ostrogradski's theorem. We obtain

$$S^0_{\alpha\beta} + \frac{2i\pi N}{m}\,\delta_{\alpha\beta} = \frac{2i\pi\Omega}{4\pi^3}\iiint_{S_F} \frac{\partial J_{k\beta}}{\partial k_\alpha}\,\mathrm{d}^3k \qquad (6\text{–}157)$$

(the integral being taken over the interior of S_F). In the general case, we can say nothing more. However, if the system is translationally invariant, J_k is given by (6–137); the above integral becomes trivial and leads to

$$S^0_{\alpha\beta} = 0 \qquad (6\text{–}158)$$

Let us emphasize that (6–158) results from the additional hypothesis of translational invariance.

6. Generalization to long-range forces

a. The notion of screening

In sections 4 and 5, we limited ourselves to short-range interactions. The Fourier transform V_q of the binary interaction then remains finite. We are going to generalize this formalism to long-range forces, taking as an example the case of an electron gas.

For a Coulomb interaction, V_q, defined by (B–15), is equal to

$$V_q = \frac{4\pi e^2}{q^2} \qquad (6\text{–}159)$$

Using (B–13), we can write the Hamiltonian of the system,

$$H = \sum_i \frac{p_i^2}{2m} + \frac{1}{2} \sum_q \frac{4\pi e^2}{\Omega q^2} \{ \rho_q \rho_{-q} - N \} \tag{6–160}$$

(the term $-N$ eliminates the interaction of an electron with itself.) In this form, the Hamiltonian makes no sense, since for $q = 0$, the interaction is infinite. This singularity expresses the physical fact that a gas of charged particles is unstable, incapable of limiting itself to a volume Ω. To make the system stable, we must neutralize the electronic charge by some positive charge—for example, ions in a metal.

We shall thus assume the electrons to be placed in a *uniform* background of positive charge, such that the total system is neutral. This positive background has the effect of removing the term $q = 0$ from the Hamiltonian [(6–160)]. We must therefore replace (6–159) by

$$V_q = \begin{cases} 4\pi e^2/q^2 & \text{if} \quad q \neq 0 \\ 0 & \text{if} \quad q = 0 \end{cases} \tag{6–161}$$

(since q takes on discrete values, this definition is perfectly clear).

Let us consider the matrix element $V(\mathbf{k}_1, \ldots \mathbf{k}_4)$ in the particular case

$$\begin{cases} \mathbf{k}_1 = \mathbf{k}' - q/2 \\ \mathbf{k}_3 = \mathbf{k} - q/2 \end{cases} \qquad \begin{matrix} \mathbf{k}_4 = \mathbf{k}' + q/2 \\ \mathbf{k}_2 = \mathbf{k} + q/2 \end{matrix}$$

According to (5–17), we have

$$V(\mathbf{k}_1, \ldots, \mathbf{k}_4) = \frac{1}{\Omega} \left[V_q - V_{k-k'}\, \delta_{\sigma,\sigma'} \right] \tag{6–162}$$

When $q \to 0$, the *direct* term V_q diverges, whereas the *exchange* term $V_{k-k'}\delta_{\sigma\sigma'}$ remains finite. It is therefore necessary to modify the diagrams of Chap. 5 in such a way as to separate explicitly direct and exchange terms. For this purpose, it is sufficient to indicate in which state, $\mathbf{k}_1$ or $\mathbf{k}_2$, the particle $\mathbf{k}_4$ is found after collision.

We therefore replace the vertices by the element indicated in Fig. 17, corresponding to the process

$$\begin{cases} \mathbf{k}_4 \to \mathbf{k}_1 \\ \mathbf{k}_3 \to \mathbf{k}_2 \end{cases}$$

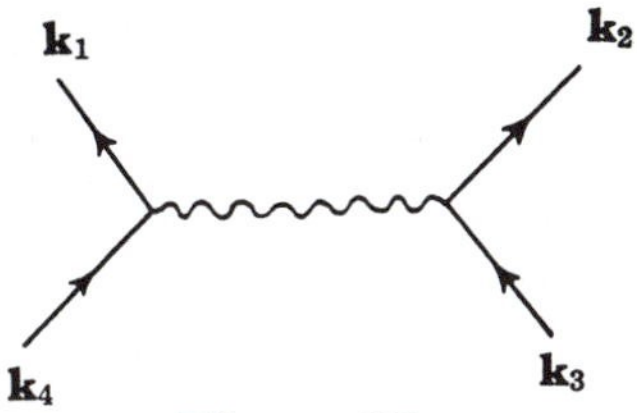

Figure 17

With this element we associate the first term of (5–17), corrected by a factor $i/2\pi$, in accordance with the rules (5–107),

$$\frac{i}{2\pi\Omega} V_{k_1-k_4}\, \delta_{\sigma_1,\sigma_4}\, \delta_{\sigma_2,\sigma_3}\, \delta_{k_1+k_2,k_3+k_4} \tag{6–163}$$

We see that along each solid line spin is conserved. The last δ function ensures conservation of total momentum. The interaction is represented by the wavy "interaction" line and corresponds to a *momentum transfer* $q = (k_1 - k_4)$. If we take care to ensure by construction the conservation of the spin of each particle and of the total wave vector, the contribution of the interaction line reduces to $iV_q/2\pi\Omega$.

To each vertex of Chap. 5 there correspond *two* elements (Fig. 17), indicated in Fig. 18. The two diagrams (a) and (b) correspond to the two terms of (5–17). We verify in Appendix F that we obtain the correct sign by keeping the old rule: each diagram carries a factor $(-1)^l$, where l is the number of closed loops formed by solid lines.

Let us now consider the diagram 19, contributing to $^{\circ}\Gamma(\mathbf{p},\mathbf{p}';\varpi)$. [Again ϖ is an abbreviated notation for (q,ε).] By construction, the central interaction line has a wave vector q. Its contribution is singular when $\varpi \to 0$; we shall say that the diagram is *improper*. On the other hand, we shall call *proper* all diagrams which cannot be split into two parts related

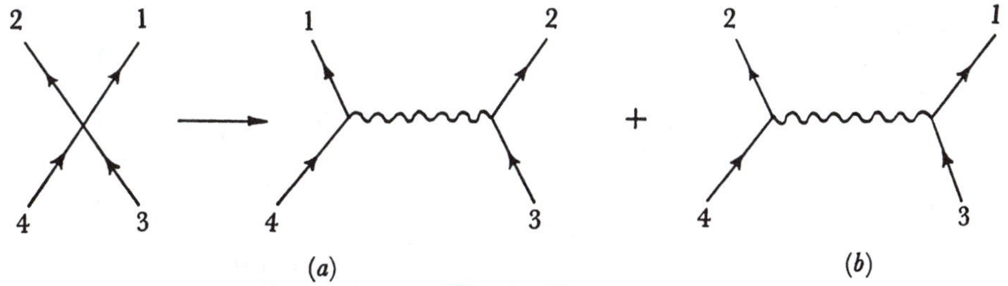

Figure 18

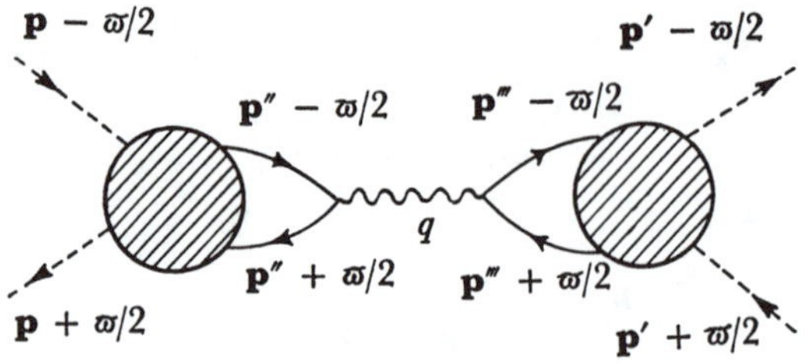

Figure 19

only by a single wavy line (Fig. 19). According to this definition, all the interaction lines of a proper diagram have variable wave vectors. We shall designate by $°\tilde{\Gamma}(\mathbf{p},\mathbf{p}';\varpi)$ the contribution from all proper diagrams.

The notion of proper and improper diagrams can easily be extended to the vertex operators $\Lambda_\mu(p;\varpi)$ and to the correlation functions $S_{\mu\nu}(\varpi)$. We indicate in Fig. 20 an example of improper diagrams for these two quantities. Let $\tilde{\Lambda}_\mu(p;\varpi)$ and $\tilde{S}_{\mu\nu}(\varpi)$ be the contribution from proper diagrams in each of these cases. We propose to calculate $°\Gamma$, Λ_μ, and $S_{\mu\nu}$ as functions of $°\tilde{\Gamma}$, $\tilde{\Lambda}_\mu$, $\tilde{S}_{\mu\nu}$.

Let us consider a diagram of $°\Gamma(\mathbf{p},\mathbf{p}';\varpi)$ containing a *single* interaction line q. If we remove this line, diagram 19 decomposes into two *proper* vertex operators $\tilde{\Lambda}_4(p;\varpi)$ and $\tilde{\Lambda}_4(p';\varpi)$. (Let us recall that we sum over $\mathbf{p}''$ and $\mathbf{p}'''$.) The contribution from diagrams of this type can thus be written as

$$\frac{-\mathrm{i}V_q}{2\pi\Omega}\tilde{\Lambda}_4(p\,;\varpi)\,\tilde{\Lambda}_4(p'\,;\varpi) \qquad (6\text{–}164)$$

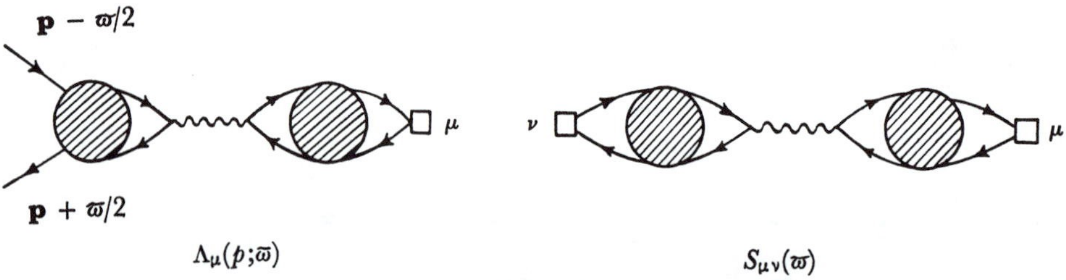

Figure 20

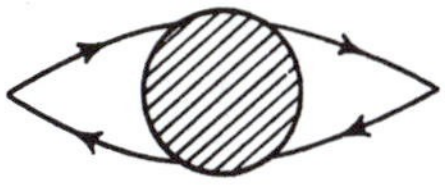

Figure 21

the minus sign arising from the fact that the line ($\mathbf{p} - \varpi/2$) terminates as ($\mathbf{p} + \varpi/2$) instead of ($\mathbf{p}' - \varpi/2$).

To obtain the most general diagram of $^{\circ}\Gamma$, we just have to insert into the interaction line q an arbitrary number of "polarization loops," made up of *proper* diagrams of the type of Fig. 21. Since we sum over $\mathbf{p}''$ and $\mathbf{p}'''$, the contribution from all polarization diagrams is just $\tilde{S}_{44}(\varpi)$. Each additional loop thus yields a factor

$$-\frac{\mathrm{i}}{2\pi\Omega} V_q \tilde{S}_{44}(\varpi) \tag{6–165}$$

Note the minus sign in (6–165), due to the fact that every polarization loop constitutes a closed loop. To obtain S, we must sum the geometric series shown in Fig. 22. We thus arrive at

$$^{0}\Gamma(\mathbf{p}, \mathbf{p}'; \varpi) = {}^{0}\tilde{\Gamma}(\mathbf{p}, \mathbf{p}'; \varpi) - \frac{\mathrm{i}V_q}{2\pi\Omega}\left(\frac{\tilde{\Lambda}_4(p; \varpi)\,\tilde{\Lambda}_4(p'; \varpi)}{1 + (\mathrm{i}V_q/2\pi\Omega)\,\tilde{S}_{44}(\varpi)}\right) \tag{6–166}$$

Repeating this argument for $\Lambda_\mu(p;\varpi)$ and $S_{\mu\nu}(\varpi)$, we easily verify that

$$\Lambda_\mu(p; \varpi) = \tilde{\Lambda}_\mu(p; \varpi) - \frac{\mathrm{i}V_q}{2\pi\Omega}\left(\frac{\tilde{\Lambda}_4(p; \varpi)\,\tilde{S}_{4\mu}(\varpi)}{1 + (\mathrm{i}V_q/2\pi\Omega)\,\tilde{S}_{44}(\varpi)}\right) \tag{6–167}$$

$$S_{\mu\nu}(\varpi) = \tilde{S}_{\mu\nu}(\varpi) - \frac{\mathrm{i}V_q}{2\pi\Omega}\left(\frac{\tilde{S}_{\mu4}(\varpi)\,\tilde{S}_{4\nu}(\varpi)}{1 + (\mathrm{i}V_q/2\pi\Omega)\,\tilde{S}_{44}(\varpi)}\right) \tag{6–168}$$

These equations relate the complete quantities $^{\circ}\Gamma$, Λ_μ, $S_{\mu\nu}$, to the proper quantities $^{\circ}\tilde{\Gamma}$, $\tilde{\Lambda}_\mu$, $\tilde{S}_{\mu\nu}$. They are very simple because the series to be summed is a geometric series. Let us point out the analogy with $G(k,\omega)$; we renormalized G by inserting into a solid line an arbitrary number of self-energy operators; we renormalize the interaction lines by inserting into them an arbitrary number of polarization loops.

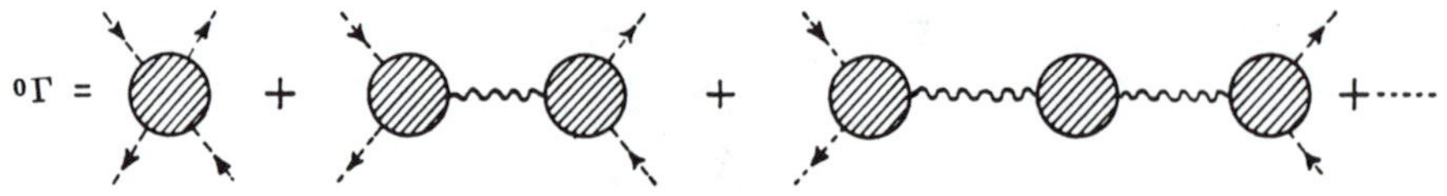

Figure 22

In the particular case $\mu = 4$, we can write (6–167) in the form

$$\Lambda_4(p\,;\varpi) = \frac{\widetilde{\Lambda}_4(p\,;\varpi)}{\epsilon(\varpi)} \qquad (6\text{–}169)$$

where the function $\epsilon(\varpi)$ is defined by

$$\epsilon(\varpi) = 1 + \frac{iV_q}{2\pi\Omega}\widetilde{S}_{44}(\varpi) \qquad (6\text{–}170)$$

Similarly, it is easily verified that

$$S_{\mu 4}(\varpi) = \frac{\widetilde{S}_{\mu 4}(\varpi)}{\epsilon(\varpi)} \qquad (6\text{–}171)$$

These relations are exact, even for finite ϖ.

$\epsilon(\varpi)$ plays the role of a dielectric constant. The *scalar* vertex operators and correlation functions (involving the charge) are reduced by a factor ϵ by the introduction of polarization loops; this simply expresses the *effect of screening* by the medium on external charges. The definition of the *polarizability*

$$\frac{iV_q}{2\pi\Omega}\widetilde{S}_{44}(\varpi)$$

is rigorous, valid to all orders in the perturbation expansion.

It is in general difficult to extend this discussion to the vector quantities Λ_α and $S_{\alpha\beta}$. For simplicity, we assume the system to be isotropic; we then have rotational symmetry about the axis q. Let us take the direction 3 (longitudinal) to be parallel to q and the directions 1 and 2 (transverse) in the plane perpendicular to q. The two transverse directions are, by symmetry, completely decoupled from the direction 3 and from the charge,

$$\tilde{S}_{14} = \tilde{S}_{24} = \tilde{S}_{13} = \tilde{S}_{23} = 0 \tag{6–172}$$

Furthermore, the directions 1 and 2 are equivalent,

$$\begin{cases} \tilde{S}_{11} = \tilde{S}_{22} = \tilde{S}_{\perp} \\ \tilde{S}_{12} = 0 \end{cases} \tag{6–173}$$

The same discussion applies to $S_{\alpha\beta}$, whose transverse components are characterized by the single function $S_{\perp}$. Referring to (6–167) and (6–168), we see that

$$\begin{cases} \Lambda_{1,2} = \tilde{\Lambda}_{1,2} \\ S_{\perp} = \tilde{S}_{\perp} \end{cases} \tag{6–174}$$

As a consequence, *there is no transverse screening effect.* This important result proceeds from our hypothesis of isotropy.

It remains to treat the longitudinal case; $\tilde{S}_{34}$ is then $\neq 0$. We shall see in section 8 that conservation of charge imposes the continuity relations

$$\begin{cases} S_{34} = \dfrac{\varepsilon}{q} S_{44} \\ S_{33} + \dfrac{2i\pi N}{m} = \dfrac{\varepsilon}{q} S_{34} \end{cases} \tag{6–175}$$

These relations are rigorous, exact even if ϖ is finite. Comparing them with (6–168), we can verify that

$$\begin{cases} \tilde{S}_{34} = \dfrac{\varepsilon}{q} \tilde{S}_{44} \\ \tilde{S}_{33} + \dfrac{2i\pi N}{m} = \dfrac{\varepsilon}{q} \tilde{S}_{34} \end{cases} \tag{6–176}$$

[We shall prove (6–176) a little later, in the limit $\varpi \to 0$, by anology with (6–152).] Comparing (6–171), (6–175), and (6–176), we obtain

$$S_{33}(\varpi) + \frac{2i\pi N}{m} = \frac{\tilde{S}_{33}(\varpi) + 2i\pi N/m}{\epsilon(\varpi)} \tag{6–177}$$

There is thus a screening effect for longitudinal fields, in accordance with the equations of continuity.

b. The limit, $\varpi \to 0$

Let us consider a proper diagram of $°\tilde{\Gamma}$; no internal line is restricted to the wave vector q. When $\varpi \to 0$, the matrix elements $V(\mathbf{k}_1, \ldots, \mathbf{k}_4)$ are therefore regular. *We find ourselves back in the case of short-ranged forces*; the only singularities arise from pairs of lines $(\mathbf{p}'' - \varpi/2)$ and $(\mathbf{p}'' + \varpi/2)$ separating two irreducible interactions.

As in section 1, we can decompose $°\tilde{\Gamma}$ into a series of *proper* and *irreducible* interaction cores, $°\tilde{I}$. Let us point out that, except for the first-order direct diagram, an irreducible diagram is always proper. As a consequence, we have

$$°\tilde{I}(\mathbf{p}, \mathbf{p}' \; ; \varpi) = °I\,(\mathbf{p}, \mathbf{p}' \; ; \varpi) - \frac{V_q}{2i\pi\Omega} \qquad \textit{(6–178)}$$

Starting with $°\tilde{I}$, we can reproduce the section on the limit $\varpi \to 0$ from beginning to end. We have only to add a tilde to all interactions $°I$ or $°\Gamma$ to the vertex operators Λ_μ and to the correlation functions $S_{\mu\nu}$. (We consider only proper diagrams.) Instead of rewriting all the results, we represent the transposed version of Eq. (6–76) to (6–96) by the symbols $(\underset{\sim}{6}–\tilde{7}6)$ to $(6\tilde{–}96)$. For example, $f(\mathbf{k},\mathbf{k}')$ and $A(\mathbf{k},\mathbf{k}')$ are defined by (6–83), obtained from (6–83) by replacing $°\Gamma$ by $°\tilde{\Gamma}$.

Let us now try to generalize the Ward identities. For this purpose, let us consider the diagram of Fig. 23, contributing to the self-energy operator $M(p)$. On the left and right of the interaction line, the total momentum is zero. The transfer in the collision is thus also zero; the line corresponds to $q = 0$. But, according to (6–161), $V_0 = 0$; the contribution from the diagram 23 is therefore zero. (This is a consequence of the continuous background of positive charge.)

Again, we can analyze the derivatives of M by the method of section 5. For example, $\partial M/\partial \omega$ is given by an expression of the type (6–98); the vertex $\mathcal{J}(\mathbf{p},\mathbf{p}')$ is obtained by removing any line of a diagram of M.

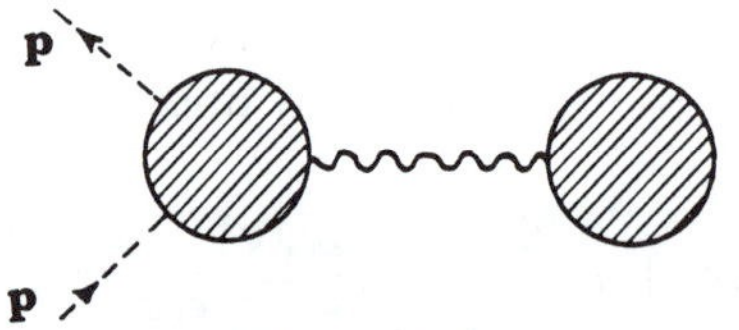

Figure 23

The absence of the diagrams of Fig. 23 guarantees that $\mathcal{J}$ contains only *proper* diagrams. As a consequence, $\mathcal{J}$ is no longer given by (6–99), but by

$$\mathfrak{J}(\mathbf{p}, \mathbf{p}') = {}^{0}\tilde{I}(\mathbf{p}, \mathbf{p}') \qquad \textit{(6–179)}$$

The difference between (6–99) and (6–179) arises from the discontinuity of V_q; $^{\circ}I(\mathbf{p},\mathbf{p}';\varpi)$ involves a q which is small, but $\neq 0$, for which V_q is divergent; in contrast $M(p)$ involves a q *identically* zero, for which $V_q = 0$.

In order to generalise the section on Ward identities, it is sufficient, as before, to add a tilde to all factors $^{\circ}I$, Λ_μ, $S_{\mu\nu}$. For example, (6–138) becomes

$$\begin{cases} z_k\tilde{\Lambda}_4^0(k) = 1 \\ z_k\tilde{\Lambda}_4^{\infty}(k) = v_k\,(\mathrm{d}k_F/\mathrm{d}\mu) \\ z_k\tilde{\Lambda}_\alpha^{\infty}(k) = v_{k\alpha} \\ z_k\tilde{\Lambda}_\alpha^0(k) = J_{k\alpha} \end{cases} \qquad \textit{(6–}\widetilde{\textit{138}}\textit{)}$$

(v_k and J_k have the same interpretation as before). Similarly, (6–144), (6–149), (6–154), and (6–158) can be written as

$$\begin{cases} \tilde{S}_{44}^0 = \tilde{S}_{4\alpha}^0 = \tilde{S}_{4\alpha}^{\infty} = \tilde{S}_{\alpha\beta}^0 = 0 \\ \tilde{S}_{\alpha\beta}^{\infty} = -\,2i\pi(N/m)\,\delta_{\alpha\beta} \\ \tilde{S}_{44}^{\infty} = -\,2i\pi\,\mathrm{d}N/\mathrm{d}\mu \end{cases} \qquad \textit{(6–180)}$$

$\tilde{S}_{\mu\nu}$ thus has for long-range forces the same properties as $S_{\mu\nu}$ for short-range ones. In particular, the continuity equation (6–152) is transposed into (6–176).

From (6–180) we can deduce the complete correlation functions $S_{\mu\nu}$. We need for that the dielectric constant ϵ^r, given by (6–170). When $r \to \infty$, it is easily verified that

$$\epsilon^{\infty} = 1 + \frac{4\pi e^2}{q^2}\frac{1}{\Omega}\frac{\mathrm{d}N}{\mathrm{d}\mu} \qquad \textit{(6–181)}$$

We thus relate the *static* dielectric constant to the compressibility of the system. For $r = 0$, the situation looks more difficult, as (6–170) is indeterminate. According to (6–176), we can write

$$\tilde{S}_{44} = \frac{q^2}{\varepsilon^2}(\tilde{S}_{33} + 2i\pi N/m) \qquad (6\text{–}182)$$

(6–182) is a *rigorous* result. As a consequence, $\tilde{S}_{44}$ is of order q^2 *whatever* ε is; there is no corrective term of order ε/E_F. This point being clarified, we replace $\tilde{S}_{33}$ by its value (6–180); we thus find

$$\begin{cases} \epsilon^0 = 1 - \omega_p^2/\varepsilon^2 \\ \omega_p^2 = 4\pi N e^2/m\Omega \end{cases} \qquad (6\text{–}183)$$

This well-known result gives the response to a *uniform* field; it can be proved by other methods.

In conclusion, our formalism can be as easily adapted to Coulomb forces as to short-range forces. We are ready to put it into practice.

7. Response to a Scalar Field. Justification of the Landau Model for short-range forces.

Rather than exhaust all possible cases, we study in detail two typical examples, frequently encountered in practice. In this section we consider short-range forces, and we calculate the response to a scalar field. We establish by means of this problem the validity of the Landau model. In the next section we shall study a system of electrons coupled to an electromagnetic field.

a. The response to a test charge

Let us introduce into our system a certain distribution of "external" particles, interacting with the particles of the system according to the same binary law as the latter do with each other. We shall call these external particles "test charges," although this name wrongly evokes an electrostatic picture. We assume the test-charge distribution to be periodic in space and time, having the form of a traveling wave. Let r_q/Ω be the amplitude of this wave; r_q is the qth Fourier component of the test-charge density. The Hamiltonian describing the interaction with the system can be written as

$$\mathcal{H}_1 = \frac{V_q}{\Omega}\left[r_q \exp(-i\varepsilon t)\, \rho_{-q} + C \cdot C\right] \qquad (6\text{–}184)$$

Again we combine in the symbol ϖ the wave vector q and the frequency

ε of the test charge. We have here a simple example of scalar coupling [see (6–40)].

The test charge distorts the density ρ of the system. Let $\langle\rho_q\rangle e^{-i\varepsilon t}$ be the average value of the qth Fourier component of ρ. We shall characterize the response of the system by the ratio $\langle\rho_q\rangle/r_q$. This quantity is given by the general theory of Chap. 2. According to (2–16), we have

$$\frac{\langle \rho_q \rangle}{r_q} = \frac{V_q}{\Omega} \chi_{\rho_q\rho_{-q}}(\varepsilon) \qquad \textit{(6–185)}$$

where $\chi_{\rho_q\,\rho_{-q}}$ is an "admittance," defined by (2–20). Instead of χ, it is preferable to use the function $\bar{\bar{S}}(q,\varepsilon)$ defined by the Fourier transform of (A–2). Comparing (2–47) and (A–9) with (2–20), we can easily verify that

$$\frac{\langle \rho_q \rangle}{r_q} = -\,\mathrm{i} V_q\, \bar{\bar{S}}(\varpi) \qquad \textit{(6–186)}$$

It is now easy to make the connection with the preceding sections. We saw in Appendix A [Eq. (A–9)] that the function $\bar{\bar{S}}(\varpi)$ is practically identical to $\bar{S}(\varpi)$, defined by (A–3), and used above,

$$\begin{cases} \mathrm{i}\bar{S}(\varpi) = \mathrm{i}\;\bar{\bar{S}}(\varpi) & \text{if} \quad \varepsilon > 0 \\ \mathrm{i}\;\bar{S}(\varpi) = [\mathrm{i}\;\bar{\bar{S}}(\varpi)]^* & \text{if} \quad \varepsilon < 0 \end{cases} \qquad \textit{(6–187)}$$

We may always assume $\varepsilon > 0$. With the help of (6–55), we can then write (6–186) in the form

$$\frac{\langle \rho_q \rangle}{r_q} = \frac{V_q}{2\pi\mathrm{i}\Omega} S_{44}(\varpi) \qquad \textit{(6–188)}$$

It can similarly be shown that the current induced by the test charge is given by

$$\frac{\langle J_{q\alpha} \rangle}{r_q} = \frac{V_q}{2\pi\mathrm{i}\Omega} S_{4\alpha}(\varpi) \qquad \textit{(6–189)}$$

These relations are very important, since they relate our perturbation formalism to measurable physical results. They are *exact*, and not restricted to the limit $\varpi \to 0$.

The total number of particles in the system must be conserved when the test charge is introduced. This imposes the continuity condition

$$\sum_\alpha q_\alpha \langle J_{q\alpha} \rangle = \varepsilon \langle \rho_q \rangle \qquad \textit{(6–190)}$$

We can deduce from this the relation

$$\sum_{\alpha} q_\alpha \, S_{4\alpha}(\varpi) = \varepsilon \, S_{44}(\varpi) \tag{6–191}$$

When $\varpi \to 0$, (6–191) goes back into (6–152). But this proof is much more powerful, since it is valid for any ϖ.

Let us now assume that ϖ is small. What error do we make by replacing $S_{\mu\nu}(\varpi)$ by the limit $S^r_{\mu\nu}$? Thanks to our detailed study of the singular terms, we have had to make no hypothesis about the ratio qv_F/ε. *The error made by replacing a function of ϖ by its limit for $\varpi \to 0$ is thus of order q/k_F or ε/E_F, whatever the ratio qv_F/ε is.* On a macroscopic scale, the passage to the limit $\varpi = 0$ is thus totally justified.

Let us neglect these errors and consider the limit $q = 0$. For short-range forces, we obtain, according to (6–144),

$$\frac{\langle \rho_q \rangle}{r_q} = 0 \tag{6–192}$$

To interpret this result we note that the test charges give rise to a force field,

$$\mathcal{F}(r, t) = -iq \frac{V_q}{\Omega} r_q \exp[i(qr - \varepsilon t)] + C.\,C. \tag{6–193}$$

Since V_0 is finite, these forces disappear when $q \to 0$.

Let us now pass to the limit $\varepsilon = 0$. According to (6–154) and (1–16), we can put the response (6–188) into the form

$$\frac{\langle \rho_q \rangle}{r_q} = -\frac{V_q}{\Omega}\frac{dN}{d\mu} = -V_q \rho^2 \chi \tag{6–194}$$

where χ is the compressibility of the system and $\rho = N/\Omega$ its density. The particles thus screen the external charge. This result can be proved directly by macroscopic considerations. The force applied per unit volume of the system is of magnitude $\rho\mathcal{F}$. Since it is longitudinal, it is equivalent to a pressure such that

$$\text{grad}\, p = -\rho\mathcal{F}$$

But in equilibrium the pressure must be uniform; the density ρ is distorted in such a way as to *compensate* this external pressure. The variation $\delta\rho$ is then given by (1–14),

$$\delta\rho = -\chi\rho p = -\frac{V_q}{\Omega} r_q \, \rho^2\chi \exp[i(qr - \varepsilon t)] + C.C. \tag{6–195}$$

By definition we have:

$$\delta\rho = \frac{1}{\Omega} \langle \rho_q \rangle \exp[i(qr - \varepsilon t)] + C.C. \tag{6–196}$$

(6–194) follows immediately.

Note: This result is not true for a Coulomb interaction; (6–194) must then be divided by the dielectric constant $\epsilon(\varpi)$. In fact, when we define the compressibility, we assume that the continuous positive background deforms with the average electronic density. Only the sum $r_q + \langle \rho_q \rangle$, acts as an external force, (6–194) being replaced by

$$\frac{\langle \rho_q \rangle}{r_q - \langle \rho_q \rangle} = - V_q \, \rho^2\chi \,, \qquad q \to 0$$

When q/ε is finite, we must calculate S^r_{44} with the help of (6–150). This requires the solution of the integral equation (6–92). It is easy to expand the solution in powers of q/ε or of ε/q. We can sometimes use the continuity relations (6–152). For example, in the neighborhood of $q = 0$, (6–152) and (6–160) imply

$$\frac{\langle \rho_q \rangle}{r_q} \sim \frac{V_q}{\Omega} \frac{Nq^2}{m\varepsilon^2} \tag{6–197}$$

Instead of exploiting this formalism directly, we shall use it to justify Landau's model. We shall thus validate all of Chap. 1.

b. The basis of the Landau Theory

We must first clearly define the notion of quasi particle. For this purpose, we reconsider one by one Landau's hypotheses:

(1) *The elementary excitations are quasi particles and quasi holes, separated by a Fermi surface S_F. The energy ε_k of a quasi particle is a continuous function when k crosses S_F, as is its gradient v_k.*

This was proved in Chap. 5, where we moreover defined the operator $A^*_{\mathbf{k}}$ creating the quasi particle **k**. A quasi particle is therefore not a myth; it is a state whose wave function we know, at least in principle.

(2) *Two elementary excitations* **k** *and* **k**′ *have an interaction energy* $f(\mathbf{k},\mathbf{k}')$ *of order* $1/\Omega$.

Let $\delta_{\mathbf{k}}M(\mathbf{k}',\omega')$ be the variation in $M(\mathbf{k}',\omega')$ when a quasi particle **k** is added. This correction appears when one of the *internal* lines of the diagram has a wave vector **k**; it is therefore of order $1/\Omega$. The energy of the quasi particle **k**′ is given by the equation:

$$\frac{k'^2}{2m} - \varepsilon_{k'} - M(k', \omega)\Big|_{\omega=\varepsilon_{k'}} = 0$$

When M changes by $\delta_{\mathbf{k}}M$, $\varepsilon_{k'}$ changes by an amount

$$f(\mathbf{k}, \mathbf{k}') = -z_{k'}\, \delta_{\mathbf{k}}M(\mathbf{k}', \varepsilon_{k'}) \qquad (6\text{–}198)$$

[see (6–128)]. We are thus led to calculate $\delta_{\mathbf{k}}M$.

Let us represent $M(\mathbf{k}',\omega')$ with the help of *skeleton* diagrams. To obtain $\delta_{\mathbf{k}}M$, we apply the method already used for the Ward identities. Repeating the discussion of section 5, we obtain an equation analogous to (6–98).

$$\delta_{\mathbf{k}}M(\mathbf{p}') = \sum_{\mathbf{p}''} {}^0I(\mathbf{p}', \mathbf{p}'')\, \delta_{\mathbf{k}}G(\mathbf{p}'') \qquad (6\text{–}199)$$

[(6–199) is valid only for short-range interactions.] When $\mathbf{k} \neq \mathbf{k}''$, $\delta_k G$ is of order $1/\Omega$; the poles of G and of $G + \delta_{\mathbf{k}}G$ are on the same side of the real axis. As a consequence, we can write

$$\delta_{\mathbf{k}}G(\mathbf{p}'') = \delta_{\mathbf{k}}M(\mathbf{p}'')\, G^2(\mathbf{p}'') \qquad (\mathbf{k} \neq \mathbf{k}'') \qquad (6\text{–}200)$$

On the other hand, if $\mathbf{k}'' = \mathbf{k}$, $\delta_{\mathbf{k}}G$ is of order 1, given by (5–213),

$$\delta_{\mathbf{k}}G(\mathbf{p}'') = -2i\pi z_k\, \delta(\omega'' - \varepsilon_k) \qquad (\mathbf{k} = \mathbf{k}'') \qquad (6\text{–}201)$$

Let us put

$$\frac{i}{2\pi z_k}\, \delta_{\mathbf{k}}M(\mathbf{p}') = Y(\mathbf{p}', \mathbf{k}) \qquad (6\text{–}202)$$

By using (6–200) and (6–201), we can put (6–199) into the form

$$Y(\mathbf{p}', \mathbf{k}) = {}^0I(\mathbf{p}', \mathbf{p})\Big|_{\omega=\varepsilon_k} + \sum_{\mathbf{p}''} {}^0I(\mathbf{p}', \mathbf{p}'')\, G^2(\mathbf{p}'')\, Y(\mathbf{p}'', \mathbf{k}) \qquad (6\text{–}203)$$

Comparing (6–203) with (6–78), we see that

$$Y(\mathbf{p}', \mathbf{k}) = {}^0\Gamma^0(\mathbf{p}', \mathbf{p}) \Big|_{\omega=\varepsilon_k} \tag{6–204}$$

Let us return to (6–198) and use the notation (6–82); finally we find

$$f(\mathbf{k}, \mathbf{k}') = 2\pi\, \mathrm{i} z_k\, z_{k'}{}^0\Gamma^0(\mathbf{k}, \mathbf{k}') \tag{6–205}$$

As we expected, (6–83) represents the interaction energy of two quasi particles. We thus rigorously justify the somewhat vague considerations of Chap. 4.

(3) *A quasi particle* **k** *contains one bare particle; it carries a current* $J_{k\alpha}$ *given by* (1–33).

The first assertion is obvious in the case of a perturbation method. As far as the current is concerned, we have defined it by translating the reference system. The relation (1–33) can then be proved with the aid of the Ward identities [see (6–139)].

(4) *When the system is compressed, the distortion* $dk_F/d\mu$ *of the Fermi surface is governed by* (1–20).

This is an immediate consequence of the Ward identities [see (6–140)].

We see that perturbation theory confirms all the "static" properties of quasi particles.

c. The dynamics of quasi particles

We shall verify the Boltzmann equation (1–78) in a particular case —the response of the system to a periodic test charge, studied at the beginning of this section. Our treatment relies on two basic approximations:

(i) The wave vector q of the test charge is $\ll k_F$, its frequency $\varepsilon \ll E_F$. This allows us to use the limit $\varpi = 0$.

(ii) The system is at absolute zero; we completely neglect collisions between quasi particles.

In other words, we shall consider the response of the system in the *ground state* to a *macroscopic* excitation.

First we must define the quasi-particle density $\delta n(\mathbf{k},r)$ induced by the test charge. We proceed by analogy with the bare particles, whose density at the point r can be written as

$$\rho(r) = \sum_\sigma \psi_\sigma^*(r)\, \psi_\sigma(r) = \sum_{\mathbf{k},q} a^*_{\mathbf{k}-q/2}\, a_{\mathbf{k}+q/2}\, \mathrm{e}^{\mathrm{i}qr} \tag{6–206}$$

(Hereafter, we shall take Ω as unit volume.) When q is small, we can interpret the quantity

$$\sum_q a^*_{\mathbf{k}-q/2}\, a_{\mathbf{k}+q/2}\, e^{iqr} \tag{6–207}$$

as a density of particles $\mathbf{k}$ at the point r. We assume the scale q^{-1} of the disturbance to be sufficiently large so that the uncertainty principle is not violated. This interpretation is equivalent to Wigner's semiclassical method.

By analogy, we introduce the quasi-particle density operator

$$\sum_q A^*_{\mathbf{k}-q/2}\, A_{\mathbf{k}+q/2}\, e^{iqr} \tag{6–208}$$

$\delta n(\mathbf{k},r,t)$ is equal to the average value of (6–208). For a periodic excitation, δn is of the form

$$\delta n(\mathbf{k}, r, t) = \delta n(\mathbf{k}) \exp[i(qr - \varepsilon t)] + C.\,C. \tag{6–209}$$

According to (6–208), $\delta n(\mathbf{k})$ is given by

$$\delta n(\mathbf{k})\, e^{-i\varepsilon t} = < \varphi \mid \mathbf{A}^*_{\mathbf{k}-q/2}(t)\, \mathbf{A}_{\mathbf{k}+q/2}(t) \mid \varphi > \tag{6–210}$$

where $|\varphi\rangle$ is the wave function perturbed by the test charge.

The average value (6–210) can be evaluated by the technique of Chap. 2. Let us apply Eq. (2–9), putting $t = 0$. We find

$$\delta n(\mathbf{k}) = i \int_{-\infty}^{0} dt < \varphi_0 \mid [\mathcal{H}_1(t), A^*_{\mathbf{k}-q/2} A_{\mathbf{k}+q/2}] \mid \varphi_0 > \tag{6–211}$$

The interaction $\mathcal{H}_1$ is given by (6–184); for reasons of translational invariance, only its first term contributes. We add a factor $e^{\eta t}$ in order to avoid the heating up of the system by the test charge. (6–211) can thus be written as

$$\delta n(\mathbf{k}) = i V_q r_q \int_{-\infty}^{0} dt\, e^{\eta t}\, e^{-i\varepsilon t} < \varphi_0 \mid [\rho_{-q}(t), A^*_{\mathbf{k}-q/2} A_{\mathbf{k}+q/2}] \mid \varphi_0 > \tag{6–212}$$

We thus obtain an explicit expression for δn.

We shall replace (6–212) by

$$\delta n(\mathbf{k}) = -\mathrm{i}V_q r_q \int_{-\infty}^{+\infty} \mathrm{d}t\, \mathrm{e}^{-\eta|t|}\, \mathrm{e}^{-\mathrm{i}\varepsilon t} < \varphi_0 \mid T\, \{\, \rho_{-q}(t)\; A^*_{\mathbf{k}-q/2}\, A_{\mathbf{k}+q/2} \,\} \mid \varphi_0 > \tag{6–213}$$

Using the method of Appendix A, we can easily verify that (6–213) is equal to (6–212) if $\varepsilon > 0$ and that it differs from it by the replacement $\eta \to -\eta$ if $\varepsilon < 0$. In order to avoid any difficulty, we shall assume that $\varepsilon > 0$. The advantage of (6–213) is that it fits naturally into our perturbation method. Let us put

$$Д_4(k, t, t' \,;\, \varpi) = \int_{-\infty}^{+\infty} \mathrm{d}t''\, \mathrm{e}^{-\mathrm{i}\varepsilon t''} < \varphi_0 \mid T\, \{\, \mathbf{A}^*_{\mathbf{k}-q/2}(t')\; \mathbf{A}_{\mathbf{k}+q/2}(t)\; \rho_{-q}(t'') \,\} \mid \varphi_0 > \tag{6–214}$$

(the factor $e^{-\eta|t''|}$, necessary if the integral is to make sense, is understood henceforth). This definition allows us to put (6–213) into the very simple form

$$\delta n(\mathbf{k}) = -\mathrm{i}V_q r_q Д_4\, (k, 0, 0 \,;\, \varpi) \tag{6–215}$$

Let us compare $Д_4$ with the function Δ_4 defined by (6–42). The only difference is the replacement of a, a^* by A, A^*. $Д_4$ is a *renormalized* version of Δ_4. It can be calculated by the method of Chap. 4, Eqs. (4–57) to (4–65). Let us decompose Δ_4 into a vertex operator Λ_4 connected to two propagators G (Fig. 8). After Fourier transformation, we have, according to (6–47),

$$\Delta_4(p \,;\, \varpi) = \Lambda_4(p \,;\, \varpi)\, G(p + \varpi/2)\, G(p - \varpi/2) \tag{6–216}$$

In order to go to $Д_4$, we need only replace the two factors G by mixed propagators g and $\tilde{g}$, defined by (4–61); more precisely,

$$Д_4(p \,;\, \varpi) = \Lambda_4(p \,;\, \varpi)\, \tilde{g}(p - \varpi/2)\, g(p + \varpi/2) \tag{6–217}$$

Let us use the expressions (4–63) for g and $\tilde{g}$ and invert the Fourier transform. We find without difficulty

$$Д_4(k, 0, 0 \,;\, \varpi) = \frac{z_k}{2\pi} \int_{-\infty}^{+\infty} \Lambda_4(p \,;\, \varpi)\, \frac{\mathrm{d}\omega}{(\varepsilon_{k+q/2} - \omega - \varepsilon/2 + \mathrm{i}\eta'')\, (\varepsilon_{k-q/2} - \omega + \varepsilon/2 + \mathrm{i}\eta')} \tag{6–218}$$

As usual, the signs of η' and η'' depend on the position of $(k \pm q/2)$ with respect to S_F. The different cases are indicated in Fig. 10.

Before calculating the integral (6–218), let us remark that the quasi particles are *true* elementary excitations of the system. We have, as a consequence,

$$\begin{cases} A_{\mathbf{k}} \mid \varphi_0 > = 0 & k > k_F \\ A_{\mathbf{k}}^* \mid \varphi_0 > = 0 & k < k_F \end{cases} \tag{6–219}$$

$\text{Д}_4(k,0,0;\varpi)$ is therefore zero in regions I and II of Fig. 10 [the matrix element of (6–214) being zero for either sign of t'']. Physically, the test charge excites a quasi particle by a momentum $\pm q$; the limitation to regions III to VI simply expresses the *exclusion principle.*

Let us return to (6–218). In regions I and II, the two poles are on the same side of the real axis. Closing the contour on the opposite side, we see that only the singularities of Λ_4 contribute; hence a result of order unity, which, in fact, is zero according to the preceding paragraph. In regions III to VI, the poles of (6–218) give a contribution of order $1/(qv - \varepsilon)$, much larger than that of order unity arising from the singularities of Λ_4. Passing to the limit $\varpi \rightarrow 0$, we easily verify that

$$\text{Д}_4(k, 0, 0 ; \varpi) = \frac{z_k}{2\pi} \Lambda_4^r(k) \frac{-2\pi i}{q \cdot v_k - \varepsilon - i\eta} q \cdot v_k \, \delta(\varepsilon_k - \mu) \tag{6–220}$$

[the factor $|q \,.\, v_k| \, \delta\,(\varepsilon_k - \mu)$ expresses the limitation to the two crescents of Fig. 10]. Going back to (6–215), we finally obtain

$$\delta n(\mathbf{k}) = - V_q r_q \frac{q \cdot v_k}{q \cdot v_k - \varepsilon - i\eta} \delta(\varepsilon_k - \mu) \, z_k \Lambda_4^r(k) \tag{6–221}$$

$\Lambda_4^r(k)$ thus acquires a very suggestive physical interpretation.

The vertex operator $\Lambda_4^r(k)$ satisfies the Bethe-Salpeter equation (6–92*a*). Using the Ward identity (6–129), we can write the latter as

$$z_k \Lambda_4^r(k) = 1 + \sum_{\mathbf{k}'} f(\mathbf{k}, \mathbf{k}') \frac{q \cdot v_{k'}}{\varepsilon - q \cdot v_{k'} + i\eta} \delta(\varepsilon_{k'} - \mu) \, z_{k'} \, \Lambda_4^r(k') \tag{6–222}$$

Let us eliminate Λ_4^r in favor of $\delta n(\mathbf{k})$. Equation (6–222) becomes

$$\begin{aligned} (\varepsilon - q \cdot v_k) \, \delta n(\mathbf{k}) = {} & V_q r_q \, q \cdot v_k \, \delta(\varepsilon_k - \mu) \\ & + q \cdot v_k \, \delta(\varepsilon_k - \mu) \sum_{\mathbf{k}'} f(\mathbf{k}, \mathbf{k}') \, \delta n(\mathbf{k}') \end{aligned} \tag{6–223}$$

Let us now introduce the amplitude $\mathscr{F}$ of the force created by the test charge [see (6–193).],

$$\mathscr{F} = -\mathrm{i}q\, V_q\, r_q \tag{6–224}$$

(6–223) then takes the final form

$$\begin{aligned} 0 = (q\cdot v_k - \varepsilon)\, \delta n(\mathbf{k}) + \mathrm{i}\, \mathscr{F}\cdot v_k\, \delta(\varepsilon_k - \mu) \\ + q\cdot v_k\, \delta(\varepsilon_k - \mu) \sum_{\mathbf{k}'} f(\mathbf{k}, \mathbf{k}')\, \delta n(\mathbf{k}') \end{aligned} \tag{6–225}$$

This is just the Boltzmann equation (1–78), governing the motion of quasi particles. We have thus *proved* Landau's kinetic equation, which is just a variation of the Bethe-Salpeter equation.

To conclude this proof, it remains to verify that the current is given by (1–79), which can be written, in the notation of this chapter, as

$$\langle J_{q\alpha} \rangle = \sum_{\mathbf{k}} \delta n(\mathbf{k})\, J_{k\alpha} \tag{6–226}$$

$\langle J_{q\alpha}\rangle$ is given by (6–189). By using (6–96*a*) in the particular case $\nu = 4$, $\mu = \alpha$, and with the help of (6–144), we can write

$$\langle J_{q\alpha} \rangle = V_q r_q \sum_{\mathbf{k}} \frac{q\cdot v_k}{\varepsilon - q\cdot v_k + \mathrm{i}\eta}\, \delta(\varepsilon_k - \mu)\, z_k\Lambda_4^r(k)\, z_k\Lambda_\alpha^0(k) \tag{6–227}$$

Let us compare (6–227), on the one hand with the definition (6–221), and on the other with the Ward identity (6–138); we immediately obtain the desired expression (6–226).

In the particular case that we have studied, the Landau model is thus completely justified. Let us emphasize once more its limitations:

(i) The ratios q/k_F and ε/E_F must be small; if not, the Bethe-Salpeter equation (6–222) would no longer be true.

(ii) Collisions must be negligible (we have neglected the lifetime of the quasi particles).

When one of these two conditions is violated, we must abandon the simplified Landau formalism and return to the complete perturbation expansion.

(6–221) constitutes a formal solution of the Boltzmann equation (1–78). By employing the formalism of this chapter, we can give various forms to this result. For example, it is easily verified that the quantity $\overline{\delta n}$, defined by (1–82), is equal to

$$\overline{\delta n}(\mathbf{k}) = V_q\, r_q\, \delta(\varepsilon_k - \mu) \left[\frac{\varepsilon}{\varepsilon - q\cdot v_k + i\eta}\, z_k \Lambda_4^r(k) - 1\right] \qquad (6\text{–}228)$$

Starting from (6–228), it is simple to prove (1–83), (1–84), by introducing $A(\mathbf{k},\mathbf{k}')$ instead of $f(\mathbf{k},\mathbf{k}')$, etc. We shall not investigate these matters further, since they introduce nothing new.

8. Response of an Electron Gas to an Electromagnetic Field—Generalization of the Landau Model.

a. Calculation of the conductivity

We shall characterize the electromagnetic field by its vector potential $A(r,t)$ (the gauge is chosen in such a way as to eliminate the scalar potential). The electric and magnetic fields $\mathcal{E}$ and $\mathcal{H}$ are given by

$$\begin{cases} \mathcal{E} = -\dfrac{1}{c}\dfrac{\partial A}{\partial t} \\ \mathcal{H} = \operatorname{rot} A \end{cases} \qquad (6\text{–}229)$$

We shall consider only the case of a traveling plane wave, such that

$$A(r, t) = \frac{A_q}{\Omega} \exp[i(qr - \varepsilon t)] + C.\,C. \qquad (6\text{–}230)$$

(With this definition, A_q is the qth Fourier component of A.) The longitudinal part of A produces only an electric field; it is equivalent to a test charge, fixed by Poisson's equation,

$$-\frac{\varepsilon}{c}\, q\cdot A_q = 4\pi\, e\, r_q \qquad (6\text{–}231)$$

In particular, a static longitudinal potential ($\varepsilon = 0$) does not produce any physical effect; this simply expresses gauge invariance.

The Hamiltonian of the system can be written as

$$H = \frac{1}{2m} \sum_i \left\{ p_i - \frac{eA(r_i, t)}{c} \right\}^2 + V \qquad (6\text{–}232)$$

We deduce from this the velocity of the ith particle,

$$\dot{r}_i = \mathrm{i}[H, r_i] = \frac{p_i}{m} - \frac{eA(r_i)}{mc} \tag{6–233}$$

The Fourier components of the *total* current can therefore be written as

$$J_q^{\text{tot}} = J_q - \frac{Ne}{mc}\frac{A_q}{\Omega} \tag{6–234}$$

The usual term J_q is given by (6–39); we call it the "paramagnetic current." The last term is called the "diamagnetic current."

The vector potential A_q produces an average current equal to

$$\langle J_{q\alpha}^{\text{tot}} \rangle = \sum_\beta K_{\alpha\beta}(\varpi)\, A_{q\beta} \tag{6–235}$$

The tensor $K_{\alpha\beta}$ characterizes the response of the system. According to (6–234), we can write

$$K_{\alpha\beta} = K_{\alpha\beta}^p - \frac{Ne}{\Omega mc}\,\delta_{\alpha\beta} \tag{6–236}$$

($K_{\alpha\beta}^p$ corresponds to just the paramagnetic current J_q). In practice, we frequently substitute for K the conductivity tensor σ, defined by

$$\langle e\, J_{q\alpha}^{\text{tot}} \rangle = \sum_\beta \sigma_{\alpha\beta}(\varpi)\, \mathcal{E}_{q\beta} \tag{6–237}$$

Using (6–229), we can verify that

$$\sigma_{\alpha\beta}(\varpi) = \frac{ec}{\mathrm{i}\varepsilon}\, K_{\alpha\beta}(\varpi) \tag{6–238}$$

To calculate $K_{\alpha\beta}^p$, we use the methods of Chap. 2. To first order, the Hamiltonian coupling the electrons to the electromagnetic field can be written

$$\mathcal{H}_1 = -\frac{e}{\Omega c}\, A_q \cdot J_q\, e^{-\mathrm{i}\varepsilon t} + C.\,C. \tag{6–239}$$

$\mathscr{H}_1$ has the same structure as that of (6–184). Following step by step the reasoning which led from (6–184) to (6–188), we easily verify that, when $\varepsilon > 0$,

$$K^{p}_{\alpha\beta} = -\frac{e}{c}\frac{1}{2\pi i\Omega} S_{\alpha\beta}(\varpi) \qquad \textit{(6–240)}$$

Similarly, the density fluctuation induced by the electromagnetic field is equal to

$$\langle \rho_q \rangle = -\sum_{\beta} \frac{e}{c}\frac{1}{2\pi i\Omega} S_{4\beta}(\varpi)\, A_{q\beta} \qquad \textit{(6–241)}$$

These results are valid for any ϖ. The continuity equation (6–190) implies the relation

$$\sum_{\alpha} q_{\alpha} \left\{ S_{\alpha\beta}(\varpi) + \frac{2\pi i N}{m} \delta_{\alpha\beta} \right\} = \varepsilon\, S_{4\beta}(\varpi) \qquad \textit{(6–242)}$$

(6–242) is *exact*, even if ϖ is finite. (We obtained it in the limit $\varpi \to 0$ by applying the results of sections 5 and 6.)

Let us compare the expression for $K_{\alpha\beta}$ with the relations (6–174) and (6–177), in the particular case of an *isotropic* system; we see that the response to a longitudinal field is reduced by the effect of screening. The *effective* field in the interior of the system is equal to $\mathscr{E}/\epsilon$. (In practice, this correction arises from the polarization charges accumulated on the external surface.) For a transverse field, on the other hand, there is no screening—$\mathscr{E}$ acts with full strength.

When the perturbation varies on a macroscopic scale, we can use the results of section 6, giving the limit $\varpi \to 0$. Let us first consider the case $r \to \infty$; according to (6–180), we then have $K_{\alpha\beta} = 0$. In other words, when $q \to 0$, a static vector potential does not create a current. If q is parallel to the direction β, this result ensures gauge invariance. Since the latter must be *rigorously* observed, we must have, *even for q finite*,

$$\sum_{\beta} K_{\alpha\beta}(q, 0)\, q_{\beta} = 0 \qquad \textit{(6–243)}$$

[this follows trivially from (6–42)]. On the other hand, if q has a direction different from β, gauge invariance no longer plays a role. Our calculations are then subject to the usual error q/k_F. Actually, for reasons of symmetry, we have

$$K_{\alpha\beta}(q, 0) \sim q^2/k_F^2 \tag{6–244}$$

For a static transverse vector potential, there thus remains a small current when q is finite. This current gives rise to the orbital diamagnetism of the gas (also called Landau diamagnetism). As soon as ε is no longer zero, a term of order ε/qv_F appears, which quickly dominates the diamagnetic term (6–244).

Note: (6–197) shows that there is no Meissner effect in a normal gas; such an effect would require $K^{\infty}_{\alpha\beta} \neq 0$ for a transverse potential.

Let us consider the other limit, $r=0$, and assume for simplicity that the system is translationally invariant. By refering to (6–174), (6–177), and (6–180), we see that in this case

$$\begin{aligned} &\text{transverse field} && K_{11} = K_{22} = -\frac{Ne}{\Omega mc} \\ &\text{longitudinal field} && K_{33} = -\frac{Ne}{\Omega mc\,\epsilon} \end{aligned} \tag{6–245}$$

the dielectric constant ϵ being given by (6–183). With the aid of (6–238), we obtain the conductivity $\sigma_{\alpha\beta}$. These results are well known and can be proved directly by choosing a reference frame moving with the traveling wave.

In the general case, the calculation of $K_{\alpha\beta}$ is more difficult. However, it is easy to make an expansion in powers of ε/q or of q/ε. By way of example, we can thus verify the relation (1–88) proved in Chap. 1 by Landau's method. We shall not investigate this point further, although it leads to numerous interesting results.

b. Generalization of the Landau model

A system of N electrons is stable only if it is associated with a uniform continuous background, of total charge $(-Ne)$. When excited states of $(N+p)$ electrons are considered, it is natural to adjust the positive background in such a way as to preserve overall neutrality. A quasi particle $\mathbf{k}$ will thus be made up of an electron of wave vector $\mathbf{k}$, clothed by its self-energy cloud, *neutralized* by a charge $(-e)$ distributed uniformly over the whole volume Ω. According to this definition, a quasi particle $\mathbf{k}$ is *neutral*. A "pure" quasi particle is distributed over the whole volume and does not produce any accumulation of charge. If, on the other hand, we make up a wave packet, we have a localized electronic charge e, neutralized by a uniform charge $-e$.

This new definition does not affect the energy ε_k of the quasi particle, which remains equal to the pole of G. On the other hand, it eliminates from $f(\mathbf{k},\mathbf{k}')$ the average Hartree interaction between the two particles $\mathbf{k}$ and $\mathbf{k}'$. There remain only the corrections due to correlations, whose range is necessarily finite. We thus come back to the hypotheses of Chap. 1.

The elementary excitations being neutral by definition, the component $q = 0$ of the Coulomb interaction remains zero in all cases, whether or not there are excited quasi particles. (This would no longer be true if the system were charged.) In other words, the addition of a quasi particle $\mathbf{k}$ does not introduce any improper diagrams into the self-energy operator $M(p)$. We can then reproduce the discussion of the preceding section [Eqs. (6–198) to (6–205)], simply replacing $^{\circ}\Gamma(\mathbf{p},\mathbf{p}')$ by the *proper* interaction $^{\circ}\tilde{\Gamma}(\mathbf{p},\mathbf{p}')$. The interaction energy $f(\mathbf{k},\mathbf{k}')$ is therefore given by

$$f(\mathbf{k}, \mathbf{k}') = 2\pi \, i \, z_k \, z_{k'} \, {}^{0}\tilde{\Gamma}^{0}(\mathbf{k}, \mathbf{k}') \tag{6–246}$$

Let us again emphasize that the difference between (6–205) and (6–246) is due to the continuous positive background.

When we compress the system macroscopically, we compress the positive charge at the same time, the combination remaining neutral at each point. $d\mu/dN$ thus involves only proper diagrams, which explains the second Ward identity (6–$\tilde{1}$38). The compressibility χ is related to $\tilde{S}_{44}^{\infty}$ instead of S_{44}^{∞} (this point has already been discussed in section 7). The current J_k is still given by (1–33) and satisfies the Ward identity (6–138). In short, the Landau theory applies directly to *neutralized* quasi particles.

It remains now to prove the Boltzmann equation in the particular case studied above: the response to a vector potential $A(r,t)$. Each particle i is subject to a force

$$\mathcal{F}_i = e\mathcal{E}(r_i) + \frac{ev_i}{c} \wedge \mathcal{H}(r_i) \tag{6–247}$$

The magnetic field $\mathcal{H}$ causes the electrons to move over a constant-energy surface; it does not change the equilibrium distribution function $n_0(\mathbf{k})$. In the Boltzmann equation, only the electric field $\mathcal{E}$ is involved, in agreement with (1–78).

Let $n(\mathbf{k},r,t)$ be the quasi-particle distribution function in the presence of the vector potential $A(r,t)$. We put

$$n(\mathbf{k}, r, t) = n_0(\mathbf{k}) + \delta n^{(p)}(\mathbf{k}, r, t) \tag{6–248}$$

$\delta n^{(p)}$ measures the variation of n_0 due to the interaction $\mathscr{H}_1$; it is a paramagnetic term. On the other hand, a quasi particle **k** located at (r,t) carries a current deduced from (6–233),

$$J_k = \frac{k}{m} - e\frac{A(r, t)}{mc}$$

(We assume the system to be translationally invariant.) The total current can thus be written as

$$J(r, t) = \sum_{\mathbf{k}} n(\mathbf{k}, r, t)\left\{\frac{k}{m} - e\frac{A(r, t)}{mc}\right\} \tag{6–249}$$

Collecting the terms linear in A, we recover the two currents, paramagnetic and diamagnetic, of the expression (6–234).

Within the framework of a Boltzmann equation, we prefer to write the total current in the form

$$J(r, t) = \sum_{\mathbf{k}} n'(\mathbf{k}, r, t)\frac{k}{m} \tag{6–250}$$

Comparing (6–249) and (6–250), we see that

$$n'(\mathbf{k}, r, t) = n\left(\mathbf{k} + \frac{eA(r, t)}{c}, r, t\right) \tag{6–251}$$

Let us put $n' = n_0 + \delta n$. According to (6–251), we can write

$$\begin{gathered} \delta n = \delta n^{(p)} + \delta n^{(d)} \\ \delta n^{(d)}(\mathbf{k}, r, t) = \frac{eA(r, t)}{c}\cdot \nabla_k n_0 = -\frac{eA(r, t)\cdot v_k}{c}\,\delta(\varepsilon_k - \mu) \end{gathered} \tag{6–252}$$

Making a Fourier tranformation [see the definition (6–209)], we obtain the Fourier component of the total current,

$$\begin{cases} \langle J_q^{\text{tot}} \rangle = \sum_{\mathbf{k}} \frac{k}{m}\,\delta n(\mathbf{k}) \\ \delta n^{(d)}(\mathbf{k}) = -e\frac{A_q\cdot v_k}{c}\,\delta(\varepsilon_k - \mu) \end{cases}$$

The diamagnetic term $\delta n^{(d)}$ arises from a *translation* of the equilibrium distribution n_0, without deformation.

The calculation of $\delta n^{(p)}$ is analogous to that carried out for short-range forces in section 7. In particular, the relation (6–211) remains valid. Replacing $\mathscr{H}_1$ by its expression (6–239), we can easily generalize (6–221) in the form

$$\delta n^{(p)}(\mathbf{k}) = \sum_{\alpha} \frac{eA_{q\alpha}}{c} \frac{q \cdot v_k}{q \cdot v_k - \varepsilon - i\eta} \delta(\varepsilon_k - \mu)\, z_k \Lambda^r_\alpha(k) \tag{6–253}$$

(we assume again $\Omega = 1$). Since the vertex operator comes in through (6–214), it is Λ_α which is involved, and not $\tilde{\Lambda}_\alpha$. Let us now add together the two contributions to $\delta n(\mathbf{k})$ and use the Ward identity (6–138) in (6–252). We find

$$\delta n(\mathbf{k}) = \sum_{\alpha} \frac{e}{c} A_{q\alpha} z_k\, \delta(\varepsilon_k - \mu) \left[\frac{q \cdot v_k}{q \cdot v_k - \varepsilon - i\eta} \Lambda^r_\alpha(k) - \tilde{\Lambda}^\infty_\alpha(k)\right] \tag{6–254}$$

Λ^r_α can be decomposed with the aid of (6–167). It is convenient to split the result into two parts by putting

$$\left\{\begin{aligned} \delta n(\mathbf{k}) &= \delta n_1(\mathbf{k}) + \delta n_2(\mathbf{k}) \\ \delta n_1(\mathbf{k}) &= \sum_{\alpha} \frac{e}{c} A_{q\alpha} z_k\, \delta(\varepsilon_k - \mu) \left[\frac{q \cdot v_k}{q \cdot v_k - \varepsilon - i\eta} \tilde{\Lambda}^r_\alpha(k) - \tilde{\Lambda}^\infty_\alpha(k)\right] \\ \delta n_2(\mathbf{k}) &= -i \frac{V_q}{2\pi} \sum_{\alpha} \frac{eA_{q\alpha}}{c} S^r_{4\alpha} z_k\, \delta(\varepsilon_k - \mu) \frac{q \cdot v_k}{q \cdot v_k - \varepsilon - i\eta} \tilde{\Lambda}^r_4(k) \end{aligned}\right. \tag{6–255}$$

[In writing δn_2 we have used the relation (6–171).] We shall study the equations satisfied by δn_1 and δn_2 separately.

To within a constant factor, δn_2 is similar to (6–221). Using the Bethe-Salpeter equation (6–92*a*) and the Ward identity (6–129), we obtain, by analogy with (6–223),

$$\begin{aligned} (\varepsilon - q \cdot v_k)\, \delta n_2(\mathbf{k}) &= \\ & ie\mathcal{E}_{qH} \cdot v_k\, \delta(\varepsilon_k - \mu) + q \cdot v_k\, \delta(\varepsilon_k - \mu) \sum_{\mathbf{k}'} f(\mathbf{k}, \mathbf{k}')\, \delta n_2(\mathbf{k}') \end{aligned} \tag{6–256}$$

where the constant vector $\mathscr{E}_{qH}$ is given by

$$e\vec{\mathcal{E}}_{qH} = \frac{V_q}{2\pi} \sum_{\alpha} \frac{e}{c} A_{q\alpha} S^r_{4\alpha} \mathbf{q} \tag{6–257}$$

Comparing this with (6–241), we see that $\mathscr{E}_{qH}$ can be written as

$$\mathcal{E}_{qH} = -\mathrm{i}\,\frac{4\pi e}{q^2}\langle \rho_q \rangle q \tag{6–258}$$

where $\langle \rho_q \rangle$ is the average charge induced by the electromagnetic field. $\mathscr{E}_{qH}$ is thus simply the Hartree electrostatic field created by the average induced charge, given by Poisson's equation,

$$\operatorname{div} \mathcal{E}_H = 4\pi\, e\, \rho \tag{6–259}$$

δn_2 measures the distortion due to this Hartree field.

Let us turn now to δn_1. Let us subtract from (6–$\tilde{9}$2*a*) the same equation taken in the particular case $r = \infty$. We find

$$\begin{aligned} z_k \tilde{\Lambda}^r_\alpha(k) &= z_k \tilde{\Lambda}^\infty_\alpha(k) \\ &+ \sum_{\mathbf{k}'} f(\mathbf{k}, \mathbf{k}')\, z_{k'}\, \delta(\mu - \varepsilon_{k'}) \left\{ \frac{q\cdot v_{k'}}{\varepsilon - q\cdot v_{k'} + \mathrm{i}\eta} \tilde{\Lambda}^r_\alpha(k') + \tilde{\Lambda}^\infty_\alpha(k') \right\} \end{aligned} \tag{6–260}$$

Comparing this result with the definition of δn_1, we can easily verify that

$$\begin{aligned} \frac{(q\cdot v_k - \varepsilon)}{q\cdot v_k}\, \delta n_1(\mathbf{k}) = \frac{\varepsilon}{q\cdot v_k}\, \delta(\varepsilon_k - \mu) \sum_\alpha \tilde{\Lambda}^\infty_\alpha(k)\, \frac{e}{c}\, A_{q\alpha} \\ - \sum_{\mathbf{k}'} \delta(\varepsilon_k - \mu)\, f(\mathbf{k}, \mathbf{k}')\, \delta n_1(\mathbf{k}') \end{aligned} \tag{6–261}$$

Let us use the Ward identity (6–$\tilde{1}$38) and the definition (6–229) of the *applied* electric field $\mathscr{E}_q$. The preceding relation takes the form

$$\begin{aligned} (\varepsilon - q\cdot v_k)\, \delta n_1(\mathbf{k}) &= \\ \mathrm{i}\, e\, \mathcal{E}_q\cdot v_k\, \delta(\varepsilon_k - \mu) &+ q\cdot v_k\, \delta(\varepsilon_k - \mu) \sum_{\mathbf{k}'} f(\mathbf{k}, \mathbf{k}')\, \delta n_1(\mathbf{k}') \end{aligned} \tag{6–262}$$

δn_1 thus measures the distortion due to the external field $\mathscr{E}_q$. Combining (6–256) and (6–262), we recover the Landau equation (1–78), the *local field* $\mathscr{E}_L$ being given by

$$\mathcal{E}_L = \mathcal{E} + \mathcal{E}_H$$

The Landau-Silin equation is thus proved.

It remains to verify the relation (6–250), giving the total current.

With the help of the Ward identities, we can write it as

$$\langle J_{q\alpha}^{\text{tot}} \rangle = \sum_{\mathbf{k}} \{ \delta n_1(\mathbf{k}) + \delta n_2(\mathbf{k}) \} z_k \Lambda_\alpha^0(k) \qquad (6\text{–}263)$$

Let us replace δn_1 and δn_2 by their expressions (6–255) and use the identities (6–96*a*). We thus obtain

$$\langle J_{q\alpha}^{\text{tot}} \rangle = -\frac{e}{c}\frac{1}{2i\pi} \sum_{\beta} A_{q\beta} \left\{ \widetilde{S}_{\alpha\beta}^{r} - \widetilde{S}_{\alpha\beta}^{\infty} + \frac{V_q}{2\pi i} S_{4\beta}^{r} \left[\widetilde{S}_{\alpha 4}^{r} - \widetilde{S}_{\alpha 4}^{0} \right] \right\} \qquad (6\text{–}264)$$

By referring to (6–168) and (6–180), we finally arrive at

$$K_{\alpha\beta}^{r} = -\frac{Ne}{mc}\delta_{\alpha\beta} - \frac{e}{c}\frac{1}{2i\pi} S_{\alpha\beta}^{r} \qquad (6\text{–}265)$$

This result is identical to (6–240); the last Landau hypothesis is confirmed.

When the system is isotropic, the screening field $\mathscr{E}_H$ appears only for a *longitudinal* disturbance. We must treat separately the case of uniform fields $q = 0$; there is then no screening effect, V_q being rigorously zero; physically, the electrons are displaced as a whole, without producing an accumulation of charge. On the other hand, when q is small, but *not zero,* there appear charge fluctuations, and hence an almost perfect screening. Let us emphasize the fundamental difference between the case $q \to 0$ and the limit $q \equiv 0$, due essentially to a change of boundary conditions; this difficulty often leads to confusion.

Chapter 7

Generalization of Perturbation Methods to Superfluids

1. Statement of the Problem

The formalism developed in Chaps. 5 and 6 applies only to normal systems. More precisely, if we begin with an unperturbed normal ground state, of Fermi surface $\overset{\circ}{S}_F$, we can generate only a normal state, with the same Fermi surface. In order to describe a superfluid system, we must thus begin with an unperturbed state which is *already superfluid.* Let H_0 be the corresponding Hamiltonian. We obtain the real ground state by adiabatically switching on the "perturbation" $H - H_0$. The result makes sense only if it is independent of the exact choice of H_0. Our study is thus comprised of three steps:

(1) Choice and diagonalization of H_0
(2) Perturbation treatment of $(H - H_0)$
(3) Elimination of H_0

As in Chap. 4, we are going to abandon conservation of the total number of particles. We must therefore replace our isolated system, containing N particles, by a "grand canonical ensemble," characterized by its chemical potential μ. The equilibrium state is obtained by minimizing the free energy,

$$\overline{H} = H - \mu N \tag{7-1}$$

We must replace H by $\bar{H}$ everywhere. In particular, we modify the Heisenberg representation, replacing the definition (5–3) by

$$\overline{\mathbf{A}}(t) = e^{i\overline{H}t} A e^{-i\overline{H}t} \tag{7-2}$$

It is easy to go from the representation (7–2) to the usual Heisenberg representation. For example, we have

$$\begin{aligned} \bar{\mathbf{a}}_{\mathbf{k}}(t) &= e^{i\mu t}\, \mathbf{a}_{\mathbf{k}}(t) \\ \bar{\mathbf{a}}^*_{\mathbf{k}}(t) &= e^{-i\mu t}\, \mathbf{a}^*_{\mathbf{k}}(t) \end{aligned} \tag{7–3}$$

In order to simplify the notation, we suppress the bars over $\bar{A}$, it being understood that H is no longer the Hamiltonian but the *free energy* (7–1).

For an isolated system μ satisfies the relation

$$\mu = \mathrm{d}E_0/\mathrm{d}N$$

According to (7–1), this implies

$$N = -\frac{\mathrm{d}}{\mathrm{d}\mu} < \varphi_0 \,|\, \overline{H} \,|\, \varphi_0 > \tag{7–4}$$

After all calculations have been completed, we shall use (7–4) to fix the value of the chemical potential μ.

Let T and V be, respectively, the kinetic and potential energies of the particles. We decompose $\overline{H}$ into two parts,

$$\left\{ \begin{aligned} \overline{H} &= \widetilde{H}_0 + \widetilde{H}_1 \\ \widetilde{H}_0 &= T - \mu N + W \\ \widetilde{H}_1 &= V - W \\ W &= \sum_{k,\sigma} \lambda_k a^*_{k\sigma} a_{k\sigma} + \sum_k \mu_k (a^*_{k,+} a^*_{-k,-} + C\,.\,C) \end{aligned} \right. \tag{7–5}$$

λ_k and μ_k are real constants, for the moment arbitrary. $\widetilde{H}_0$ is our unperturbed Hamiltonian, which we shall diagonalize directly. The *perturbation* $\widetilde{H}_1$ will be switched on adiabatically (see section 3). Note that $\widetilde{H}_0$ and $\widetilde{H}_1$ do not commute with N, in contrast to the complete Hamiltonian $\overline{H}$. We have thus artificially abandoned conservation of particle number; it is for this reason that we work with a grand canonical ensemble.

2. The Unperturbed Ground State

The Hamiltonian $\widetilde{H}_0$ can easily be diagonalized by Bogoliubov's method. Let us introduce the operators

$$\alpha^*_{k\sigma} = u_k\, a^*_{k\sigma} + \sigma v_k\, a_{-k,-\sigma} \tag{7–6}$$

($\sigma = \pm 1$), the constants u_k and v_k being defined by

$$\begin{cases} u_k = \sqrt{\dfrac{1}{2} + \dfrac{\xi_k}{2E_k}}, & v_k = \sqrt{\dfrac{1}{2} - \dfrac{\xi_k}{2E_k}} \quad (7\text{–}7) \\ \xi_k = \dfrac{k^2}{2m} - \mu + \lambda_k\,, & E_k = \sqrt{\xi_k^2 + \mu_k^2} \quad (7\text{–}8) \end{cases}$$

It is easily verified that

$$\begin{cases} u_k^2 + v_k^2 = 1 \\ u_k^2 - v_k^2 = \xi_k/E_k \end{cases} \qquad 2u_k v_k = \mu_k/E_k \tag{7–9}$$

The transformation which takes $a_{k\sigma}$ into $\alpha_{k\sigma}$ is *canonical* (it preserves commutation relations). The inverse transformation is given by

$$a^*_{k\sigma} = u_k\ \alpha^*_{k\sigma} - \sigma v_k\ \alpha_{-k,-\sigma} \tag{7–10}$$

Inserting expression (7–10) into the Hamiltonian $\tilde{H}_0$, we obtain

$$\begin{cases} \tilde{H}_0 = U + \sum\limits_{k\sigma} E_k \alpha^*_{k\sigma}\alpha_{k\sigma} \\ U = \sum\limits_{k\sigma} \xi_k v_k^2 \end{cases} \tag{7–11}$$

In this form, $\tilde{H}_0$ is diagonal.

The elementary excitations of $\tilde{H}_0$ are quasi particles, created by the operator $\alpha^*_{k\sigma}$. These excitations are neither holes nor particles; they belong to both types at the same time (see Chap. 4). The ground state $|0\rangle$ is the "vacuum" for these quasi particles, so that

$$\alpha_{k\sigma} \,|\, 0 > = 0 \tag{7–12}$$

The ground-state energy is equal to U. Let $|vac\rangle$ be the "vacuum" state for *bare* particles, defined by

$$a_{k\sigma} \,|\, vac > = 0 \tag{7–13}$$

It is easily verified that the ground state $|0\rangle$ is given by

$$|\, 0 > = \prod_k (u_k - v_k a^*_{k,+} a^*_{-k,-}) \,|\, vac > \tag{7–14}$$

The state (7–14), discovered by Bardeen, Cooper, and Schrieffer, exhibits the characteristics of superfluidity. It can therefore be taken as the unperturbed ground state. It has all the properties studied in Chap. 4, which we shall not review here.

Let us introduce the single-particle Green's function characteristic of $\tilde{H}_0$, defined with the notation (4–116),

$$G_0(kt\sigma\rho, k't'\sigma'\rho') = i < 0 \mid T\{ \mathbf{a}^{\rho}_{k\sigma}(t)\, \mathbf{a}^{\rho'}_{k'\sigma'}(t') \} \mid 0 > \qquad (7\text{–}15)$$

(7–15) is identical to the definition (4–115), the exponential factor being incorporated into our modified Heisenberg representation. In order to calculate G_0, it is convenient to go through the intermediary of the one-*quasi-particle* Green's function,

$$\mathcal{G}_0(kt\sigma\rho, k't'\sigma'\rho') = i < 0 \mid T\{ \boldsymbol{\alpha}^{\rho}_{k\sigma}(t)\, \boldsymbol{\alpha}^{\rho'}_{k'\sigma'}(t') \} \mid 0 > \qquad (7\text{–}16)$$

G_0 and $\mathcal{G}_0$ both have the general structure (4–119); they are each characterized by two functions: F_{01} and F_{02}, $\mathcal{F}_{01}$ and $\mathcal{F}_{02}$. Using (7–12) we see that

$$\begin{cases} \mathcal{F}_{01}(k, t) = i \exp(-iE_k t)\, \theta(t) \\ \mathcal{F}_{02}(k, t) = 0 \end{cases} \qquad (7\text{–}17)$$

where $\theta(t)$ is the step function,

$$\theta(t) = \begin{cases} 1 & \text{if} \quad t > 0 \\ 0 & \text{if} \quad t < 0 \end{cases} \qquad (7\text{–}18)$$

With the aid of (7–10), we can easily go back from $\mathcal{G}_0$ to G_0; we thus obtain

$$\begin{cases} F_{01}(k, t) = i\, u_k^2 \exp(-iE_k t)\, \theta(t) - i v_k^2 \exp(iE_k t)\, \theta(-t) \\ F_{02}(k, t) = i u_k v_k \exp(-iE_k \mid t \mid) \end{cases} \qquad (7\text{–}19)$$

Note that $\mathcal{F}_{01}$ is unidirectional; a quasi particle cannot move backward in time. Conversely, a bare particle can propagate either forward or backward in time.

The functions G_0 and $\mathcal{G}_0$ are going to appear at the heart of the perturbation expansions. We define their Fourier transforms with the help of (4–114) and (4–117). These transforms have the structure (4–120), the functions F_1 and F_2 being given, respectively, by

$$\begin{cases} \mathcal{F}_{01}(p) = \dfrac{1}{E_k - \omega - i\delta} \\ \mathcal{F}_{02}(p) = 0 \end{cases}$$

$$\begin{cases} F_{01}(p) = \dfrac{u_k^2}{E_k - \omega - i\delta} - \dfrac{v_k^2}{E_k + \omega - i\delta} \\ F_{02}(p) = \dfrac{u_k v_k}{E_k - \omega - i\delta} + \dfrac{u_k v_k}{E_k + \omega - i\delta} \end{cases} \tag{7-20}$$

u_k^2 and v_k^2 thus appear as the residues $z_\pm$ of F_{01} at the two poles $\pm\, E_k$. The "unperturbed" state $|0\rangle$ is thus completely known; we are prepared to look into the perturbation treatment.

3. The Perturbation Formalism

a. General equations

In order to generate the ground state of the real system, we switch on $\tilde{H}_1$ *adiabatically*. Let $g = e^{\eta t}$ be the coupling constant, increasing slowly from 0 to 1 between the times $-\infty$ and 0. We write the interaction at time t in the form

$$\tilde{H}_1(g) = gV - \sum_{k\sigma} \lambda_k(g)\, a_{k\sigma}^* a_{k\sigma} - \sum_k \mu_k(g)\,(a_{k,+}^* a_{-k,-}^* + C.C) \tag{7-21}$$

The functions $\lambda_k(g)$ and $\mu_k(g)$ are arbitrary, subject only to the conditions

$$\begin{cases} \mu_k(0) = \lambda_k(0) = 0 \\ \lambda_k(1) = \lambda_k\,,\ \mu_k(1) = \mu_k \end{cases} \tag{7-22}$$

[We shall give later some indication as to how the choice of $\lambda_k(g)$ and $\mu_k(g)$ is to be made.]

As for a normal system, we go over to the interaction representation, defined by (5–4), where we replace H_0 by $\tilde{H}_0$. The time development of the system is then characterized by an operator $U_\eta(t, t')$, defined by (5–7). U_η satisfies a "dynamical" equation analogous to (5–11),

$$\begin{cases} U_\eta(t', t) = 1 - i\displaystyle\int_t^{t'} \tilde{H}_1(t'')\, U_\eta(t'', t)\, dt'' \\ \tilde{H}_1(t'') = \exp(i\tilde{H}_0 t'')\, \tilde{H}_1(g'')\, \exp(-\, i\tilde{H}_0 t'') \end{cases} \tag{7-23}$$

This integral equation can be solved by iteration. The solution is given by (5–14), on condition that we add a ~ over all operators.

By analogy with Chap. 5, we expect that the real ground state $|\varphi_0\rangle$ will be given by

$$|\varphi_0> = \lim_{\eta\to 0} U_\eta(0, -\infty)\,|\,0> \qquad (7\text{–}24)$$

(after extraction of a singular phase factor). We shall thus be led to calculate matrix elements of the type

$$<0\,|\,\alpha_1 \ldots \alpha_n U_\eta\,|\,0> \qquad (7\text{–}25)$$

between two eigenstates of $\tilde{H}_0$.

Again we must decompose a chronological product of p factors $\tilde{H}_1$. Two approaches are possible:

(a) We can eliminate the bare particles a, a^* in favor of the quasi particles α, α^*. This simplifies the calculation of the matrix elements (7–25). The propagators are simple—completely diagonal. In contrast, the interaction V becomes very complex and contains terms $\alpha^*\alpha^*\alpha^*\alpha^*$, $\alpha\,\alpha^*\alpha^*\alpha^*$, etc.

(b) We can retain the bare particles a, a^*. The interaction V contains only terms a^*a^*aa. On the other hand, the propagators become nondiagonal.

We shall choose the second method, which is simpler to use and can be connected directly with the physics (the quasi particles α_k are intermediaries in the calculation, without any real existence).

b. Generalization of Wick's theorem

We want to calculate matrix elements of a product of operators a, a^* between states characterized by α^*. The usual Wick's theorem is not sufficient; however, it is easy to generalize it. Let us consider two operators U and V, which can be of the type a or the type a^*. We retain the definition (5–20) of the chronological product $T(UV)$. On the other hand, we change that of the normal product $N(UV)$, in such a way as to maintain the property

$$<0\,|\,N(U\,V)\,|\,0> = 0 \qquad (7\text{–}26)$$

To define $N(UV)$, we first express U and V as a linear combination of α and α^*; we then move all the α^* to the left; finally, we re-express the result in terms of a and a^*. Using (7–10), we can easily verify that

$$\begin{cases} N(a^*_{k\sigma}\, a_{k\sigma}) = a^*_{k\sigma}\, a_{k\sigma} - v_k^2 \\ N(a_{k\sigma}\, a_{-k,-\sigma}) = a_{k\sigma}\, a_{-k,-\sigma} - \sigma u_k v_k \end{cases} \qquad (7\text{–}27)$$

Finally, we introduce the "contraction" of U and V,

$$\overbrace{U\, V} = T(UV) - N(UV)$$

The contraction is a *number* which, according to (7–26), is given by

$$\overbrace{U\, V} = \langle 0 \,|\, T(UV) \,|\, 0 \rangle \qquad (7\text{–}28)$$

We can therefore write, using the notation (4–116),

$$\overbrace{a^{\rho}_{k\sigma}(t)\; a^{\rho'}_{k'\sigma'}(t')} = -\mathrm{i}\, G_0(kt\sigma\rho,\, k't'\sigma'\rho') \qquad (7\text{–}29)$$

Let us point out the existence of contractions $\widehat{aa}$, absent for a normal gas.

Once these definitions have been made, Wick's theorem can be expressed just as for a normal gas; a chronological product $T(U_1 \ldots U_p)$ can be decomposed into:

(a) A normal product $N(U_1 \ldots U_p)$

(b) $p[(p-1)/2]$ normal products in which any two factors U_i have been contracted

(c) Etc., until all factors in the product have been contracted

As usual, each permutation of operators introduces a sign change. The proof of the theorem is identical to that of Appendix E. Superfluidity is simply expressed by a certain number of new contractions.

Let us consider the average value $\langle 0 | U_\eta | 0 \rangle$; only the completely contracted terms of the Wick expansion contribute here. For other matrix elements, the situation is more complex. Let us consider, for example, the element $\langle 0 | \alpha_1 \ldots \alpha_n\, U_\eta(0, -\infty) | 0 \rangle$; there appear mixed contractions

$$\begin{cases} \overbrace{\alpha_{k\sigma}\; a^*_{k\sigma}(t)} = \mathrm{i} u_k \exp(\mathrm{i} E_k t) \\ \overbrace{\alpha_{k\sigma}\; a_{-k,-\sigma}(t)} = \mathrm{i} \sigma v_k \exp(\mathrm{i} E_k t) \end{cases} \qquad (7\text{–}30)$$

(with $t < 0$). There is no difficulty in principle, only a notational complication.

c. Graphical representation of the expansion of $\langle 0|U_\eta|0\rangle$

With each operator $a^\rho_{k\sigma}$ we associate half an oriented line, according to the scheme of Fig. 1. A contraction between two operators $kt\sigma\rho$ and $k't'\sigma'\rho'$ is represented by a line connecting the two corresponding end points. Note that the sense of propagation can be reversed from one end of the line to the other (see Fig. 2). This new aspect is characteristic of superfluidity; it expresses the emission or absorption of a pair by the condensed phase. We provisionally associate the contraction (7–29) with each line.

There exist two types of vertices, arising, respectively, from V and $(-W)$. Let us first take the usual vertices, coming from V. They involve two incoming and two outgoing lines (Fig. 3). If we take the precaution of aligning line 1 with line 4, line 2 with line 3, the vertex yields a contribution $V(\mathbf{k}_1,\mathbf{k}_2,\mathbf{k}_3,\mathbf{k}_4)$, given by (5–17) [supplemented by a factor $-i$ arising from the general expansion (5–13)]. It is sometimes convenient to characterize a vertex by the four lines (k_i,σ_i,ρ_i) which end there. The corresponding contribution $V(k_i,\sigma_i,\rho_i)$ exists only if

$$\begin{cases} \sum_i \rho_i = 0 \\ \sum_i \rho_i\sigma_i = 0 \\ \sum_i \rho_i k_i = 0 \end{cases} \tag{7–31}$$

These relations express, respectively, conservation of particle number, of total spin, and of total wave vector.

Let us now consider the vertices coming from $(-W)$. There are three types of them, indicated in Fig. 4. With a type (a) vertex there is associated the contribution $i\lambda_k$. To within a sign, the vertices (b) and (c) give a contribution $i\mu_k$ (this sign is still uncertain, since it depends on the conventions which we are going to choose).

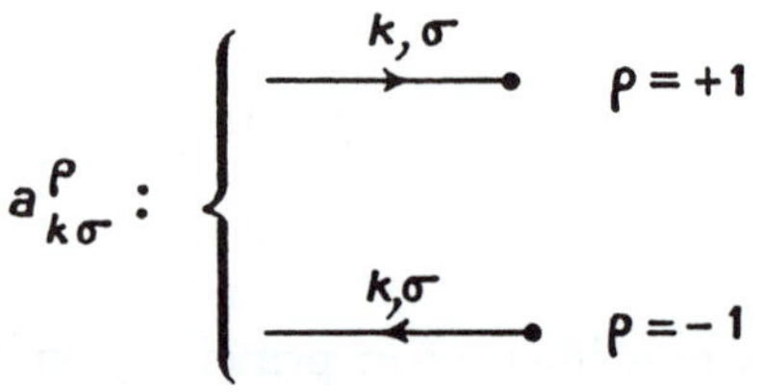

Figure 1

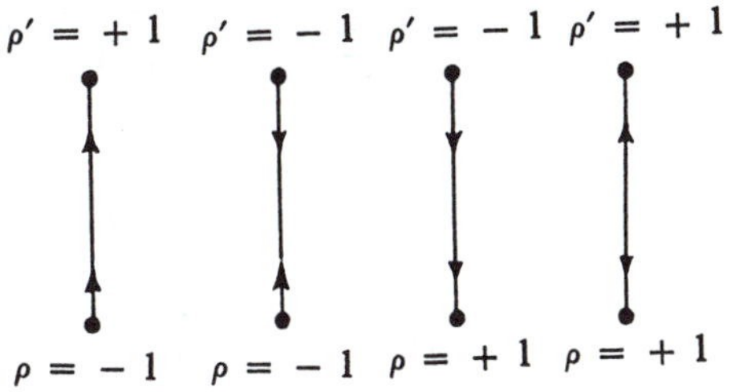

Figure 2

To obtain the matrix element $\langle 0|U_\eta|0\rangle$, we contract *all* the factors of U_η; this amounts to considering only *unlinked* diagrams. It remains only to settle the delicate question of sign. Let us take a line of the diagram and follow it continuously from vertex to vertex (each vertex corresponds to the *crossing* of two lines); we thus trace out a *closed loop*. Let us first assume that this loop contains only the vertices of Fig. 3 or 4a. Let us follow it in either direction and number the vertices from 1 to p in the order in which they appear. With the ith vertex there is associated a pair of factors $a_i{}^*a_i$. We can put these pairs side by side without changing the sign,

$$| \overset{*}{a}_1 a_1 | \overset{*}{a}_2 a_2 | \ldots\ldots\ldots\ldots | \overset{*}{a}_p a_p |$$

We must contract an element of pair 1 with an element of pair 2, etc., until we return to pair 1.

Let us assume that a_1 is contracted with pair 2, $\overset{*}{a}_1$ with pair p (we can always come back to this case by reversing the tracing out of the loop). There is a change of sign if a_1 is contracted with a_2 and none if a_1 is contracted with $\overset{*}{a}_2$. In the two cases, the remaining operator 2 must be contracted with one of the operators of pair 3; this returns us to just the situation encountered with a_1, with respect to pair 2. This is repeated

Figure 3

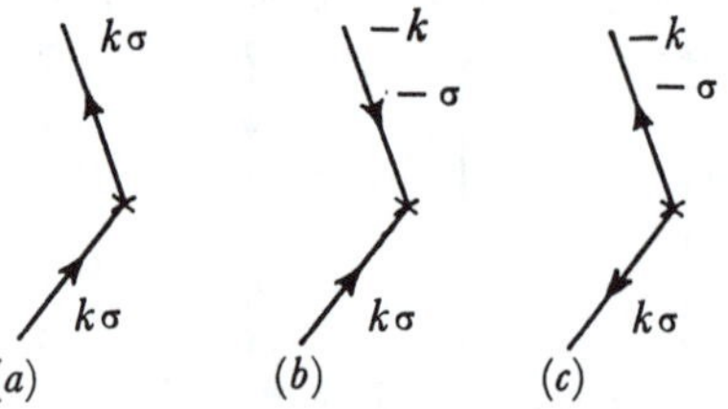

Figure 4

from pair to pair, until the last step, where the remaining operator of pair p is contracted with $\overset{*}{a}_1$; in this case we automatically have a minus sign. We can condense this discussion into a rule:

(7–32)
Trace out each closed loop in either direction. With each line starting from $kt\sigma\rho$ and ending at $k't'\sigma'\rho'$ associate a factor

$$-i\tilde{G}_0(k\sigma\rho t, k'\sigma'\rho't') = i\rho' G_0(k\sigma\rho t, k'\sigma'\rho't')$$

Associate a minus sign with the whole closed loop.

The result does not depend on the direction in which the line is traversed.

Let us now assume that our oriented path crosses a vertex of the type shown in Fig. 4b, arriving in the state (k,σ) and leaving in the state $(-k,-\sigma)$. The set of operators to be contracted has the form

$$a^*a \mid a_-a_+ \mid a^*a$$

For the rules (7–32) to remain valid, it is necessary to attribute to the vertex (b) the contribution $i\sigma\mu_k$. This conclusion extends to the vertices (c), for analogous reasons. Let us emphasize the *absolute* necessity of a *consistent* sign convention throughout the calculation.

Until now, we have considered labeled diagrams. It is easy to replace these by unlabeled diagrams. As in Chap. 5, we must not count the same term more than once, and we must pay attention to equivalent lines and to permutations of vertices leaving labeled diagrams invariant. In the present case, there appears a third risk of duplication, due to "symmetric loops" of the type

or

In the contribution a^*a^*aa from the corresponding vertex, there is actually only a single way of contracting the two a or the two a^*.

Let us consider an unlabeled unlinked diagram containing p vertices V and q vertices W. It contains $(2p + q)$ internal lines. We can calculate the contribution of this diagram by the following rules:

(7–33)

(a) Decompose the lines into closed loops, traced out in an arbitrary, but well-defined, direction. Each line thus has an *origin* and an *end point.*

(b) For each line going from $kt\sigma\rho$ to $k't'\sigma'\rho'$, introduce a factor

$$\tilde{G}_0(kt\sigma\rho,\ k't'\sigma'\rho')$$

(c) Associate the factor $V(k_i, \sigma_i, \rho_i)$ with each V vertex.

(d) Associate the contributions λ_k and $\sigma\mu_k$ with the W vertices.

(e) Add to this the factor

$$\frac{(\mathrm{i})^p(-1)^l}{r2^{m+n}}$$

where p = number of V vertices
l = number of closed loops
r = number of permutations of vertices leaving the diagram invariant
m = number of pairs of equivalent lines
n = number of symmetric loops

Note that all the factors i have been collected. The rules (7–33) generalize the rules (5–35) established for a normal gas. (The propagators $\tilde{G}_0$ are now matrix, not scalar quantities.) To obtain $\langle 0|U_\eta|0\rangle$, we add the contributions of all distinct unlinked diagrams, after summation over all internal variables.

It is easy to extend this discussion to the nondiagonal elements of U_η, such as (7–25). We make "external points" of the diagram correspond to the operators $\alpha_1 \dots \alpha_n$; with the lines connecting these external points with the vertices, we associate the mixed contractions (7–30). The only difficulty is in the assignment of the overall sign; we shall not take up this question, which, simple in principle, leads to a complicated discussion.

d. Elimination of disconnected diagrams and Fourier transformation of the expansion

As in Chap. 5, we can carry out a partial summation of the disconnected diagrams. The time-development operator thus takes the form (5–45). This allows us to pass to the limit $\eta \to 0$. The conclusions are the same as for a normal gas.

Let us choose $\lambda_k(g)$ and $\mu_k(g)$ so that, for each value of g, $|0\rangle$ generates the ground state of the corresponding system. With this reservation, $U_{\eta L}$ has a well-defined limit when $\eta \to 0$, although $U_{\eta 0c}$ has an imaginary part of order η. We can thus define without ambiguity the ground state $|\varphi_0\rangle$ of the real system by the relation (5–63). The ground-state energy E_0 can be written as

$$E_0 = U + \Delta E_0 \tag{7-34}$$

where U is given by (7–11) and ΔE_0 by (5–80). To obtain ΔE_0, it is sufficient to consider the set of connected unlinked diagrams having a vertex fixed at the origin. The reader can refer to Chap. 5 for a detailed discussion of all these problems.

In its actual form, the diagrammatic expansion is quite complex, since $\tilde{G}_0$ depends on the four indices $\rho\sigma$, $\rho'\sigma'$. This difficulty disappears if we make a Fourier transformation with respect to time. $\tilde{G}_0$ is then given by (4–122) and depends only on the indices σ and σ'; we can represent it by a 2×2 matrix.

Therefore, let us replace all factors

$$\tilde{G}_0(k\rho\sigma t, k'\rho'\sigma' t')$$

by their Fourier expansions (4–117),

$$\tilde{G}_0(k\rho\sigma t, k'\rho'\sigma' t') = \frac{1}{2\pi}\int d\omega\, d\omega'\, \tilde{G}_0(k\rho\sigma\omega, k'\rho'\sigma'\omega') \exp[-\,i(\rho\omega t + \rho'\omega' t')] \tag{7-35}$$

Each end of a line is thus characterized by four quantities k, ρ, σ, ω, which we group together under the single symbol ξ [see (4–118)]. Let us consider a V vertex where the four lines $\xi_1 \ldots \xi_4$ come together; it carries a phase factor

$$\exp[-\,i(\rho_1\omega_1 + \ldots + \rho_4\omega_4)\, t] \tag{7-36}$$

When we integrate over the time t, we find a contribution

$$2\pi\delta(\rho_1\omega_1 \ldots + \rho_4\omega_4) \tag{7-37}$$

Therefore, there is conservation of energy at each vertex. The same conclusion applies to a W vertex. This is to be combined with conservation of momentum and spin.

With each line starting from ξ and ending at ξ', we associate the quantity

$$\frac{1}{2\pi}\,\tilde{G}_0(\xi,\xi')$$

According to (4–122), $\tilde{G}_0(\xi,\xi')$ contains a factor $\delta(\xi,\xi')$. Energy, spin, and momentum are thus conserved from one end of a line to the other, just as at each vertex. It is preferable to ensure these conservation laws by construction and to omit all δ-function factors.

The "local" conservation of (k,σ,ω) implies overall conservation from one end to the other of the diagram. For an unlinked connected diagram, the conservation relations are therefore *automatically* satisfied at any one vertex. In calculating ΔE_0, we shall choose precisely this vertex to be fixed; this amounts to omitting the corresponding factor (7–37).

In conclusion, we collect all these results in the form of a practical set of rules for calculation of ΔE_0. Let us consider an unlinked connected diagram containing p vertices V and q vertices W—that is to say, $(2p+q)$ internal lines. Let us collect all factors 2π, taking account of the "fixed" vertex. The contribution of this diagram to the expression (5–80) is obtained by the following rules:

(7–38)

(a) Ensure by construction the conservation of k, σ, ω; orient each closed loop.
(b) Associate with each V vertex the factor $V(\xi_1 \ldots \xi_4)$, with the W vertices the factors λ_k or $\sigma\mu_k$ (σ being the spin before *entering* the vertex).
(c) To characterize a line starting from ξ, ending at ξ', it is sufficient to give the spins σ and σ' and the four-vector $p=(k,\omega)$. With this line associate the factor $\tilde{G}_{\sigma\sigma'}(p)$ defined by (4–122).
(d) Add an over-all factor $\left(\frac{i}{2\pi}\right)^{p+1}\frac{(-1)^l}{r2^{m+n}}$.

These rules generalize the set (5–108) to superfluid systems. They are more convenient than (7–33). In applying them, one must pay careful attention to questions of sign.

4. Calculation of the Green's Functions

a. *Graphical representation*

Let us first consider the complete one-particle Green's function,

$$\tilde{G}(k\rho\sigma t, k'\rho'\sigma' t') = -i\rho' < \varphi_0 \mid T\{\, a^{\rho}_{k\sigma}(t)\, a^{\rho'}_{k'\sigma'}(t')\,\} \mid \varphi_0 > \qquad \textbf{\textit{(7–39)}}$$

We can repeat the analysis made for the normal gas and put $\tilde{G}$ in a form analogous to (5–97):

$$\tilde{G}(k\rho\sigma t, k'\rho'\sigma' t') = -\frac{i\rho' < 0 \mid T\{ S_\eta a_{k\sigma}^{\rho}(t)\, a_{k'\sigma'}^{\rho'}(t') \} \mid 0 >}{< 0 \mid S_\eta \mid 0 >} \qquad (7\text{–}40)$$

In this expression, $S_\eta = U_\eta(+\infty, -\infty)$ is the "S matrix" of the system; the operators a are expressed in the interaction representation. The formal chronological product indicates that in the perturbation expansion of S_η, the two operators a must be inserted in their chronological positions.

As in Chap. 5, the numerator of (7–40) can be represented with the aid of diagrams having two external lines coming from the points $k\rho\sigma t$ and $k'\rho'\sigma' t'$. The denominator of (7–40) has the effect of eliminating all disconnected diagrams. $\tilde{G}$ is therefore given by the set of *connected* diagrams indicated in Fig. 5. In addition to a certain number of closed loops, these diagrams contain an open line relating the external points. It is easily verified that $\tilde{G}$ is obtained by application of the rules (7–33), supplemented by the following rule:

(7–41) (f) The open line must be traversed from the point $k\rho\sigma$ to the point $k'\rho'\sigma'$. It does *not* contribute the factor (-1) characteristic of a closed loop.

To zeroth order, $\tilde{G}$ is equal to $\tilde{G}_0$, which confirms the accuracy of our sign rules.

In general, we prefer to calculate the Fourier transform $\tilde{G}(k\rho\sigma\omega, k'\rho'\sigma'\omega')$, whose reduced form $\tilde{G}_{\sigma\sigma'}(p)$ is given by (4.122). For this

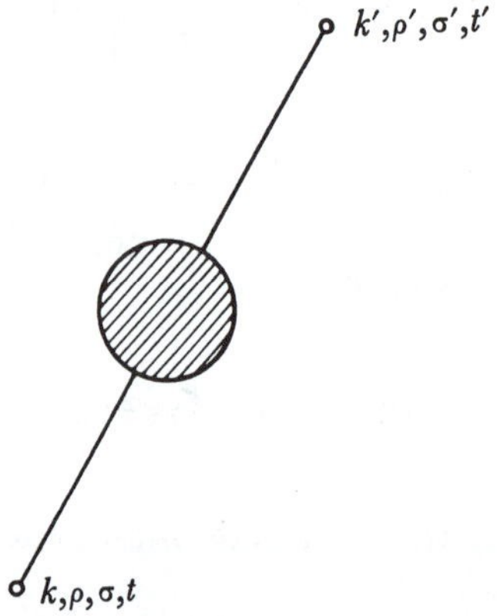

Figure 5

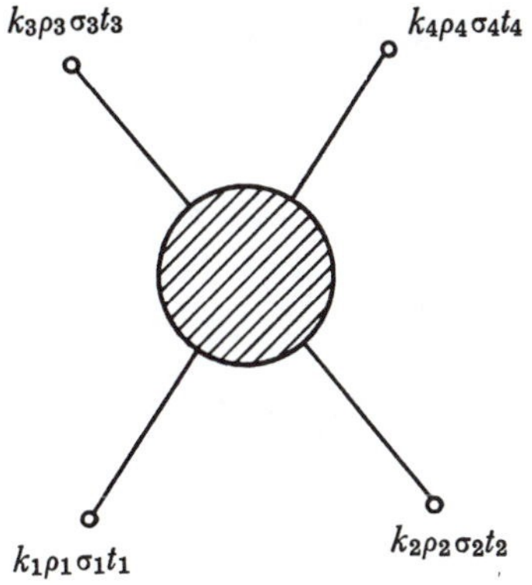

Figure 6

purpose, we assign to the two ends of the diagram 7-5 the indices $k\rho\sigma\omega$ and $k'\rho'\sigma'\omega'$ and apply rules analogous to (7–38):

(7–42) $\begin{cases} \text{Retain the rules (7–38) (a), (b), and (c), as well as (7–41).} \\ \text{Add a numerical factor } (i/2\pi)^p(-1)^l/(r2^{m+n}). \end{cases}$

We leave to the reader the task of verifying this result.

Similarly we can calculate the two-particle Green's function $K(k_i\rho_i\sigma_it_i)$, defined by (4–128) and (4–133). It is given by diagrams having four external lines, of the type indicated in Fig. 6. Here again, the only difficulty lies in questions of sign. Let us replace K by the quantity

$$\widetilde{K}(1, 2, 3, 4) = \rho_3\rho_4K(1, 2, 3, 4) \qquad (7\text{–}43)$$

and draw the diagram so that line 1 ends at 3 and line 2 at 4. We can then verify that the rules (7–33) and (7–41) remain valid. The Fourier transform $K(p_i\rho_i\sigma_i)$, defined by (4–133), is still given by (7–42). If for any reason we must permute two end points, we introduce an additional minus sign. Note that, in contrast to $\widetilde{G}$, $\widetilde{K}$ depends on the indices ρ_i or, more exactly, on their relative values.

Among the diagrams of Fig. 6, there are some which are made up of two disconnected parts (Fig. 7). These diagrams correspond to the free part K_L given by (4–130). The bound part ΔK comes only from connected diagrams. As in Chap. 4, we separate off the four external propagators G; the remaining central core then corresponds to an "interaction". Let us set

$$\Delta\widetilde{K}(\xi_1\xi_2\xi_3\xi_4) = \sum_{\xi'_i} \widetilde{\gamma}(\xi'_1\xi'_2\xi'_3\xi'_4) \prod_{i=1}^{4} \widetilde{G}(\xi_i, \xi'_i) \qquad (7\text{–}44)$$

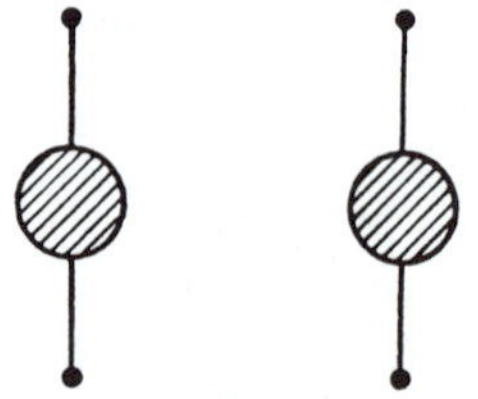

Figure 7

Comparing this with the definition (4–136), we see that $\tilde{\gamma}$ and the interaction operator of Chap. 4 are related by

$$\tilde{\gamma}(1\ 2\ 3\ 4) = \rho_1\rho_2\,\gamma(1\ 2\ 3\ 4) \tag{7–45}$$

[Note the difference from (7–43).] The operator $\tilde{\gamma}$ is evaluated according to the set of rules (7–42) (with the same conventions as for $\tilde{K}$).

Very generally, we see that the perturbation expansion gives the quantities $\tilde{G}$, $\tilde{K}$, and $\tilde{\gamma}$ in a natural way. We must obtain them as an intermediate step for calculating G, K, and γ.

The practical calculation of $\tilde{\gamma}$ quickly becomes very complicated. In first order, $\tilde{\gamma}$ reduces to a single vertex, which ensures conservation of particle number. Starting with the second order, many possibilities appear, as sketched in Fig. 8. We shall try to avoid the explicit calculation of these quantities.

b. Reduction of the single-particle Green's function

We call all diagrams of G having the structure indicated in Fig. 9 "reducible." An *irreducible* diagram is not separable into two independent parts by the cutting of a single line. Let $\bar{M}(\xi,\xi')$ be the contribution from

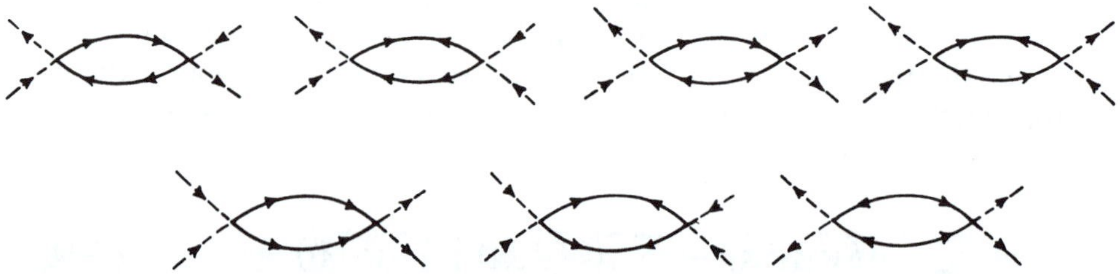

Figure 8

all the diagonal irreducible diagrams, defined as for a normal gas. To obtain $\tilde{G}$, we insert into a line $\tilde{G}_0$ an arbitrary number of self-energy parts $\bar{M}$. Instead of having a simple algebraic product, as for a normal gas, we now have a matrix product.

Before examining this question, we must know the symmetry properties of $\tilde{G}_0$, $\tilde{G}$, and $\bar{M}$. We know that $\tilde{G}_0$ has the simple form (4–122), F_{02} being an even function of p. Using the notation of Chap. 4, we write the "matrix" $\tilde{G}_0$ in the form (4–125),

$$\tilde{\mathbf{G}}_0(p) = F_{01}(p)\,\mathbf{1} + F_{02}(p)\,\boldsymbol{\beta}_1 \tag{7–46}$$

where the matrices $\mathbf{1}$ and $\boldsymbol{\beta}_1$ are defined by (4–124). Within the limits of a perturbation expansion, $\tilde{G}$ and $\bar{M}$ *necessarily* have the same symmetries as $\tilde{G}_0$:

Reversing all the arrows in a diagram does not change its contribution. Thus, $\tilde{G}(\xi,\xi')$ and $\bar{M}(\xi,\xi')$ do not depend on ρ and ρ'.

Upon reversal of all the spins of a diagram, each factor μ_k or F_{02} is multiplied by (-1). These factors always create or destroy a pair. We must therefore have an even number of them if $\rho = -\rho'$ $(\sigma = \sigma')$ and an odd number if $\rho = \rho'$ $(\sigma = -\sigma')$; $\tilde{G}$ and $\bar{M}$ thus have the structure (7–46),

$$\begin{cases} \tilde{\mathbf{G}}(p) = F_1(p)\,\mathbf{1} + F_2(p)\,\boldsymbol{\beta}_1 \\ \bar{\mathbf{M}}(p) = \bar{M}_1(p)\,\mathbf{1} + \bar{M}_2(p)\,\boldsymbol{\beta}_1 \end{cases} \tag{7–47}$$

Finally, if in a diagram with end points (ρ,σ,p) and $(\rho,-\sigma,-p)$ we flip all the spins, this amounts to changing p into $-p$. The change of sign is compensated by the fact that the beginning and end of the open line are interchanged; thus $\bar{M}_2$ and F_2 are even in p.

We thus directly confirm the deductions made in Chap. 4.

Let us now return to the reduction of $\tilde{\mathbf{G}}$. Isolating the first self-energy

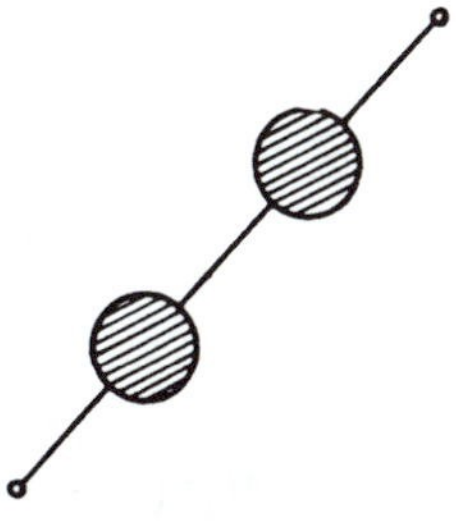

Figure 9

part $\overline{\mathbf{M}}$ from the bottom, we can analyze $\tilde{\mathbf{G}}$ according to the scheme of Fig. 10 (a double line representing a propagator $\tilde{\mathbf{G}}$). Because of the relations (7–46) and (7–47), it is easy to express this figure in mathematical terms. Using the identity $\boldsymbol{\beta}_1 \cdot \boldsymbol{\beta}_1 = -1$, we obtain

$$\left\{\begin{aligned} F_1(p) &= F_{01}(p) + F_{01}(p)\,\overline{M}_1(p)\,F_1(p) - F_{02}(p)\,\overline{M}_1(-p)\,F_2(p) \\ &\qquad - F_{01}(p)\,\overline{M}_2(p)\,F_2(p) - F_{02}(p)\,\overline{M}_2(p)\,F_1(p) \\ F_2(p) &= F_{02}(p) + F_{01}(p)\,\overline{M}_1(p)\,F_2(p) + F_{01}(p)\,\overline{M}_2(p)\,F_1(-p) \\ &\qquad + F_{02}(p)\,\overline{M}_1(-p)\,F_1(-p) - F_{02}(p)\,\overline{M}_2(p)\,F_2(p) \end{aligned}\right. \tag{7–48}$$

This allows us to calculate the Green's functions from the self-energy operators.

The solution of Eqs. (7–48) is tedious but simple. It is easily verified that

$$\left\{\begin{aligned} F_1 &= \frac{F_{01} - \overline{M}_1^-(F_{01}\,F_{01}^- + F_{02}^2)}{\Delta} \\ F_2 &= \frac{F_{02} + \overline{M}_2(F_{01}\,F_{01}^- + F_{02}^2)}{\Delta} \\ \Delta &= 1 - F_{01}\overline{M}_1 - F_{01}^-\overline{M}_1^- + 2\,F_{02}\overline{M}_2 + (\overline{M}_2^2 + \overline{M}_1\overline{M}_1^-)\,(F_{02}^2 + F_{01}F_{01}^-) \end{aligned}\right. \tag{7–49}$$

[where F_{01}^-, $\overline{M}_1^-$ are abbreviations for $F_{01}\,(-p)$, $\overline{M}_1\,(-p)$]. These results are simplified considerably by specifying F_{01} and F_{02}; according to (7–9) and (7–20) we can write

$$\left\{\begin{aligned} F_{01}(p) &= \frac{\xi_k + \omega}{\xi_k^2 + \mu_k^2 - \omega^2} \\ F_{02}(p) &= \frac{\mu_k}{\xi_k^2 + \mu_k^2 - \omega^2} \\ (F_{01}F_{01}^- + F_{02}^2) &= \frac{1}{\xi_k^2 + \mu_k^2 - \omega^2} \end{aligned}\right. \tag{7–50}$$

Inserting these expressions into (7–49), we find

$$\left\{\begin{aligned} F_1 &= \frac{\xi_k + \omega - \overline{M}_1^-}{D} \\ F_2 &= \frac{\mu_k + \overline{M}_2}{D} \\ D &= (\xi_k - \omega - \overline{M}_1)\,(\xi_k + \omega - \overline{M}_1^-) + (\mu_k + \overline{M}_2)^2 \end{aligned}\right. \tag{7–51}$$

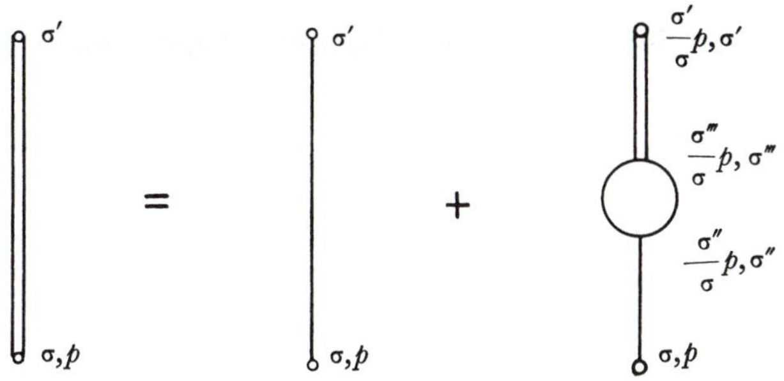

Figure 10

The reduction of the Green's functions is thus achieved in a very simple form.

This "algebraic" step having been carried out, let us analyze more in detail the structure of the self-energy operators $\overline{\mathbf{M}}$. There is one diagram which plays a separate role—the one made up of a single W vertex (Fig. 11a). Comparing (7–47) with the rules (7–38), we see that the corresponding contributions to M_1 and M_2 are, respectively, λ_k and $-\mu_k$. The differences

$$\begin{cases} M_1 = \overline{M}_1 - \lambda_k \\ M_2 = \overline{M}_2 + \mu_k \end{cases} \tag{7–52}$$

come entirely from diagrams of the type of Fig. 11b, whose beginning and end are made up of V vertices (which does not exclude the possibility of

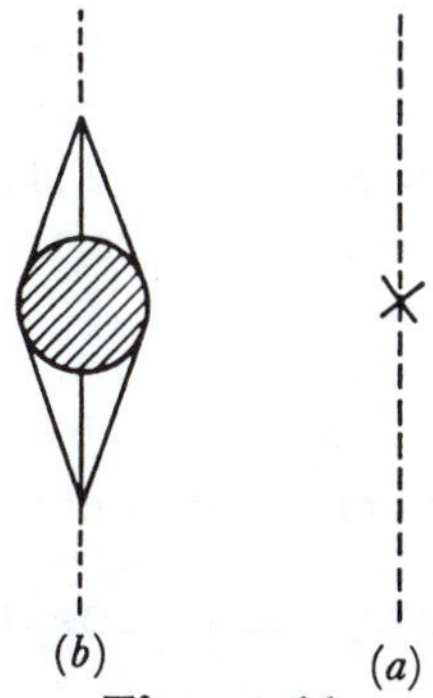

Figure 11

W vertices on the interior of the diagram). Combining (7–51) and (7–52), we see that

$$\begin{cases} F_1 = \dfrac{k^2/2m - \mu + \omega - M_1^-}{D} \\ F_2 = M_2/D \\ D = (k^2/2m - \mu + \omega - M_1^-)\,(k^2/2m - \mu - \omega - M_1) + M_2^2 \end{cases} \qquad (7\text{–}53)$$

Let us compare (7–53) with (4–152). The functions M_1 and M_2 are identical to those introduced in Chap. 4. This is not at all surprising; in fact, the diagrams 11b can be decomposed into a single vertex followed by a two-particle Green's function; it is just in this way that we introduced the self-energy operator **M** in Chap. 4. As for a normal gas, **M** thus has a very simple graphical interpretation.

c. *Passage to skeleton diagrams; elimination of* λ_k *and* μ_k

Let us consider a diagram (Fig. 11b) and remove all the diagonal insertions; we obtain a *skeleton* diagram. In the calculation of **M** we can limit ourselves to skeleton diagrams on condition that we associate a factor $\tilde{G}$ instead of $\tilde{G}_0$ with each line; we have thus renormalized the propagators. **M** becomes a functional of $\tilde{\mathbf{G}}$, denoted by $\mathbf{M}[\tilde{\mathbf{G}}]$. This functional does not involve W vertices, since these necessarily constitute a diagonal insertion.

Knowing **M**, we go back to $\tilde{\mathbf{G}}$ by means of the relations (7–53), which no longer involve either λ_k or μ_k; **M** and $\tilde{\mathbf{G}}$ are therefore determined by two coupled equations, whose solution is, in principle, possible. We note that the constants λ_k and μ_k, artificially introduced at the beginning of the problem, have disappeared. They constitute an intermediate step necessary for developing our perturbation expansion, but they have no effect on the final answer. This property is essential; without it, our formalism would make no physical sense.

The equations determining **M** are implicit. In addition to an eventual "superfluid" solution, they will always have a "normal" solution, corresponding to $M_2 = 0$. In the end, we shall have to decide which of these solutions is stable. In general, it is the superfluid solution which prevails—when it exists, of course.

The simplest approximation consists in taking for **M** the first-order skeleton diagram, indicated in Fig. 12. The interaction being instantaneous, M_1 and M_2 are independent of the energy ω. Let us take

$$\zeta_k = k^2/2m - \mu - M_1(k) \qquad (7\text{–}54)$$

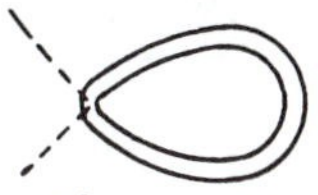

Figure 12

M_2 is given by the equations

$$\left\{\begin{aligned} M_2(k) &= \frac{i}{2\pi}\frac{1}{2}2\sum_{k'}\int d\omega' V(k-k')\, F_2(k', \omega') \\ F_2(k, \omega) &= \frac{M_2(k)}{\zeta_k^2 + M_2^2 - \omega^2 - i\delta} \end{aligned}\right. \tag{7-55}$$

(the factor $\frac{1}{2}$ arises from the symmetric loop, the factor 2 from the summation over spins). It is easy to integrate over ω'; we thus obtain an integral equation for M_2,

$$M_2(k) = -\tfrac{1}{2}\sum_{k'} \frac{V(k-k')\, M_2(k')}{\sqrt{\zeta_{k'}^2 + M_2(k')^2}} \tag{7-56}$$

This equation can have a solution only if $V(k - k') < 0$, that is to say, for an attraction. (7–56) is just the famous equation of Bardeen, Cooper, and Schrieffer, giving the width of the "forbidden gap" in the excitation spectrum. It takes account of superfluidity qualitatively, but it constitutes a rough approximation (equivalent to the Hartree-Fock approximation for a normal gas).

λ_k and μ_k do not explicitly enter into the final result. We can therefore choose them more or less arbitrarily. This latitude is limited, nevertheless, by conditions of *convergence* and *stability*. Let us consider the diagram of Fig. 13 and move the core M_2 from one end of the line to the other. The corresponding contribution is of order M_2/E_k, where E_k is given by (7–8). When we insert an increasing number of cores M_2 into a line, we construct a geometric series of ratio M_2/E_k. For this series to converge, it is necessary that

$$M_2 < E_k \tag{7-57}$$

If we had started from a normal system, E_k would have vanished at the Fermi surface, giving a highly divergent series. To ensure convergence,

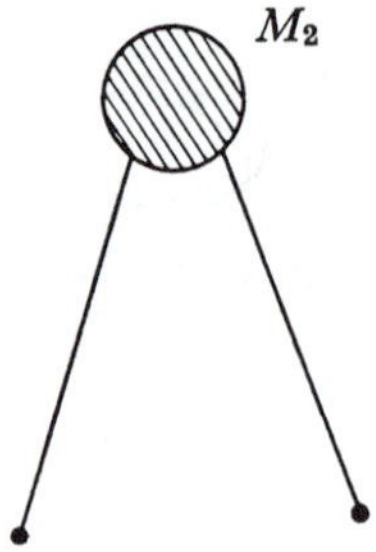

Figure 13

we must choose μ_k so that in the neighborhood of the Fermi surface,

$$M_2 < \mu_k \tag{7–58}$$

(7–58) emphasizes the *necessity* of the artifice used in this chapter.

Questions of convergence aside, the adiabatic switching on must generate the ground state of the real system. We saw in Chap. 5 that, for a normal system, this leads to choosing a "good" Fermi surface from the start. For a superfluid gas, the limitations are less clear, since we hardly know what remains constant during the adiabatic switching on (besides the chemical potential μ). To justify our method, it is sufficient to know that there *exist* certain functions $\lambda_k(g)$ and $\mu_k(g)$ which lead to the ground state sought; of this there is no doubt. However, we do not know how to formulate the corresponding conditions precisely. Intuitively, we feel that $|\varphi_0(g)\rangle$ must remain as close as possible to the real state $|\varphi_0\rangle$ at all times. One interesting possibility consists in choosing $\lambda_k(g)$ and $\mu_k(g)$ so that the ratio z_+/z_- of the residues of the Green's function F_1 is independent of g for each value of k (equal, in particular, to u_k^2/v_k^2). This is only a conjecture. We shall not examine this obscure point further; it is of academic interest only.

5. The Interaction of Two Elementary Excitations

a. Bethe-Salpeter equation

To conclude, we are going to generalize to superfluid gases the considerations of Chap. 6. Let us consider the diagram of Fig. 14, contributing to the interaction operator $\tilde{\gamma}(\xi_1\xi_2\xi_3\xi_4)$ [again, we use the notation (4–118)]. By analogy with a normal system, we shall say that it is "reducible" with respect to the collision 1, 2 → 3, 4 (we must specify which are the two incoming and which the two outgoing lines). Let $I(\xi_1\xi_2\xi_3\xi_4)$ be the contribution to $\tilde{\gamma}$ from the irreducible diagrams; we

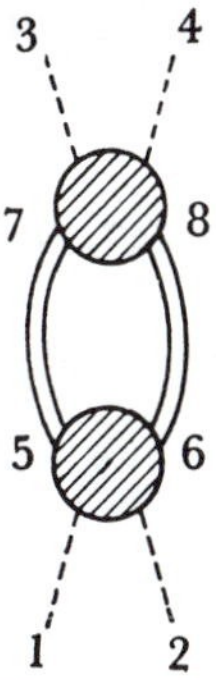

Figure 14

call it the "irreducible interaction operator." To obtain a diagram of $\tilde{\gamma}$, we line up a certain number of cores I. Repeating the proof of (6–1), we can verify that

$$\tilde{\gamma}(\xi_1\xi_2\xi_3\xi_4) = I(\xi_1\xi_2\xi_3\xi_4) + \frac{1}{2} \sum_{\xi_5\xi_6\xi_7\xi_8} I(\xi_1\xi_2\xi_5\xi_6)\, \tilde{G}(\xi_5\xi_7)\, \tilde{G}(\xi_6\xi_8)\, \tilde{\gamma}(\xi_7\xi_8\xi_3\xi_4) \tag{7–59}$$

(the factor $\frac{1}{2}$ is introduced because the intermediate lines of Fig. 14 are equivalent). (7–59) is the Bethe-Salpeter equation for a superfluid gas. It describes the multiple scattering of two elementary excitations (each of which is both a particle and a hole). In the special case of a normal system (7–59) contains the two equations (6–1) and (6–11).

In the form (7–59), the Bethe-Salpeter equation is so condensed that it is of no great use. By a judicious choice of notation, we are going to make it more explicit; we proceed by analogy with the analysis made in Chap. 6. First, let us consider the total spin of the pair (1,2); two cases should be considered:

(a) $\rho_1\sigma_1 = \rho_2\sigma_2 = \sigma$. The total spin of the pair is ± 1. The conservation relations fix all the spins of (7–59),

$$\begin{cases} \rho_3\sigma_3 = \rho_4\sigma_4 = -\sigma \\ \rho_5\sigma_5 = \rho_6\sigma_6 = -\sigma \\ \rho_7\sigma_7 = \rho_8\sigma_8 = +\sigma \end{cases} \tag{7–60}$$

We can limit ourselves to the case $\sigma > 0$. The opposite case may be deduced from it with the aid of the symmetry relation (4–129*a*), which can be written as

$$\tilde{\gamma}(p_i, \sigma_i, \rho_i) = \sigma_1\sigma_2\sigma_3\sigma_4\, \tilde{\gamma}(p_i, -\sigma_i, \rho_i) \tag{7–61}$$

(b) $\rho_1\sigma_1 = -\rho_2\sigma_2 = \sigma$. The total spin of the pair is zero. The conservation relations imply

$$\begin{cases} \rho_3\sigma_3 = -\rho_4\sigma_4 \\ \rho_5\sigma_5 = -\rho_7\sigma_7 = -\rho_6\sigma_6 = \rho_8\sigma_8 \end{cases} \tag{7–62}$$

There remains a summation over one of the internal spins—for example, over σ_5. We shall limit ourselves to the case $\rho_1\sigma_1 = \rho_3\sigma_3 = +1$; we can always come back to it by using either the antisymmetry of $\tilde{\gamma}$ or the relations (7–61). These same considerations show that the integral (7–59) is independent of the value of σ_5. We can therefore take

$$\rho_1\sigma_1 = -\rho_5\sigma_5 = \rho_6\sigma_6 = -\rho_3\sigma_3 \tag{7–63}$$

and eliminate the factor $\frac{1}{2}$ from (7–59). If we draw the diagram lining up the points 1, 5, 7, 3 on one hand, and 2, 6, 8, 4 on the other, the spin $\pm\rho\sigma$ is constant along each of these lines.

Having thus clarified the problem of the spins, we introduce the notation [see (6–14)].

$$\begin{cases} \rho_1 p_1 = \varpi/2 + p \\ \rho_2 p_2 = \varpi/2 - p \end{cases} \qquad \begin{cases} \rho_3 p_3 = -\varpi/2 - p' \\ \rho_4 p_4 = -\varpi/2 + p' \end{cases} \tag{7–64}$$

ϖ measures the total wave vector and energy of the pair (1,2). We then define the following functions for the two cases a and b:

Case a

$$\begin{aligned} &\rho_1\sigma_1 = \rho_2\sigma_2 = -\rho_3\sigma_3 = -\rho_4\sigma_4 = +1 \\ &I(p_1\sigma_1\rho_1 \ldots p_4\sigma_4\rho_4) = {}^1I^{\rho_3\rho_4}_{\rho_1\rho_2}(p, p'; \varpi) \end{aligned} \tag{7–65}$$

Case b

$$\begin{aligned} &\rho_1\sigma_1 = -\rho_2\sigma_2 = \rho_4\sigma_4 = -\rho_3\sigma_3 = +1 \\ &I(p_1\sigma_1\rho_1 \ldots p_4\sigma_4\rho_4) = {}^0I^{\rho_3\rho_4}_{\rho_1\rho_2}(p, p'; \varpi) \end{aligned} \tag{7–66}$$

${}^0I$ and ${}^1I$ are 4×4 matrices, functions of the three variables p, p', and ϖ. We define ${}^0\tilde{\Gamma}$ and ${}^1\tilde{\Gamma}$ in the same way. These relations are to be compared with the definitions (6–15) and (6–16) given for a normal system. Note that ${}^0I$ does not involve spin variables, in contrast with (6–15); the corresponding degree of freedom is obtained by changing ρ_3, ρ_4 into $-\rho_3$, $-\rho_4$.

To rewrite the Bethe-Salpeter equation, it remains to specify the propagators $\widetilde{G}$ of (7–59). For this purpose we introduce the quantity

$$ {}_\sigma G_{\rho_1}^{\rho_2}(p) = \widetilde{G}(-\rho_1 p, -\rho_1\sigma, \rho_1\,;\, \rho_2 p, \rho_2\sigma, \rho_2) \tag{7–67} $$

This matrix is to be added to the large collection of various forms taken for the Green's function G; like each of the preceding forms, it has the merit of simplifying the writing of a special problem. Referring to (4–122), we see that

$$ {}_\sigma G_{\rho_1}^{\rho_2}(p) = F_1(-\rho_1 p)\,\delta_{\rho_1,-\rho_2} + \rho_1\sigma F_2(p)\,\delta_{\rho_1,\rho_2} \tag{7–68} $$

It is instructive to write G in the matrix form

$$ {}_\sigma G_{\rho_1}^{\rho_2}(p) = \begin{array}{|cc} \rho_2 = 1 & \rho_2 = -1 \\ \hline \sigma F_2(p) & F_1(-p) \\ F_1(p) & -\sigma F_2(p) \end{array} \tag{7–69} $$

It is important not to confuse this matrix with that of Eq. (4–123).

We are now prepared to rewrite the Bethe-Salpeter equation. The latter can be decomposed into two parts,

$$ {}^0\widetilde{\Gamma}_{\rho_1\rho_2}^{\rho_3\rho_4}(p, p'\,;\,\varpi) = {}^0I_{\rho_1\rho_2}^{\rho_3\rho_4}(p, p'\,;\,\varpi) $$
$$ + \sum_{\substack{\rho_5,\rho_6\\ \rho_7,\rho_8}} \sum_{p''} {}^0I_{\rho_1\rho_2}^{\rho_5\rho_6}(p, p''\,;\,\varpi)_+ G_{\rho_5}^{\rho_7}(p'' + \varpi/2)_- G_{\rho_6}^{\rho_8}(-p'' + \varpi/2)\; {}^0\widetilde{\Gamma}_{\rho_7\rho_8}^{\rho_3\rho_4}(p'', p'\,;\,\varpi) \tag{7–70} $$

$$ {}^1\widetilde{\Gamma}_{\rho_1\rho_2}^{\rho_3\rho_4}(p, p'\,;\,\varpi) = {}^1I_{\rho_1\rho_2}^{\rho_3\rho_4}(p, p'\,;\,\varpi) $$
$$ + \tfrac{1}{2} \sum_{\substack{\rho_5,\rho_6\\ \rho_7,\rho_8}} \sum_{p''} {}^1I_{\rho_1\rho_2}^{\rho_5\rho_6}(p, p''\,;\,\varpi)_+ G_{\rho_5}^{\rho_7}(p'' + \varpi/2)_+ G_{\rho_6}^{\rho_8}(-p'' + \varpi/2)\; {}^1\widetilde{\Gamma}_{\rho_7\rho_8}^{\rho_3\rho_4}(p'', p'\,;\,\varpi) \tag{7–71} $$

These two equations, of somewhat appalling appearance, are actually much simpler than (7–59), since they involve only 4×4 matrices, instead of 16×16 matrices. Let us recall that the indices ρ fix the orientations of the lines at each point of the diagram. Let us also point out the

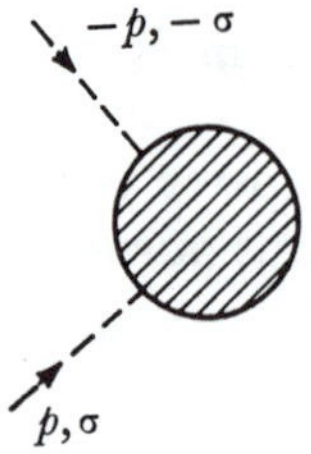

Figure 15

factor $\frac{1}{2}$ in (7–71), which is due to the fact that the two intermediate lines have the same total spin.

These equations generalize (6–18) and (6–19). From them we could study collective modes, bound states, the response to an external field, etc. We shall not investigate these questions in this book.

b. Gap equation; the limit, $\varpi \to 0$

Let us consider the skeleton diagram of Fig. 15, contributing to $-\sigma M_2(p)$. (We follow it from bottom to top.) It corresponds to the destruction of one pair. Let us assume that this diagram contains a number $n \geqslant 0$ of lines of the type •—←——→—•; it must necessarily contain $(n+1)$ lines of the type •—→——←—• in order to ensure the overall balance of the number of particles. Let us select any one of these $(n+1)$ lines; we can draw the diagram in the form of Fig. 16*a*. Since we started from a

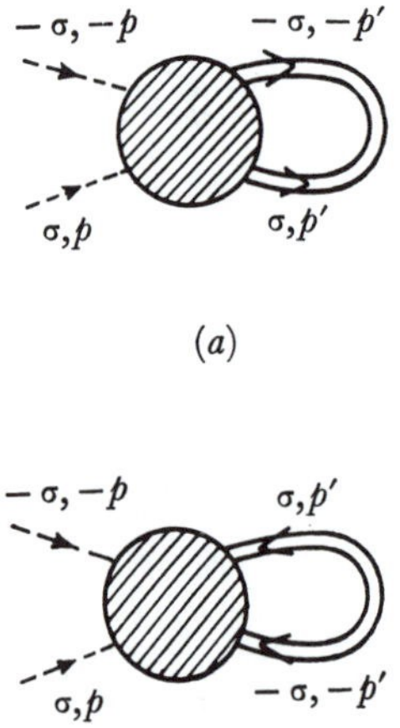

(*a*)

(*b*)

Figure 16

skeleton diagram, the shaded core is necessarily irreducible. If we sum over all possible cores, each diagram 15 is obtained $n + 1$ times, once for each line •→——←• selected. If we carry out the same operation on the diagrams 16*b*, each diagram 15 appears n times. Taking the difference of the diagrams 16*a* and 16*b*, we shall thus reproduce $-\sigma M_2$. We thus obtain an integral equation for M_2.

Let us consider the diagram 16*a*. The shaded core, to within a sign, is a diagram of ${}^0I^{--}_{++}(p,p';0)$; the diagram being irreducible, the passage to the limit $\varpi \to 0$ poses no problem. Similarly, Fig. 16*b* involves the core ${}^0I^{++}_{++}(p,p';0)$. In order to abbreviate the writing, we suppress the index $\varpi = 0$ in these expressions.

Let us now look closer at the question of signs. Let us draw the diagram 16*a*, for example, so that the incoming line follows the path indicated in Fig. 17, from A to D (in conformity with the rules for calculation of $-\sigma M_2$). The external line contributes a factor $-\sigma F_2(p')$. As for the core, it differs from ${}^0I$ because the section CD is traversed in the wrong direction. In the case of Fig. 17 this is of no importance, since the section DC involves an *even* number of lines F_2. On the other hand, for the diagrams 16*b*, this inversion of CD introduces a minus sign [the line p' giving a factor $\sigma F_2(p')$].

This discussion allows us to write the equation

$$M_2(p) = \sum_{p'} \{ {}^0I^{--}_{++}(p,p') - {}^0I^{++}_{++}(p,p') \} F_2(p') \tag{7–72}$$

Inverting the two arrows in Fig. 15, we can similarly show that

$$M_2(p) = \sum_{p'} \{ {}^0I^{++}_{--}(p,p') - {}^0I^{--}_{--}(p,p') \} F_2(p') \tag{7–73}$$

F_2 is given by (7–53); we obtain a nonlinear integral equation, allowing us to calculate M_2. (7–73) is just the *dynamical* equation determining the gap.

In general, the solution of (7–73) is difficult. We can simplify it when the gap is very small compared with the Fermi energy (weak

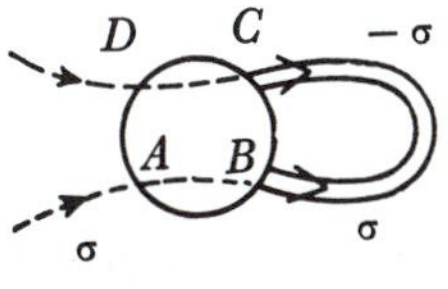

Figure 17

coupling). The factors $F_2(k')$ are then important only when k' is in the immediate neighborhood of the Fermi surface, in a region of width M_2/v_F; the sum over k' thus extends over a narrow region. We are interested in keeping the number of factors F_2 as small as possible; we thus replace ${}^0I$ and M_1 by their values for a normal gas, in the absence of a condensed phase,

$$\begin{cases} {}^0I^{++}_{--}(p, p') = {}^0I(p, p') \\ {}^0I^{--}_{--}(p, p') = 0 \end{cases} \tag{7-74}$$

In this approximation (7–73) reduces to

$$M_2(p) = \sum_{p'} {}^0I(p, p')\, F_2(p') \tag{7-75}$$

In order to go over to (7–55), we must replace ${}^0I$ by the single first-order diagram; this second approximation is not justified. When the gap is important, (7–75) is no longer valid; we must return to the general equation (7–73).

To conclude, we shall relate (7–72) and (7–73) to the Bethe-Salpeter equation (7–70). Let us introduce a "vector,"

$$V_{\rho_1\rho_2}(p) = \rho_1\, M_2(p)\, \delta_{\rho_1\rho_2} \tag{7-76}$$

Using (7–68) and (7–53), we easily verify that

$$\sum_{\rho_3\rho_4} {}_{+}G^{\rho_3}_{\rho_1}(p)\; {}_{-}G^{\rho_4}_{\rho_2}(-p)\; V_{\rho_3\rho_4}(p) = -\,\rho_1 F_2(p)\, \delta_{\rho_1\rho_2} \tag{7-77}$$

Furthermore, by using (7–61) and the symmetry of ${}^0I$ when $\varpi = 0$ on the one hand, (4–129*b*) on the other, we can prove that

$$\begin{cases} {}^0I^{++}_{+-}(p, p') = -\,{}^0I^{++}_{-+}(-p, -p') = {}^0I^{--}_{+-}(p, p') \\ {}^0I^{++}_{-+}(p, p') = \cdots = {}^0I^{--}_{-+}(p, p') \end{cases} \tag{7-78}$$

By making use of (7–77) and (7–78), we can put Eqs. (7–72) and (7–73) into the form

$$V_{\rho_1\rho_2}(p) = \sum_{\rho_5,\rho_6,\rho_7,\rho_8} \sum_{p'} {}^0I^{\rho_5\rho_6}_{\rho_1\rho_2}(p, p')\; {}_{+}G^{\rho_7}_{\rho_5}(p')\; {}_{-}G^{\rho_8}_{\rho_6}(-p')\; V_{\rho_7\rho_8}(p') \tag{7-79}$$

$V_{\rho_1\rho_2}(p)$ is thus a solution of the *homogeneous* Bethe-Salpeter equation (7–70) (with the term ${}^0I$ removed) in the case $\varpi = 0$.

This remark contains much information. First, we know that the solutions of the homogeneous Bethe-Salpeter equation correspond to the bound states of the system; we thus explicitly see the appearance of the bound pairs whose existence we have assumed for a long time. On the other hand, the existence of a solution of the homogeneous equation implies a singularity of ${}^0\tilde{\Gamma}(p,p';\varpi)$ in the neighborhood of $\varpi = 0$; ${}^0\tilde{\Gamma}$ cannot be expanded in powers of ϖ. This singularity makes possible the existence of different limits according to the direction of ϖ with respect to p and p'. In particular, the current correlation functions can depend on the direction of the wave vector q. While retaining gauge invariance for longitudinal fields, we can have a Meissner effect for transverse fields. All this would be impossible if $\tilde{\Gamma}$ were an analytic function of ϖ in the neighborhood of $\varpi = 0$. The existence of this singularity is therefore essential.

As in Chap. 6, the interaction operator ${}^0\tilde{\Gamma}$ is singular when $\varpi \to 0$. But the reasons for this singularity are completely different. For a normal gas, it arises from the discontinuity of the distribution function at the Fermi surface, the product of two factors G being singular. For a superfluid system, all the factors in (7–70) are well-behaved when $\varpi \to 0$, but the *equation* is singular, the associated homogeneous equation having a nonzero solution.

The development of these results seems promising. Unfortunately, it leads to difficult mathematical problems, and the work remains to be done.

Appendix A

Different Forms of the Correlation Function

We have already emphasized the close relationship between S and the dielectric constant ε. We shall now study the question in more detail.

The correlation function $S(q,t)$ is defined by (2–51),

$$S(q, t) = \frac{1}{\Omega} < \varphi_0 \mid \rho_q(t) \; \rho_{-q} \mid \varphi_0 > \tag{A–1}$$

Let us introduce two other functions, $\bar{\bar{S}}(q,t)$ and $\bar{S}(q,t)$, defined by

$$\bar{\bar{S}}(q, t) = \begin{cases} 0 & \text{for } t < 0 \\ 1/\Omega < \varphi_0 \mid [\rho_q(t), \rho_{-q}] \mid \varphi_0 > & \text{for } t > 0 \end{cases} \tag{A–2}$$

$$\bar{S}(q, t) = \begin{cases} 1/\Omega < \varphi_0 \mid \rho_{-q} \; \rho_q(t) \mid \varphi_0 > & \text{for } t < 0 \\ 1/\Omega < \varphi_0 \mid \rho_q(t) \; \rho_{-q} \mid \varphi_0 > & \text{for } t > 0 \end{cases} \tag{A–3}$$

The Fourier transform $\bar{\bar{S}}(q,\omega)$ is closely connected to the dielectric constant. Comparing (2–10) with (A–2), we see that $\bar{\bar{S}}$ is just the response function

$$\frac{i}{\Omega} \varphi_{\rho_q, \rho_{-q}}(t)$$

Consequently,

$$\frac{1}{\varepsilon(q, \omega)} - 1 = -\frac{4i\pi e^2}{q^2} \bar{\bar{S}}(q, \omega) \tag{A–4}$$

The function $\bar{S}(q,t)$ will, on the other hand, appear naturally in Chap. 3, in the discussion of Green's functions.

We now define the functions $S_{\pm}(q,t)$,

$$
\begin{aligned}
S_+(q, t) &= \begin{cases} S(q, t) & \text{if } t > 0 \\ 0 & \text{if } t < 0 \end{cases} \\
S_-(q, t) &= \begin{cases} 0 & \text{if } t > 0 \\ S(q, t) & \text{if } t < 0 \end{cases}
\end{aligned}
\tag{A-5}
$$

We see at once that

$$
\begin{cases}
S(q, t) = S_+(q, t) + S_-(q, t) \\
\bar{\bar{S}}(q, t) = S_+(q, t) - S_-(q, -t) \\
\bar{S}(q, t) = S_+(q, t) + S_+(q, -t)
\end{cases}
\tag{A-6}
$$

which leads to

$$
\bar{S}(q, t) = \bar{\bar{S}}(q, t) + S(q, -t) \tag{A-7}
$$

Now consider the Fourier transforms with respect to t. We know that $S(q,\omega)$ is a real function which is zero when $\omega < 0$. We can easily check that

$$
\begin{aligned}
S_+(q, \omega) &= -\frac{1}{2i\pi}\int_{-\infty}^{+\infty} d\omega' \, \frac{S(q, \omega')}{\omega - \omega' + i\eta} \\
S_-(q, \omega) &= \frac{1}{2i\pi}\int_{-\infty}^{+\infty} d\omega' \, \frac{S(q, \omega')}{\omega - \omega' - i\eta}
\end{aligned}
\tag{A-8}
$$

(where η is a positive infinitesimal). Using (A–6) and (A–8), we finally obtain

$$
\begin{cases}
\bar{\bar{S}}(q, \omega) = -\dfrac{1}{2i\pi}\displaystyle\int_{-\infty}^{+\infty} d\omega' \, S(q, \omega') \left\{ \frac{1}{\omega - \omega' + i\eta} - \frac{1}{\omega + \omega' + i\eta} \right\} \\
\bar{S}(q, \omega) = -\dfrac{1}{2i\pi}\displaystyle\int_{-\infty}^{+\infty} d\omega' \, S(q, \omega') \left\{ \frac{1}{\omega - \omega' + i\eta} - \frac{1}{\omega + \omega' - i\eta} \right\}
\end{cases}
\tag{A-9}
$$

Comparing this with (2–47) and (A–4), we again find (2–37). Our treatment is thus consistent.

$\bar{\bar{S}}(q,\omega)$ and $\bar{S}(q,\omega)$ are very similar. Their imaginary parts are the same; their real parts are equal when $\omega > 0$, opposite when $\omega < 0$. They both arise from the *real* function $S(q,\omega)$, defined on the real positive axis, which contains *all* the available physical information. The introduction of $\bar{\bar{S}}$ and $\bar{S}$ is just a convenient mathematical trick, with no physical content whatsoever. Let us point out that these relations can be generalized to nonzero temperatures. The results are slightly more complicated, since $S(q,\omega)$ now extends onto the negative real axis.

Appendix B

Second Quantization

Let us consider a collection of N particles, contained in a box of volume Ω, on which we impose periodic boundary conditions. To describe the state of such a system, we ordinarily use the wave function $\psi(r_1 \ldots r_N)$, depending on the $3N$ coordinates of the particles in "configuration" space. Such a representation allows a natural generalization of methods applied to a single particle. It is not the only possible representation; there are others, much more practical.

Let us consider a complete set of *single-particle states*—for example, plane waves of wave vector k (we ignore spin for the moment). One can specify the state of a system by indicating the number n_k of particles found in the "box" k. We are thus led to choose as a basis of the space of *states of the total system* the vectors

$$|n_1, \ldots, n_k, \ldots >$$

where each of the n_k can take on any positive integral value. (Note that the total number of particles is no longer conserved.) Such a state is simply a product (eventually symmetrized or antisymmetrized) of plane waves, one for each particle.

We can define a "creation" operator a_k^* by the relation

$$a_k^* \,|\ldots n_k \ldots > = \sqrt{n_k + 1}\,|\ldots n_k + 1 \ldots > \qquad \textit{(B-1)}$$

In the present representation, known as the "occupation-number" representation, a_k^* has a single nonzero matrix element. Its complex conjugate, or "destruction" operator, is defined by

$$a_k \,|\, \ldots n_k \ldots > \; = \sqrt{n_k} \,|\, \ldots n_k - 1 \ldots > \tag{B–2}$$

Our objective is to express all the properties of the system in terms of these creation and destruction operators.

Let us first take a boson gas, and let us consider the product

$$\overset{*}{a_k} \, \overset{*}{a_{k'}} \,|>$$

We have created a particle k', then another, k. The same result would be obtained by inverting the order, as the total wave function is invariant with respect to permutation of two particles. Hence we have the relation

$$[\overset{*}{a_k}, \overset{*}{a_{k'}}] = 0 \tag{B–3}$$

By the same method it is easily shown that

$$[a_k, a_{k'}] = 0 \,, \qquad [a_k, \overset{*}{a_{k'}}] = \delta_{kk'} \tag{B–4}$$

The operator

$$N_k = \overset{*}{a_k} a_k \tag{B–5}$$

measures the number of particles in the state k (since $N_k | \ldots n_k \ldots \rangle = n_k | \ldots n_k \ldots \rangle$).

Let us turn now to the fermion gas: the interchange of two particles must produce a change of sign. The commutation relations become anticommutation relations,

$$[\overset{*}{a_k}, \overset{*}{a_{k'}}]_+ = [a_k, a_{k'}]_+ = 0 \,, \qquad [\overset{*}{a_k}, a_{k'}]_+ = \delta_{kk'} \tag{B–6}$$

It can be seen from (B–6) that $a_k{}^2 = 0$; a state can accommodate only a single particle. Moreover, the occupation-number operator

$$N_k = \overset{*}{a_k} a_k$$

is such that $N_k{}^2 = N_k$ [as can be verified with the help of Eqs. (B–6)]. Therefore, N_k can equal only 0 or 1.

Actually, fermions have a nonzero spin. Each state is characterized by a wave vector k and a spin index σ. The creation and destruction operators thus become $a^*_{k\sigma}$ and $a_{k\sigma}$. They satisfy anticommutation relations similar to (B–6),

$$\begin{aligned} &[a_{k\sigma}, a_{k'\sigma'}]_+ = [a^*_{k\sigma}, a^*_{k'\sigma'}]_+ = 0 \\ &[a^*_{k\sigma}, a_{k'\sigma'}]_+ = \delta_{kk'}\,\delta_{\sigma\sigma'} \end{aligned} \tag{B–6a}$$

The normalized wave function of a plane wave of wave vector k is

$$\frac{1}{\sqrt{\Omega}}\,\mathrm{e}^{ikr}$$

The probability amplitude for the destruction of the particle k taking place at the point r is thus represented by the operator

$$\frac{1}{\sqrt{\Omega}}\,\mathrm{e}^{ikr}\,a_k$$

This leads us to introduce the destruction operator at the point r, $\psi(r)$, defined by

$$\begin{cases} \psi(r) = \dfrac{1}{\sqrt{\Omega}} \displaystyle\sum_k a_k\,\mathrm{e}^{ikr} \\ a_k = \dfrac{1}{\sqrt{\Omega}} \displaystyle\int \mathrm{d}^3r\,\psi(r)\,\mathrm{e}^{-ikr} \end{cases} \tag{B–7}$$

$\psi(r)$ describes the destruction at r of a particle of any momentum whatever. Similarly, the creation operator $\psi^*(r)$ is defined by

$$\psi^*(r) = \frac{1}{\sqrt{\Omega}} \sum_k a^*_k\,\mathrm{e}^{-ikr}$$

The commutation relations which ψ and ψ^* satisfy are easily deduced from (B–3) and (B–4) or from (B–6). They can be written

$$\begin{aligned} &\textit{Bosons} \quad \begin{cases} [\psi(r), \psi(r')] = [\psi^*(r), \psi^*(r')] = 0 \\ [\psi(r), \psi^*(r')] = \delta(r - r') \end{cases} \\ &\textit{Fermions} \quad \begin{cases} [\psi_\sigma(r), \psi_{\sigma'}(r')]_+ = [\psi^*_\sigma(r), \psi^*_{\sigma'}(r')]_+ = 0 \\ [\psi_\sigma(r), \psi^*_{\sigma'}(r')]_+ = \delta(r - r')\,\delta_{\sigma,\sigma'} \end{cases} \end{aligned} \tag{B–8}$$

In the case of fermions we have explicitly introduced spin coordinates. It is important not to confuse the *operators* ψ, ψ^* with the state *vectors* $|\varphi\rangle$.

For the sake of simplicity, we shall consider only Fermi gases. Just as the operator $N_k = a_k^* a_k$ represents the number of particles in the state k, the operator

$$\rho(r, \sigma) = \psi_\sigma^*(r)\, \psi_\sigma(r) \tag{B–9}$$

represents the density of particles at the point r. The total number of particles in the system is given by

$$N = \sum_\sigma \int d^3r\, \psi_\sigma^*(r)\, \psi_\sigma(r) = \sum_{k\sigma} a_{k\sigma}^* a_{k\sigma} \tag{B–10}$$

These results are extended without difficulty to the study of other physical properties of the system. For example, the kinetic energy

$$T = \sum_i \frac{p_i^2}{2m}$$

can be written in the form

$$T = \sum_{k,\sigma} \frac{\hbar^2 k^2}{2m} N_{k\sigma} = \sum_{k,\sigma} \frac{\hbar^2 k^2}{2m} a_{k\sigma}^* a_{k\sigma} \tag{B–11}$$

By a Fourier transformation, we can go back to ψ, ψ^*. (B–11) then becomes

$$T = -\frac{\hbar^2}{2m} \sum_\sigma \int \psi_\sigma^*(r)\, \nabla^2 \psi_\sigma(r)\, d^3r = \frac{\hbar^2}{2m} \sum_\sigma \int \nabla\psi_\sigma^*(r) \cdot \nabla\psi_\sigma(r)\, d^3r \tag{B–12}$$

Let us turn now to the study of the potential energy V, which we assume results from a binary interaction, depending only on position,

$$V = \tfrac{1}{2} \sum_{i \neq j} V(r_i - r_j)$$

We can rewrite this interaction in the form

$$\begin{cases} V = \tfrac{1}{2} \iint d^3r\, d^3r'\, V(r - r') \left[\rho(r)\, \rho(r') - \rho(r)\, \delta(r - r')\right] \\ \rho(r) = \sum_\sigma \rho(r, \sigma) \end{cases} \tag{B–13}$$

where the second term eliminates the interaction of a charge with itself. By using (B–9) and the commutation relations, we can transform (B–13) into

$$V = \tfrac{1}{2} \sum_{\sigma,\sigma'} \iint d^3r\, d^3r'\, V(r - r')\, \psi^*_\sigma(r)\, \psi^*_{\sigma'}(r')\, \psi_{\sigma'}(r')\, \psi_\sigma(r) \qquad \textit{(B–14)}$$

In general we shall prefer an expression using the operators a_k rather than the operators $\psi(r)$. Let us introduce the Fourier transform of the binary potential,

$$V_q = \int d^3r\, e^{-iqr}\, V(r) \qquad \textit{(B–15)}$$

(B–14) can then be written as

$$V = \frac{1}{2\Omega} \sum_{\substack{kk'q \\ \sigma,\sigma'}} V_q\, a^*_{k+q,\sigma}\, a^*_{k'-q,\sigma'}\, a_{k',\sigma'}\, a_{k,\sigma} \qquad \textit{(B–16)}$$

By combining (B–11) and (B–16), we obtain the Hamiltonian operator

$$H = T + V$$

If we go into the Heisenberg representation, the dynamics of any operator A are governed by the equation

$$i\hbar \frac{d\mathbf{A}}{dt} = [H, \mathbf{A}] \qquad \textit{(B–17)}$$

This relation applies in particular to the creation and destruction operators.

It is easy to calculate in the same way the expression for any physical quantity whatever in this representation (current, etc.). For example, let us consider the Fourier component of the density,

$$\rho_q = \int d^3r\, \rho(r)\, e^{-iqr} \qquad \textit{(B–18)}$$

Because of (B–7) and (B–9), this can be written as

$$\rho_q = \sum_{k,\sigma} a^*_{k,\sigma}\, a_{k+q,\sigma} \tag{B–19}$$

To conclude, let us emphasize that we have discovered nothing new in this appendix: we have just chosen a particularly convenient representation for treating the problems of interest to us.

Appendix C

Some Properties of the Single-Particle Green's Function

Let us again take the Lehmann representation of the Green's function (3–23),

$$G(k, \omega) = \int_0^\infty d\omega' \left\{ \frac{A_+(k, \omega')}{\omega' - \omega + \mu - i\eta} - \frac{A_-(k, \omega')}{\omega' + \omega - \mu - i\eta} \right\} \qquad (C\text{-}1)$$

By using the standard expression

$$\frac{1}{x + i\eta} = P(1/x) - i\pi\,\delta(x)$$

we can separate the real and imaginary parts of G (for real ω),

$$\operatorname{Re} G = \int_0^\infty d\omega' \left\{ A_+(k, \omega')\, P\left(\frac{1}{\omega' - \omega + \mu}\right) - A_-(k, \omega')\, P\left(\frac{1}{\omega' + \omega - \mu}\right) \right\} \qquad (C\text{-}2)$$

$$\operatorname{Im} G = \begin{cases} i\pi\, A_+(k, \omega - \mu) & \text{if} \quad \omega > \mu \\ -\, i\pi\, A_-(k, \mu - \omega) & \text{if} \quad \omega < \mu \end{cases}$$

The imaginary part of G is thus simply proportional to A_+ and A_-.

Equation (C–1) allows us to continue the function $G(k, \omega)$ into the complex plane. Let us put $\omega = \omega_1 + i\omega_2$: if ω_2 is different from zero, we can neglect the terms $i\eta$ in the denominator. Let us remark at once that, since A_+ and A_- are real functions, we have

$$G(k, \omega^*) = [G(k, \omega)]^* \tag{C-3}$$

(C–3) is valid in the whole complex plane, *except* for the real axis. On the other hand, let us calculate the imaginary part of G,

$$\operatorname{Im} G(k, \omega) = \int_0^\infty d\omega' \, i\omega_2 \left\{ \frac{A_+(k,\omega')}{(\omega' - \omega_1 + \mu)^2 + \omega_2^2} + \frac{A_-(k,\omega')}{(\omega' + \omega_1 - \mu)^2 + \omega_2^2} \right\} \tag{C-4}$$

The quantity within braces is always positive, and it will not always be zero. As a consequence

$$\operatorname{Im} G \neq 0 \qquad \text{if} \qquad \operatorname{Im} \omega \neq 0 \tag{C-5}$$

This important conclusion gives the assurance that G has no zeros in the complex plane. The only zeros possible must be found on the real axis. The singularities of $1/G$ are therefore all on the real axis.

Let us turn now to the analytic properties of G. By a theorem due to Cauchy, the integral (C–1) is analytic with respect to ω, except for values of ω which make the integrand infinite at some point in the interval $0 < \omega' < \infty$. There are then two cases to consider:

(a) The functions A_+ and A_- are discrete, that is to say, are made up of a certain number of δ functions. In a noninteracting gas, for example, the set (A_+, A_-) is reduced to a single δ function, corresponding to $\omega' = |\overset{\circ}{\varepsilon}_k - \mu|$. The calculation of G is then trivial, and we obtain the expression (3–18). We see that G has only a single pole, for $\omega = \overset{\circ}{\varepsilon}_k$ If the set (A_+, A_-) is composed of several δ functions, G has several discrete poles, well separated.

(b) Actually, as we have seen, such a structure is known to arise only from an idealized approximation. In reality, A_+ and A_- are continuous functions, which in general extend from $\omega' = 0$ to $\omega' = \infty$. Under these conditions G is singular when

$$\begin{cases} \omega = \mu + \omega' - i\eta \\ \omega = \mu - \omega' + i\eta \end{cases} \qquad \omega' > 0$$

This singularity can be represented as a continuous series of poles, which gives rise to a *branch cut*, separating two different domains of G. G is *discontinuous* across this branch cut, which is represented in Fig. C–1. The point $\omega = \mu$ plays the part of a branch point; we expect G to have a

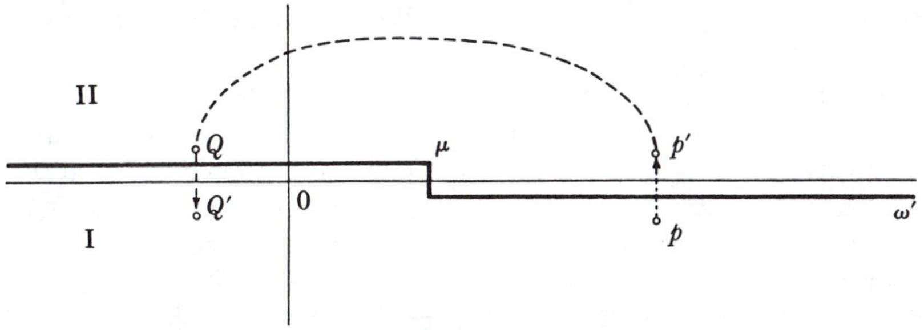

Figure 1

logarithmic singularity in the neighborhood of this point. On each side of this cut G is analytic in the whole half-plane, corresponding to a function G_{I} in domain I, G_{II} in domain II. For $Im\omega \neq 0$, G_{I} and G_{II} are related by (C–3). The value of the discontinuity across the cut follows from (C–2) and (C–3),

$$G_{\mathrm{II}} - G_{\mathrm{I}} = \begin{cases} 2 \operatorname{Im} G = 2\, i\pi\, A_+(k, \omega - \mu) & \text{if } \omega > \mu \\ -2 \operatorname{Im} G = 2i\pi\, A_-(k, -\omega + \mu) & \text{if } \omega < \mu \end{cases} \tag{C-6}$$

Note that the real part of G is continuous across the cut.

Let us now try to continue analytically G_{I} and G_{II} beyond the cut; we come to a new definition of the function G, corresponding to another Riemann sheet. For example, let us continue G_{I} from the point P to the point P', in region II (see Fig. C–1): we go over to a new definition G' of G, given by (C–6),

$$G' = G - 2i\pi\, A_+(k, \omega - \mu) \tag{C-7}$$

We can go farther and continue G' from P' to Q: we always remain on the same Riemann sheet. If, on the contrary, we go on to Q', we again change definition upon crossing the cut and we go over to

$$G'' = G - 2i\pi\, A_+(k, \omega - \mu) + 2i\pi\, A_-(k, \mu - \omega)$$

If we had gone directly from P to Q' by passing through region I, we would have retained the definition G instead of going over to G''. As a consequence, the analytic continuation of G_{I} generates the family of functions

$$\begin{cases} G_I + 2i\pi\, pJ \\ J(k, \omega) = A_+(k, \omega - \mu) - A_-(k, \mu - \omega) \end{cases} \tag{C-8}$$

where p is any integer. Similarly, the analytic continuation of G_{II} generates the family $G_{II} + 2\pi i p J$.

We have seen that the physically defined G was separately analytic in domains I and II. If we turn to other definitions of G, the functions obtained are no longer analytic: they can have poles which, according to (C–8), are automatically those of A_+ and A_-. In particular, if we carry out this continuation by making the real axis turn slightly about the point μ, in a clockwise sense (Fig. C–2), we shall obtain poles of G in the shaded region, the right half corresponding to poles of A_+ and the left half to poles of A_-. Of course, G has other singularities, but the latter are located on sheets further removed from the "physical" sheet.

In order to be able fully to utilize these results concerning the analytic structure of G, we need to know the asymptotic form of G when $\omega \to \infty$. By going back to (C–1), we see that

$$G(k,\omega) \underset{\omega\to\infty}{\to} -\frac{1}{\omega}\int_0^\infty d\omega'[A_+(k,\omega') + A_-(k,\omega')] \tag{C-9}$$

But from (3–21)

$$\begin{cases} \displaystyle\int_0^\infty A_+(k,\omega')\,d\omega' = \sum_n \left| < \varphi_n \mid a_{\mathbf{k}}^* \mid \varphi_0 > \right|^2 = < \varphi_0 \mid a_{\mathbf{k}} a_{\mathbf{k}}^* \mid \varphi_0 > \\ \qquad = (1 - m_k) \\ \displaystyle\int_0^\infty A_-(k,\omega')\,d\omega' = \sum_n \left| < \varphi_n \mid a_{\mathbf{k}} \mid \varphi_0 > \right|^2 = < \varphi_0 \mid a_{\mathbf{k}}^* a_{\mathbf{k}} \mid \varphi_0 > = m_k \end{cases} \tag{C-10}$$

Hence

$$\int_0^\infty (A_+ + A_-)\,d\omega' = 1 \tag{C-11}$$

This important result arises from the commutation relations and just expresses the discontinuity of $G(k,t)$ at $t = 0$. (Its physical origin is simple: the sum of the norms of $a_{\mathbf{k}}|\varphi_0\rangle$ and $a_{\mathbf{k}}^*|\varphi_0\rangle$ must be 1, because the particle $\mathbf{k}$ can only be created or destroyed.) By putting (C–11) into (C–9), we see that

$$G(k,\omega) \to -1/\omega \quad \text{if } \omega \to \infty \tag{C-12}$$

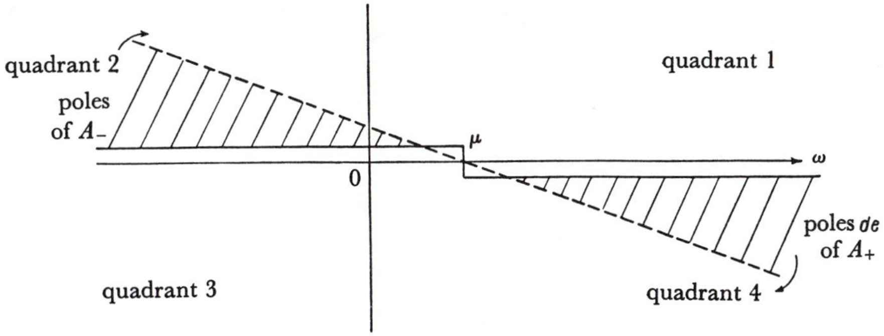

Figure 2

The asymptotic form of G does not depend on the dynamics of the system.

Let us consider the integral

$$\int_C G(k, \omega)\, d\omega$$

the contour C being indicated in Fig. C–3. From (3–16) we see that this integral is equal to

$$2\pi[G(k, t)]_{t=-0} = -2i\pi m_k$$

Similarly, the equivalent integral taken over the contour C' is equal to

$$2\pi[G(k, t)]_{t=+0} = 2i\pi(1 - m_k)$$

(C–12) allows us to evaluate the contributions from the infinite semi-circles, which leads us to

$$\int_{-\infty}^{+\infty} G(k, \omega)\, d\omega = i\pi(1 - 2m_k) \qquad \textit{(C–13)}$$

(C–13) is an example of a relation arising from the structure of G and from its asymptotic form (C–12).

We can pursue the study of the asymptotic form further and calculate the terms of order $1/\omega^2$. From (C–1) we obtain

$$G(k, \omega) + 1/\omega \underset{\omega\to\infty}{\to} (1/\omega^2) \int_0^\infty d\omega' \{ A_-(k, \omega') [\omega' - \mu] - A_+(k, \omega') [\omega' + \mu] \} \qquad \textit{(C–14)}$$

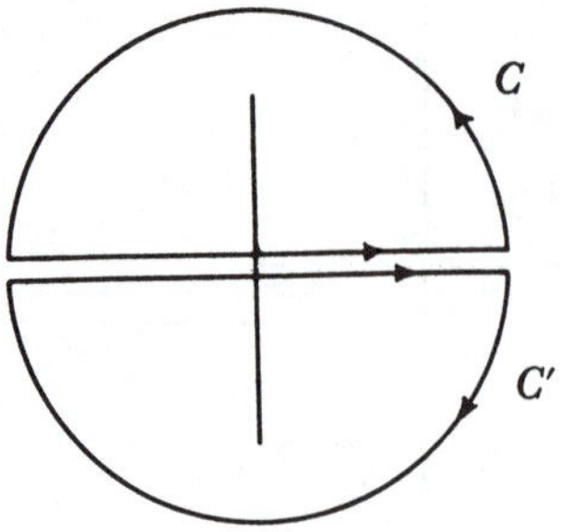

Figure 3

By returning to the definition (3–21) of the functions $A_{\pm}$, it is easy to show that

$$G(k, \omega) + 1/\omega \rightarrow -(1/\omega^2)\{<\varphi_0 \mid [a_{\mathbf{k}}, [H, a^*_{\mathbf{k}}]]_+ \mid \varphi_0>\} \qquad \textit{(C–15)}$$

It can be verified that the expression between braces is just the Hartree-Fock energy of a *bare* particle of wave vector **k** (that is to say, kinetic energy plus an average potential). This result, a priori quite surprising, becomes understandable if it is noted that the term of G in $1/\omega^2$ is intimately related to the discontinuity of $[dG(k,t)/dt]$ for $t = 0$. In fact we have, from (3–19),

$$\left(\frac{dG}{dt}\right)_{+0} - \left(\frac{dG}{dt}\right)_{-0} = <\varphi_0 \mid [a_{\mathbf{k}}, [H, a^*_{\mathbf{k}}]]_+ \mid \varphi_0> \qquad \textit{(C–16)}$$

Let us consider $G(t)$ for t a positive infinitesimal: we can write

$$G(k, t) \sim i(1 - m_k)\exp(-iE_k t) \qquad \textit{(C–17)}$$

where E_k is the *instantaneous* energy of the particle **k** which we have introduced at time $t = 0$. If t is small, the system does not have time to "respond" to the additional particle, and E_k is simply equal to the "static" energy of the particle k, that is to say, to its Hartree-Fock energy. Similarly, for t a negative infinitesimal, we have

$$G(k, t) \sim -im_k \exp[-i(-E_k)(-t)] \qquad \textit{(C–18)}$$

where $-E_k$ is the energy of the hole which we have introduced at time $-t$. By combining (C–17) and (C–18) we find

$$\left(\frac{\mathrm{d}G}{\mathrm{d}t}\right)_{+0} - \left(\frac{\mathrm{d}G}{\mathrm{d}t}\right)_{-0} = E_k \qquad (C\text{–}19)$$

which explains the result (C–16), and consequently (C–15). Although the *instantaneous* response to an additional particle is an interesting problem, we shall not explore this question.

We shall end this investigation of the properties of G with some important relations, direct consequences of the analytic structure. Let us define an operator X, acting on functions of ω (ω real) through the relation

$$Xf(\omega) = \begin{cases} f(\omega) & \text{if} \quad \omega > \mu \\ f^*(\omega) & \text{if} \quad \omega < \mu \end{cases} \qquad (C\text{–}20)$$

The function $XG(k,\omega)$ is then analytic in the upper half-plane, including the real axis (just like the dielectric constant in Chap. 2.) It is readily deduced from this that

$$\int_{-\infty}^{+\infty} XG(k, \omega')\, P\left(\frac{1}{\omega - \omega'}\right) \mathrm{d}\omega' = \mathrm{i}\pi\, X\, G(k, \omega) \qquad (C\text{–}21)$$

(C–21) is analogous to the Kramers-Kronig relations which are obtained from causality considerations. Similarly, one can establish that

$$\int_{-\infty}^{+\infty} XG(k, \omega')\, \mathrm{d}\omega' = \mathrm{i}\pi \qquad (C\text{–}22)$$

by making use of asymptotic form (C–12). [(C–22) does nothing but reexpress (C–11) and (C–13).]

In this appendix we have tried to outline the principal properties of G. The methods we have used allow us to establish others, if needed, and furnish us with a convenient "machinery". Let us emphasize that all the properties established are rigorous, valid for any system whatever (normal gas, superconductor, etc). In particular cases one can establish more restrictive properties.

Appendix D

The Analytic Properties of $K(\mathbf{k}_i, \omega_i)$

Let us consider, for example, the time ordering

$$t_1 > t_2 > t_3 > t_4$$

The corresponding part of $K(\mathbf{k}_i, \omega_i)$ is easily constructed,

$$K(\mathbf{k}_i, \omega_i) = \frac{i}{2\pi} \delta(\omega_1 + \omega_2 - \omega_3 - \omega_4) \times \sum_{mns} \frac{(a_{\mathbf{k}_1})_{0m} (a_{\mathbf{k}_2})_{ms} (a^*_{\mathbf{k}_3})_{sn} (a^*_{\mathbf{k}_4})_{n0}}{(-\omega_4 + \omega_{n0} - i\eta)(\omega_1 - \omega_{m0} + i\eta)(\omega_1 + \omega_2 - \omega_{s0} + i\eta)} \qquad \textit{(D-1)}$$

where η is a positive infinitesimal. In this expression, the states n and m contain $(N + 1)$ particles; the intermediate state s contains $(N + 2)$. Note that the δ function assures conservation of energy. Let us disregard this singular factor. It is then found that the singularities of K arise for

$$\begin{cases} \omega_4 = \omega_{n0} - i\eta \\ \omega_1 = \omega_{m0} - i\eta \\ \omega_1 + \omega_2 = \omega_{s0} - i\eta \end{cases} \qquad \textit{(D-2)}$$

ω_{n0} and ω_{m0} vary continuously: the corresponding singularities give rise to cuts across which $K(\mathbf{k}_i, \omega_i)$ is discontinuous. Similarly, ω_{s0} can take on a continuous set of values, corresponding to pairs of *independent* particles. This gives a continuous set of poles, which leads to a cut with respect to the variable $(\omega_1 + \omega_2)$.

In addition to these cuts, $K(\mathbf{k}_i, \omega_i)$ can possess poles. Those which involve separately ω_1 and ω_4 correspond to the energies of the quasi particles $\mathbf{k}_1$ and $\mathbf{k}_4$. It is in the hope of eliminating these singularities that we take out a factor $G_1G_2G_3G_4$ from δK. For a normal system, the interaction $\gamma(\mathbf{k}_i, \omega_i)$ which results from this no longer contains the singularities relative to a single quasi particle. On the other hand, the poles relative to the variable $(\omega_1 + \omega_2)$ always exist in $\gamma(\mathbf{k}_i, \omega_i)$ and $\Gamma(\mathbf{k}_i, \omega_i)$; it is these which give the bound states of two particles, already studied (see the second section of Chap. 3). In general the energies of these poles are complex, which leads to damping of the corresponding states.

To sum up, (D–1) exhibits:

(a) Cuts with respect to the variables ω_1 and ω_4.

(b) Poles with respect to these same variables, located at the energies of the corresponding quasi particles

(c) A cut with respect to $(\omega_1 + \omega_2)$ arising from the continuum of free pairs

(d) Possibly, poles with respect to $(\omega_1 + \omega_2)$, corresponding to bound pairs

The preceding discussion can be repeated for any order of the time variables. In this way the different terms of $K(\mathbf{k}_i, \omega_i)$ are generated. Unfortunately, there are 24 possible combinations, which makes the calculations very tedious. In order to express the result in compact form, let us consider the quantity

$$< \varphi_0 \mid \mathbf{A}^{\pm}(t_1)\, \mathbf{B}^{\pm}(t_2)\, \mathbf{C}^{\pm}(t_3)\, \mathbf{D}^{\pm}(t_4) \mid \varphi_0 > \qquad t_1 > t_2 > t_3 > t_4 \qquad \textit{(D–3)}$$

where the $\pm$ signs signify, respectively, destruction operator or creation operator. The Fourier transform of (D–3) is given by

$$\frac{i}{2\pi}\, \delta(\pm\, \omega_1 \pm \omega_2 \pm \omega_3 \pm \omega_4)$$
$$\times \frac{\sum_{mns} A^{\pm}_{0m}\, B^{\pm}_{ms}\, C^{\pm}_{sn}\, D^{\pm}_{n0}}{(\pm\, \omega_4 + \omega_{n0} - i\eta)\, (\pm\, \omega_1 - \omega_{m0} + i\eta)\, (\pm\, \omega_1 \pm \omega_2 - \omega_{s0} + i\eta)} \qquad \textit{(D–4)}$$

The general expression (D–4) allows for the generation of all the terms of $K(\mathbf{k}_i, \omega_i)$.

It is seen from (D–4) that the bound states of two particles give rise to poles for

$$\omega_1 + \omega_2 = \omega_{s0} - i\eta \qquad \textit{(D–5)}$$

whereas the states of two holes correspond to

$$\omega_1 + \omega_2 = -\,\omega_{s0} + i\eta \qquad \textit{(D-6)}$$

The states containing a particle and a hole give a whole set of singularities, located at

$$\begin{cases} \omega_1 - \omega_3 = \pm\,(\omega_{s0} - i\eta) \\ \omega_1 - \omega_4 = \pm\,(\omega_{s0} - i\eta) \end{cases} \qquad \textit{(D-7)}$$

Let us point out that, in the last case, the poles are grouped in pairs of opposite values. We thus confirm the results stated in Chap. 3.

These results remain valid if the bound states are subject to damping. The poles of K are then complex:

(a) The poles with respect to $(\omega_1 + \omega_2)$ located in the lower half plane correspond to pairs of particles.

(b) The poles of the same type located in the upper half plane correspond to pairs of holes.

(c) The poles with respect to $(\omega_1 + \omega_3)$ and $(\omega_1 - \omega_4)$ occur in pairs of opposite values and give states containing a particle and a hole.

Appendix E

Wick's Theorem

To establish Wick's theorem (5–30), we first prove the following lemma.

Lemma. Let us consider a *normal* product $U_1 \ldots U_{p-1}$, and let U_p be an operator *prior* to all the factors $U_1 \ldots U_{p-1}$. We have the relation

$$U_1 \ldots U_{p-1} U_p = N(U_1 \ldots U_{p-1} U_p) + \sum_{i=1}^{p-1} \lambda_i \, \widehat{U_i U_p} \, N_i(U_1 \ldots U_{p-1}) \qquad \textbf{(E-1)}$$

As in (5–30), $\lambda_i = (-1)^{p-i-1}$ has the sign of the permutation which brings U_i next to U_p. $N_i(U_1 \ldots U_{p-1})$ is the normal product which remains after removal of the factor U_i.

U_p can be of the type b_k or b_k^*; we treat these two cases separately.

(a) U_p is a destruction operator b_k. We then have

$$\widehat{U_i U_p} = 0 \qquad \textbf{(E-2)}$$

Furthermore, the product $U_1 \ldots U_p$ is normal, and the relation (E–1) is thus trivially satisfied.

(b) U_p is a creation operator b_k^*. We can easily verify that in this case

$$\widehat{U_i U_p} = [U_i, U_p]_+ \qquad \textbf{(E-3)}$$

Furthermore, the normal product $N(U_1 \ldots U_p)$ can be written as

$$N(U_1 \ldots U_p) = (-1)^{p-1} U_p U_1 \ldots U_{p-1} \tag{E-4}$$

In (E–4) let us displace U_p progressively toward the right. Each permutation introduces a correction proportional to a *scalar* anticommutator; we find

$$\begin{aligned} U_p U_1 \ldots U_{p-1} &= -U_1 U_p U_2 \ldots U_{p-1} + [U_p, U_1]_+ U_2 \ldots U_{p-1} \\ &= \ldots\ldots\ldots\ldots \\ &= (-1)^{p-1} U_1 \ldots U_{p-1} U_p + \sum_{i=1}^{p-1} [U_i, U_p]_+ (-1)^{i-1} N_i(U_1 \ldots U_{p-1}) \end{aligned} \tag{E-5}$$

Combining (E–3), (E–4), and (E–5), we recover the desired lemma (E–1).

Let us now turn to Wick's theorem. We shall prove it by an induction method. This theorem is obvious for a product of two factors, since (5–30) then reduces to the definition (5–27). Let us assume the theorem to be true for a product of $(p-1)$ factors, and consider the chronological product of p factors,

$$T(U_1 \ldots U_p)$$

Let U_m be the factor corresponding to the earliest time. According to the definition of the chronological product, we can write

$$\begin{aligned} T(U_1 \ldots U_{m-1} U_m U_{m+1} \ldots U_p) &= (-1)^{m-p}\, T(U_1 \ldots U_{m-1} U_{m+1} \ldots U_p U_m) \\ &= (-1)^{m-p}\, T_m(U_1 \ldots U_p)\, U_m \end{aligned} \tag{E-6}$$

where T_m is the chronological product without the factor U_m. Let us apply Wick's theorem to T_m—the theorem has been assumed exact for $(p-1)$ factors—and let us put the result back into (E–6). Each of the terms of this expansion is subject to the lemma (E–1). We thus prove Wick's theorem for the product of p factors,

$$T(U_1 \ldots U_{m-1} U_{m+1} \ldots U_p U_m)$$

The theorem remains true if we return U_m to its normal place; in fact, this permutation amounts to multiplying all the factors $\lambda_{i1 \ldots ip, j1 \ldots jp}$ appearing in (5–30) by $(-1)^{m-p}$. We have thus proved Wick's theorem in its most general form.

Appendix F

Some Properties of Diagrams involving Interaction Lines

We mentioned in Chap. 6 that a usual interaction *vertex* gives rise to two interaction diagrams, drawn in Fig. F–1. With diagram (a) there is associated for, example, the contribution (6–163); there is no sign correction, since line 4 continues to 1, just as in the original vertex. In diagram (b), on the contrary, line 4 continues to 2; this permutation changes by 1 the number of closed loops, and hence there is an additional minus sign. Adding these two contributions, we recover the matrix element (5–17).

While permitting a more detailed analysis, diagrams with interaction lines are clearly more cumbersome. In particular, it becomes difficult to evaluate the number of permutations which leave a given diagram invariant, and it is then preferable to return to the old diagrams of Chap. 5.

Let us consider a solid line of any diagram and follow it from interaction to interaction; it can form an open or closed path. In all cases, the spin remains constant along the whole line. The spins of two different

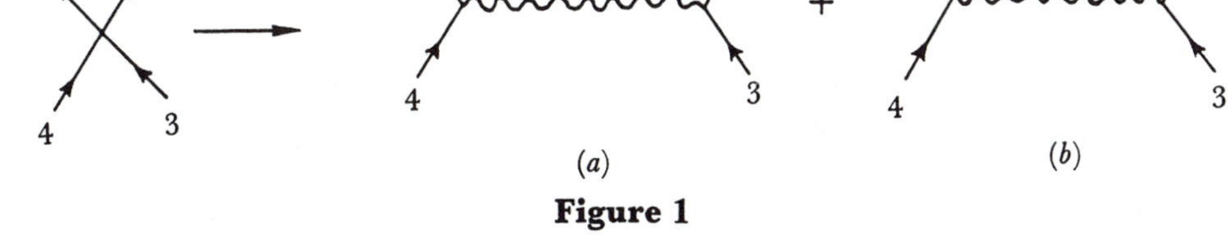

Figure 1

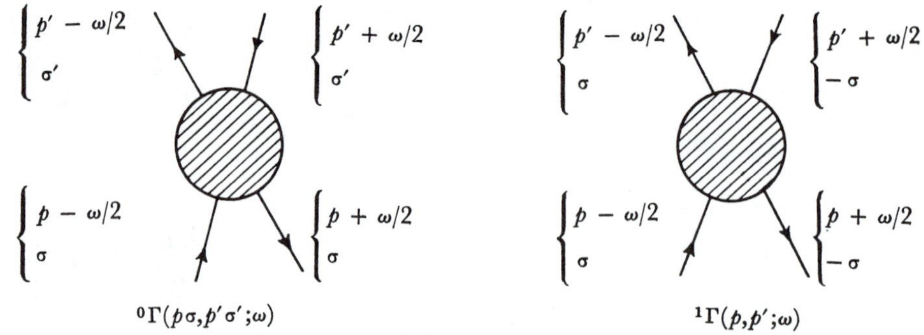

Figure 2

lines are completely independent and are not involved in the contribution from the diagram.

In the light of these remarks, let us consider the vertex operators $^{\circ}\Gamma(p\sigma,p'\sigma';\varpi)$ and $^{1}\Gamma(p,p';\varpi)$ defined in Eq. (6–16). The corresponding diagrams are indicated in Fig. F–2. The line $(p - \varpi/2)$ can end up as $(p' - \varpi/2)$ or as $(p + \varpi/2)$. These two types are shown in Fig. F–3. For each, the contribution from the diagram does not depend on the spin variables. Along a single line, the spin must remain constant; as a consequence:

$^{1}\Gamma$ contains only diagrams of type (a).

$^{\circ}\Gamma$ contains only diagrams of type (b) if $\sigma' = -\sigma$.

When $\sigma' = \sigma$, $^{\circ}\Gamma$ contains both diagrams (a) and (b).

We deduce from this

$$^{0}\Gamma(p\sigma, p'\sigma\,;\,\varpi) = {}^{1}\Gamma(p, p'\,;\,\varpi) + {}^{0}\Gamma(p\sigma, p'\,{-\sigma}\,;\,\varpi) \qquad \textit{(F–1)}$$

We thus prove the relation (6–22), ensuring the equivalence of the different orientations of the triplet state.

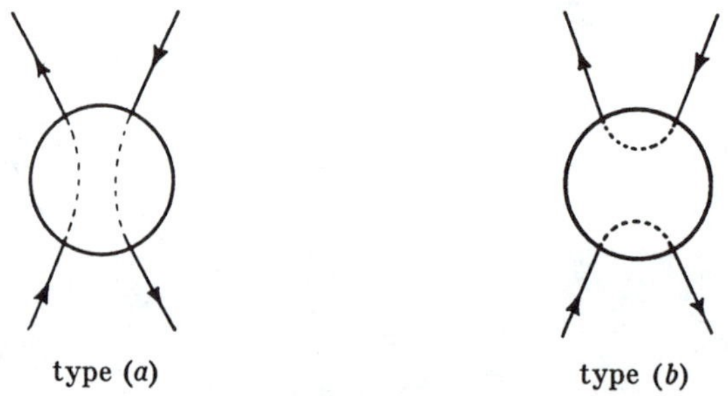

Figure 3

Notation

General Conventions

r	space coordinate
k	particle wave vector
t	time
ω	particle energy
σ	spin (= 1)
χ	(r, t)
p	(k, ω)
$\mathbf{x}, \mathbf{r}$	$(x, \sigma), (r, \sigma)$
$\mathbf{p}, \mathbf{k}$	$(p, \sigma), (k, \sigma)$
q	total wave vector of a particle hole pair
ϵ	total energy of a particle hole pair
ϖ	(q, ϵ)
α, ρ	indices indicating destruction ($\alpha = +1$) and creation ($\alpha = -1$) operators

Representations

$\|\varphi\rangle$, $\mathbf{A}$	Heisenberg representation
$\|\varphi_S\rangle$, A_S	Schrödinger representation
$\|\varphi\rangle$, A	interaction representation (Chapters 5 to 7)

In Chapter 7, the Heisenberg and interaction representations are defined with the Hamiltonian H replaced by the free energy $(H - \mu N)$

Main Notations

(The numbers refer to pages on which the various quantities are defined.)

$a_{\mathbf{k}}, a_{\mathbf{k}}^*$ destruction and creation operators for a bare particle with wave vector $\mathbf{k}$ 339

$A_{\mathbf{k}}, A_{\mathbf{k}}^{*}$	destruction and creation operators for a quasi particle with wave vector **k** 99
$A_{\pm}(k, \omega)$	spectral densities of the Green's function $G(k, \omega)$ 66
$A(\mathbf{k}, \mathbf{k}')$	forward scattering amplitude of two quasi particles on the Fermi surface 264
$b_{\mathbf{k}}, b_{\mathbf{k}}^{*}$	destruction and creation operators for an elementary excitation of the noninteracting gas 152
$d\gamma$	solid angle differential element
E_n, E°	energies of the eigenstates for the real and unperturbed systems, respectively 146
$f(\mathbf{k}, \mathbf{k}')$	interaction energy of two quasi particles 5
$F_1(p), F_2(p)$	components of the single-particle Green's function in a superfluid gas 116
g	coupling constant 171
$G(p), G_0(p)$	single-particle Green's function for the real and unperturbed system, respectively 59
$\mathscr{G}(p),\ g(p),\ \tilde{g}(p)$	single-quasi particle and mixed Green's function 105
$\bar{G}, \hat{G}, \check{G}$	single-particle Green's functions for a superfluid system 113
$\tilde{\mathbf{G}},(p)$	matrix form of the Green's function, appearing in a perturbation treatment 120
H	Hamiltonian 343
H_0, H_1	unperturbed Hamiltonian, perturbation 146
$\mathscr{H}_1$	coupling Hamiltonian with an external field 250
${}^0I, {}^1I$	irreducible particle-hole interaction operators, for total spin $\sigma_z = 0$ or 1 245
J_q	Fourier component of the total current density 251
J_k	current carried by a quasi particle **k** 12
k_F	Fermi wave vector 1
$K(\mathbf{x}_i)$	two-particle Green's function 73
$\mathscr{K}(\mathbf{x}_i)$	two-quasi-particle Green's function 104
m	bare particle mass
m^*	effective mass of a quasi particle 6
m_k	distribution of bare particles in the ground state 63
$M(p)$	self-energy operator 88
M_1, M_2	components of M for a superfluid system 127
$n(\mathbf{k})$	distribution function for quasi particles 3
$\lvert n\rangle$	eigenstate of the unperturbed system 146
N	number of particles in the system
r	limiting value of the ratio q/ϵ (Chapter 6) 260
r_q	density of test charge 288
$R(p; \varpi), \tilde{R}(p; \varpi)$	singular term of the product $G(p + \varpi/2)\ G(p - \varpi/2)$ 261
S_F	Fermi surface 1
$S(q, t)$	density correlation function 54
$\bar{S}, \bar{\bar{S}}$	other forms of the correlation function S 336

S_η	S-matrix in the presence of the interaction $H_1 e^{-\eta t}$	179
$\underset{\sim}{S}_{\mu\nu}(\varpi)$	generalized correlation functions	255
$\widetilde{S}_{\mu\nu}(\varpi)$	contribution of the proper graphs to S	282
$T\{\ldots\}$	time-ordered product for fermions	59
$U(t, t')$	time-development operator	148
$U_\eta(t, t')$	time-development operator in the presence of the interaction $H_1 e^{-\eta t}$	169
U_L, U_{oc}	contribution of the linked and unlinked corrected graphs to U	166
V_k	velocity of the quasi particle $\mathbf{k}$	6
Vq	Fourier component of the binary interaction potential	342
$V(\mathbf{k}_i)$	matrix element of the interaction	150
z_k	residue of $G(k, \omega)$	72
z^1, z	residues of the Green's functions for a superfluid gas	128
$\gamma(\mathbf{x}_i)$	interaction operator	75
$\gamma R(\mathbf{p}_i)$	renormalized interaction operator	105
$F(\mathbf{p}_i)$	interaction operator without its δ function	77
${}^0\Gamma$, ${}^1\Gamma$	components of Γ with total spin σ_z equal to 0 or 1	246
${}^0\widetilde{\Gamma}$	contribution of the proper graphs to ${}^0\Gamma$	282
δK, δM	bound parts of K, M (normal systems)	75
$\delta_{\mathbf{k}}A$	correction to A when a quasi particle $\mathbf{k}$ is added	96
ΔK, ΔM	bound parts of K, M (superfluid systems)	122
ΔE_0	shift of the ground-state energy due to the interaction H_1	177
ϵ_k	energy of the quasi particle $\mathbf{k}$	3
$\epsilon(\varpi)$	dielectric constant	284
η	adiabatic switching constant	168
$\lambda\mu(k)$	zeroth-order vertex parts	254
$\underset{\sim}{\Lambda}\mu(p; \varpi)$	vertex parts	254
$\widetilde{\Lambda}\mu(p; \varpi)$	proper vertex parts	282
$\Lambda_\mu(k)$	value of Λ_μ on the Fermi surface ($\omega = \epsilon_k = \mu$)	266
μ	chemical potential	65
ξ_k	free energy of a quasi particle	66
ρ	average density of the system	
ρ_q	Fourier component of the density operator	343
$\|\varphi_n\rangle$	eigenstates of the real system	146
$\varphi_{BA}(t)$	response function to an impulse excitation	38
χ	compressibility	7
$\chi_{BA}(\omega)$	admittance	41
$\psi_\sigma(r)$, $\psi_\sigma^*(r)$	destruction and creation operators of a quasi particle at point r	342
ω_{n0}	excitation energy of the state $\|\varphi_n\rangle$	42
Ω	volume of the system	

References and Literature

There is such a large number of articles concerned with the many-body problem that it is nearly impossible to establish an exhaustive list of references (which indeed would soon become obsolete). We shall only quote the references from which this book was derived, as well as those that constitute a natural extension of our treatment. Of course this choice is purely subjective; its only purpose is to keep the same language in the text and in the references. In addition to the original papers, we list a number of books that provide a general survey of this subject. If needed, the reader will find much more detailed bibliographies in references 7 and 9.

General

1. 1959, *Introduction to the Physics of Many Body Systems*, D. Ter Haar, Interscience, New York.
2. 1959, *The Many Body Problem: Lecture Notes of the Les Houches Summer School*, Dunod, Paris.
3. 1960, *Proceedings of the Utrecht International Congress on Many Particle Problems* (supplement to *Physica*, Vol. 26).
4. 1960, *Lectures on the Many Body Problem: Lecture Notes of the Naples Spring School*, Academic, New York.
5. 1961, *The Quantum Mechanics of Many Body Systems*, D. J. Thouless, Academic, New York.
6. 1961, *The Many-Body Problem: Lecture Notes from the First Bergen International School of Physics*—1961, Benjamin, New York.
7. 1961, *The Many-Body Problem*, D. Pines, Benjamin, New York.
8. 1961, *Quantum Theory of Many-Particle Systems*, L. Van Hove, N. M. Hugenholtz, and L. P. Howland, Benjamin, New York.

9. 1962, *The Green's Function Method in Statistical Mechanics*, V. L. Bonch Bruevich and S. V. Tyablikov (original edition in Russian, Moscow, 1960); English translation by D. Ter Haar, North Holland, Amsterdam.
10. 1962, *Quantum Statistical Mechanics*, L. Kadanoff and G. Baym, Benjamin, New York.
11. 1963, *1962 Cargèse Lectures in Theoretical Physics*, Benjamin, New York.

References 1, 5, 7, 9, and 10 give a full exposition of the many-body problem. The other books are either conference proceedings, or lecture notes together with reprinted original articles.

References

Chapter 1

Theory of Fermi liquids

L. D. Landau, *J.E.T.P.*, **30**, 1058 (1956) (*Sov. Phys.*, **3**, 920); **32**, 59 (1957) (*Sov. Phys.*, **5**, 101).
J. Ia. Pomeranchuk, *J.E.T.P.*, **35**, 524 (1958) (*Sov. Phys.*, **8**, 361).

Application to ^{3}He

I. M. Khalatnikov and A. A. Abrikosov, *Repts. Progr. Physics*, **XXII**, 329 (1959).
V. P. Silin, *J.E.T.P.*, **33**, 1227 (1957) (*Sov. Phys.*, **6**, 945).
I. L. Bekarevich and I. M. Khalatnikov, *J.E.T.P.*, **39**, 1699 (1960) (*Sov. Phys.*, **12**, 1187).
A. A. Abrikosov and I. M. Khalatnikov, *J.E.T.P.*, **41**, 544 (1961) (*Sov. Phys.*, **14**, 389).
A. J. Akhiezer, I. A. Akhiezer, and I. Ia. Pomeranchuk, *J.E.T.P.*, **41**, 478 (1961) (*Sov. Phys.*, **14**, 343); *Nucl. Phys.*, **40**, 139 (1963).
D. Hone, *Phys. Rev.*, **121**, 669 (1961); *ibid.*, **125**, 1494 (1962).

Application to Electrons

V. P. Silin, *J.E.T.P.*, **33**, 495 (1957) (*Sov. Phys.*, **6**, 387); *ibid.*, **33**, 1282 (1957) (*Sov. Phys.*, **6**, 985); *ibid.*, **34**, 707 (1958) (*Sov. Phys.*, **7**, 486); *ibid.*, **34**, 781 (1958) (*Sov. Phys.*, **7**, 538); *ibid.*, **35**, 1243 (1958) (*Sov. Phys.*, **8**, 870); *ibid.*, **37**, 273 (1959) (*Sov. Phys.*, **10**, 192).
M. Ia. Azbel, *J.E.T.P.*, **39**, 1138 (1960) (*Sov. Phys.*, **12**, 793).

Application to ferromagnets

A. A. Abrikosov and I. E. Dzialoshinskii, *J.E.T.P.*, **35**, 771 (1958) (*Sov. Phys.*, **8**, 535).
A. J. Akhiezer and I. Ia. Pomeranchuk, *J.E.T.P.*, **36**, 859 (1959) (*Sov. Phys.*, **9**, 605).

Chapter 2

Admittances

R. Kubo, *J. Phys. Soc.* (*Japan*), **9**, 888 (1954); *ibid.*, **12**, 570, 1203 (1957); *ibid.*, **14**, 56 (1959).

Diffusion in Born approximation
L. Van Hove, *Phys. Rev.*, **95**, 249, 1374 (1954).

Dielectric constant
J. Lindhard, *Kgl. Danske Videnskab. Selskab, Mat-fys. Medd*, **28**, 8 (1954).
U. Fano, *Phys. Rev.* **103**, 1202 (1956).
P. Nozières and D. Pines, *Nuovo Cimento*, **9**, 470 (1958).

Chapters 3 to 7

Zero-temperature formulation

Field theoretical methods
J. Hubbard, *Proc. Roy. Soc. (London)*, **A240**, 539 (1957); *ibid.*, **A243**, 336 (1957).
A. B. Migdal, *J.E.T.P.*, **32**, 399 (1957) (*Sov. Phys.*, **5**, 333).
V. M. Galitskii and A. B. Migdal, *J.E.T.P.*, **34**, 139 (1958) (*Sov. Phys.*, **7**, 96).
A. Klein and R. Prange, *Phys., Rev.* **112**, 994 (1958).
A. Klein, *Phys. Rev.* **121**, 950, 957 (1961).

Time-independent formulation
J. Goldstone, *Proc. Roy. Soc. (London)*, **A239**, 267 (1957).
N. M. Hugenholtz and L. Van Hove, *Physica*, **23**, 363 (1958).
N. M. Hugenholtz, *Physica*, **23**, 481 (1958).
C. Bloch, *Nucl. Phys.* **7**, 451 (1958).

Extension to finite temperatures
L. D. Landau, *J.E.T.P.*, **34**, 262 (1958) (*Sov. Phys.*, **7**, 182).
A. A. Abrikosov, L. P. Gor'kov, and I. E. Dzialoshinskii, *J.E.T.P.*, **36**, 900 (1959) (*Sov. Phys.*, **9**, 636).
E. S. Fradkin, *J.E.T.P.*, **36**, 1286 (1959) (*Sov. Phys.*, **9**, 912).
J. M. Luttinger and J. C. Ward, *Phys., Rev.* **118**, 1417 (1960).
J. M. Luttinger, *Phys. Rev.*, **119**, 1153 (1960); *ibid.*, **121**, 942 (1961).
A. I. Alekseev, *Usp. Fiz. Nauk.*, **73**, 41 (1961) (*Sov. Phys., Usp.*, **4**, 23).
S. V. Maleev, *J.E.T.P.*, **41**, 1675 (1961) (*Sov. Phys.*, **14**, 1191).

The formalism used in the preceding references is a natural extension of that of the present book. Other methods, of a rather different nature, have been proposed by:

E. Montroll and J. C. Ward, *Phys. Fluids*, **1**, 55 (1958).
P. C. Martin and J. Schwinger, *Phys. Rev.*, **115**, 1342 (1959) (this approach is used in General Reference 10).
E. Glassgold, W. Heckrotte, and K. M. Watson, *Phys. Rev.*, **115**, 1374 (1959).
C. De Dominicis and C. Bloch, *Nucl. Phys.*, **7**, 459 (1959); *ibid.*, **10**, 509 (1959).
R. Balian and C. de Dominicis, *Nucl. Phys.*, **16**, 502 (1960).
C. de Dominicis, Thèse de Doctorat, Paris, 1961.
V. L. Bonch Bruevich, *J.E.T.P.*, **31**, 522 (1956) (*Sov. Phys.*, **4**, 456) (this approach is used in General Reference 9).
D. N. Zubarev, *Usp. Fiz. Nauk.*, **71**, 71 (1960) (*Sov. Phys. Usp.*, **3**, 320).

Normal fermion systems
M. Gell-Mann and K. A. Brueckner, *Phys. Rev.*, **106**, 364 (1957).

D. V. Du Bois, *Ann. Phys. N.Y.*, **8**, 24 (1959).
V. M. Galitskii, *J.E.T.P.*, **34**, 151 (1958) (*Sov. Phys.*, **7**, 104); *ibid.*, **34**, 1011 (1958) (*Sov. Phys.*, **7**, 698).
L. D. Landau, *J.E.T.P.*, **35**, 97 (1958) (*Sov. Phys.*, **8**, 70).
A. I. Larkin, *J.E.T.P.*, **37**, 264 (1959) (*Sov. Phys.*, **10**, 186).
J. Goldstone and K. Gottfried, *Nuovo Cimento*, **13**, 849 (1959).
E. Daniel and S. Vosko, *Phys. Rev.*, **120**, 2041 (1960).
J. S. Langer and S. Vosko, *J. Phys. Chem. Solids*, **12**, 196 (1960).
J. S. Langer, *Phys. Rev.*, **120**, 714 (1960); *ibid.*, **124**, 997, 1003 (1961); *ibid.*, **127**, 5 (1962).
I. A. Akhiezer and S. V. Peletminskii, *J.E.T.P.*, **38**, 1829 (1960) (*Sov. Phys.*, **11**, 1316); *J.E.T.P.*, **39**, 1308 (1960) (*Sov. Phys.*, **12**, 913).
J. M. Luttinger, *Phys. Rev.*, **121**, 1251 (1961).
Yu. A. Bychkov, *J.E.T.P.*, **39**, 1401 (1960) (*Sov. Phys.*, **12**, 977).
Yu. A. Bychkov and L. P. Gor'kov, *J.E.T.P.*, **41**, 1592 (1961) (*Sov. Phys.*, **14**, 1132).
V. N. Tsytovich, *J.E.T.P.*, **42**, 457 (1962) (*Sov. Phys.*, **15**, 320); *ibid*, **42**, 803 (1962) (*Sov. Phys.*, **15**, 561).
V. I. Perel and G. M. Eliashberg, *J.E.T.P.*, **41**, 886 (1961) (*Sov. Phys.*, **14**, 633).
G. M. Eliashberg, *J.E.T.P.*, **41**, 1241 (1961) (*Sov. Phys.*, **14**, 886).
J. M. Luttinger and P. Nozières, *Phys. Rev.*, **127**, 1423, 1431 (1962).

Electron-phonon interaction

A. B. Migdal, *J.E.T.P.*, **34**, 1438 (1958) (*Sov. Phys.*, **7**, 996).
S. V. Tyablikov and N. N. Tolmachev, *J.E.T.P.*, **34**, 1254 (1958) (*Sov. Phys.*, **7**, 867).
G. M. Eliashberg, *J.E.T.P.*, **38**, 966 (1960) (*Sov. Phys.*, **11**, 696).

Superfluid Fermi systems

Superconductors

L. N. Cooper, *Phys. Rev.*, **104**, 1189 (1956).
J. Bardeen, L. N. Cooper, and J. R. Schrieffer, *Phys. Rev.*, **108**, 1175 (1957).
P. W. Anderson, *Phys. Rev.*, **112**, 1900 (1958).
N. N. Bogoliubov, *J.E.T.P.*, **34**, 58 (1958) (*Sov. Phys.*, **7**, 41); *J.E.T.P.*, **34**, 73 (1958) (*Sov. Phys.*, **7**, 51).
N. N. Bogoliubov, N. N. Tolmachev, and D. V. Shirkov, *A New Method in the Theory of Superconductivity* (English translation), Consultants Bureau, New York, 1959.
J. Bardeen and G. Rickayzen, *Phys. Rev.*, **118**, 936 (1960).
L. P. Gor'kov, *J.E.T.P.*, **34**, 735 (1958) (*Sov. Phys.*, **7**, 505); *J.E.T.P.*, **36**, 1918 (1959) (*Sov. Phys.*, **9**, 1364); *J.E.T.P.*, **37**, 833 (1959) (*Sov. Phys.*, **10**, 593); *J.E.T.P.*, **37**, 1407 (1959) (*Sov. Phys.*, **10**, 998).
A. A. Abrikosov and L. P. Gor'kov, *J.E.T.P.*, **35**, 1558 (1958) (*Sov. Phys.*, **8**, 1090); *J.E.T.P.*, **36**, 319 (1959) (*Sov. Phys.*, **9**, 220).
A. A. Abrikosov, L. P. Gor'kov, and I. M. Khalatnikov, *J.E.T.P.*, **35**, 265 (1958) (*Sov. Phys.*, **8**, 182); *J.E.T.P.*, **37**, 187 (1959) (*Sov. Phys.*, **10**, 132).
Y. Nambu, *Phys. Rev.*, **117**, 648 (1960).

L. Kadanoff and P. C. Martin, *Phys. Rev.*, **124**, 670 (1961).
D. Thouless, *Ann. Phys. N.Y.*, **10**, 553 (1960).
P. W. Anderson and P. Morel, *Phys. Rev.*, **125**, 1263 (1962).
G. M. Eliashberg, *J.E.T.P.*, **39**, 1437 (1960) (*Sov. Phys.*, **12**, 1000).
V. L. Pokrovskii, *J.E.T.P.*, **40**, 143 (1961) (*Sov. Phys.*, **13**, 100).
L. P. Gor'kov and T. K. Melik Barkhudarov, *J.E.T.P.*, **40**, 1452 (1961) (*Sov. Phys.*, **13**, 1018).
V. G. Vaks, V. M. Galitskii, and A. I. Larkin, *J.E.T.P.*, **41**, 1655 (1961) (*Sov. Phys.*, **14**, 1177).
A. B. Migdal, *Nucl. Phys.*, **30**, 239 (1962).

Superfluidity of ^{3}He

K. A. Brueckner, T. Soda, P. W. Anderson, and P. Morel, *Phys. Rev.*, **118**, 1442 (1960).
V. J. Emery and A. M. Sessler, *Phys. Rev.*, **119**, 43 (1960).
P. W. Anderson and P. Morel, *Phys. Rev.*, **123**, 1911 (1961).
L. P. Pitaevskii, *J.E.T.P.*, **37**, 1794 (1959) (*Sov. Phys.*, **10**, 1267).
L. P. Gor'kov and V. M. Galitskii, *J.E.T.P.*, **40**, 1124 (1961) (*Sov. Phys.*, **13**, 792).
L. P. Gor'kov and L. P. Pitaevskii, *J.E.T.P.*, **42**, 600 (1962) (*Sov. Phys.*, **15**, 417).
P. Morel and P. Nozières, *Phys. Rev.*, **126**, 1909 (1962).

Boson systems

N. Bogoliubov, *J. Phys. U.R.S.S.*, **11**, 23 (1947).
T. D. Lee, K. Huang, and C. N. Yang, *Phys. Rev.*, **106**, 1135 (1957).
S. T. Beliaev, *J.E.T.P.*, **34**, 417 (1958) (*Sov. Phys.*, **7**, 289); *J.E.T.P.*, **34**, 433 (1958) (*Sov. Phys.*, **7**, 299).
N. M. Hugenholtz and D. Pines, *Phys.*, *Rev.*, **116**, 489 (1959).
L. P. Pitaevskii, *J.E.T.P.*, **36**, 1168 (1959) (*Sov. Phys.*, **9**, 830); *J.E.T.P.*, **40**, 646 (1961) (*Sov. Phys.*, **13**, 451).

Index